T0310301

SOLVENT MICROEXTRACTION

SOLVENT MICROEXTRACTION
Theory and Practice

JOHN M. KOKOSA
MDRC Consulting/Mott Community College, Flint, MI

ANDRZEJ PRZYJAZNY
Kettering University, Flint, MI

MICHAEL A. JEANNOT
St. Cloud State University, St. Cloud, MN

WILEY

A JOHN WILEY & SONS, INC., PUBLICATION

Library of Congress Cataloging-in-publication Data:

Kokosa, John M., 1945–
    Solvent microextraction: theory and practice / John M. Kokosa, Andrzej Przyjazny,
Michael A. Jeannot.
        p. cm.
    Includes bibliographical references and index.
    ISBN 978-0-470-27859-8 (cloth)
1. Solvent extraction. 2.   Chemistry, Analytic–Technique. I. Przyjazny, Andrzej J. II.
Jeannot, Michael A., 1970– III.   Title.
    QD63.E88k65 2009
    660'.284248–dc22   2009009768

Printed in the United States of America

10 9 8 7 6 5 4 3 2 1

We dedicate this book to our parents, wives, and children, and thank them for their love, patience, and understanding:

to Cora, Michael, and Sarah

to Anna, Marcin, and Katarzyna

and to Lori, Erin, and Ben

CONTENTS

PREFACE

From its inception as a curious and possibly useful academic research technique, solvent microextraction (SME) has, in the last dozen or so years, blossomed and matured into an important tool not only for research, but also for practicing industrial, forensic, clinical, and environmental analysts. SME has been used in the analysis of metabolic products in biological fluids, in contaminants in drinking and wastewater, in soils, in foods, in plastics, and in pharmaceuticals. It is a unique methodology, requiring little more than a manual microsyringe and a few microliters of solvent, in its simplest forms, for the extraction and concentration of ultratrace to percent levels of volatile, semivolatile, polar, or ionized organic and inorganic analytes from liquids and solids. The number of application papers in the field has grown to well over 600, an indication of its usefulness. However, SME has also suffered from two major deficiencies: the inability to automate the procedure and a single source the analyst could use to find a useful compilation of reference material, a guide to proper experimental procedures, and a better understanding of how and why the technique works. Recently, the ability to completely automate the single-drop microextraction mode of SME has been made available, which allows the unattended sampling and instrumental analysis of samples. Removing the second deficiency, making SME a truly practical analysis tool, is the purpose of this book, which we hope will find its place on the shelf with other useful manuals available to the practicing analyst.

Acknowledgments

We wish to acknowledge the following individuals and organizations whose help was invaluable in completing this work. Students who have worked with us over the years to develop the SME technique include Aaron Theis, Philip Tourand, Susan Hansen, Jenna Bungert, Adam Waldack, Jenna Schwartz, and Charlene Schnobrich from St. Cloud State University, and Joel Austin, Andrew Essenmacher, Ryan Jones, Casey Essary, Donald Harris, Rashad Simmons, Sydney Forrester, and Nicole Booms from Kettering University. In addition, we wish to thank Bruce Deitz from the Kettering University library for his help in obtaining and compiling references, Ingo Christ and Chromsys LLC for the research collaboration and the single magnet mixer used in this work, and finally, CTC Analytics AG for making available the PAL autosampler instrumentation used in developing the experimental procedures described in this book.

JOHN M. KOKOSA

ANDRZEJ PRZYJAZNY

MICHAEL A. JEANNOT

CHAPTER 1

SOLVENT MICROEXTRACTION: COMPARISON WITH OTHER POPULAR SAMPLE PREPARATION METHODS

1.1 INTRODUCTION

One of the most important steps in any analytical procedure is sample preparation. Most analyses are carried out on samples containing complex mixtures of very small amounts of the chemicals that need to be identified and/or quantified. At the same time, most sample matrices, such as soils or wastewater, are also very complex. Thus, a successful sample preparation method typically has three major objectives: (1) sample matrix simplification and/or replacement, (2) analyte enrichment, and (3) sample cleanup.

Many useful sample preparation methods have been developed over the years to address specific needs for analyzing waste and drinking water, foods, medicinals, soil, and air. As an example, many of the most important methods were codified for the analysis of waste and drinking water samples, as exemplified by the U.S. Environmental Protection Agency's (EPA's) 500 and 600 methods. However, these sample preparation and analysis methodologies, which originally involved traditional laboratory liquid–liquid, liquid–solid, and gaseous extractions, suffered from a number of limitations, including the requirement for significant time-consuming manual labor or in some cases, the use of large quantities of hazardous extracting solvents. As a result, there has been a continual search for improved sample preparation procedures with the following goals: (1) reduction in the number of steps

Solvent Microextraction: Theory and Practice, By John M. Kokosa, Andrzej Przyjazny, and Michael A. Jeannot
Copyright © 2009 John Wiley & Sons, Inc.

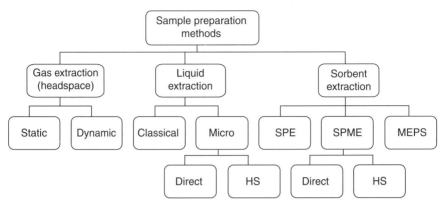

FIGURE 1.1. Sample preparation methods for aqueous and solid samples.

required for the procedure, (2) reduction or total elimination of solvents required for extraction, (3) adaptability to field sampling, and (4) automation.

The more commonly used sample preparation methods for water and solid matrix samples are represented schematically in Figure 1.1. *Solvent microextraction* (SME) is a fairly recent development in sample preparation that has the potential for meeting all four goals cited above. SME, which, in its most commonly used modifications, has also been referred to as *liquid-phase microextraction* (LPME), *single-drop microextraction* (SDME), or *dispersive liquid–liquid microextraction* (DLLME), can integrate sampling, analyte extraction and concentration, and sample introduction into a single step. Because it is the most comprehensive descriptive term available, we have decided to use the term SME originally coined by Jeannot and Cantwell[1] to cover all of the variations of this method. SME is compatible and has been used successfully with most common analytical instrumentation, including gas chromatography (GC), high-performance liquid chromatography (HPLC), capillary electrophoresis (CE), and atomic absorption spectroscopy (AAS). In its simplest and earliest practical variation, SME consists of a single drop of extracting solvent suspended from a GC syringe needle that is immersed in an aqueous sample solution. The same GC syringe is then used to introduce the solvent drop with extracted analytes into the GC (see Figure 1.2).

SME has its origins in traditional sample preparation methodologies, including liquid–liquid extraction, and in this chapter we not only give a brief history of the method, but also provide a general comparison with other commonly used sample preparation techniques. We include the advantages and disadvantages of each method, allowing the reader to gauge whether SME would be an appropriate method for their sample preparation needs.

1.2 COMPARISON OF SAMPLE PREPARATION METHODS

As indicated above, many of the most useful sample preparation methods have been rigorously detailed and codified into methods adopted by the U.S. EPA and other

FIGURE 1.2. A 1-µL drop of octane suspended in aqueous solution on the tip of a Hamilton 701 GC syringe needle.

national and state agencies. The most commonly used procedures include variants of liquid–liquid extraction, sorbent extraction [solid-phase extraction (SPE)] and headspace extraction [including purge-and-trap (PT)], all based on classical chemical laboratory analysis techniques. More recently, solid-phase microextraction (SPME) has been developed and used for many environmental, forensic, and food analysis applications.[2,3] SME, in its most commonly implemented format, is very similar to SPME, using 1 to 2 µL of solvent at the end of a syringe needle to extract a sample rather than a liquid or porous polymer coating on a fused silica or metal fiber. Each of these techniques has advantages and disadvantages, and no one technique can be or should be used for all analyses. A number of other techniques, such as supercritical fluid extraction, cryogenic trapping, microextraction in packed syringe, and stir bar sorptive extraction are important in their own right, but are not discussed here. In addition, although SME and other techniques discussed here may be used for atmospheric sampling, we limit our discussion to extraction from liquid and solid matrices.

1.2.1 Liquid–Liquid Extraction

Traditional liquid–liquid extraction methodology was taken directly from standard techniques used for the purification of chemicals prepared in the laboratory. Thus, a solvent such as dichloromethane, pentane, or ether is used to extract analytes from water using a separatory funnel or continuous extractor, and from solids such as soil or plant material using a Soxhlet extractor. Specific examples include EPA method 625 for municipal wastewater, method 525.2 for drinking water, and method 551.1 for chlorination disinfection by-products in drinking water.

1.2.1.1 *EPA Method 625: Liquid–Liquid Extraction*   EPA method 625 is used for the analysis of 54 specific chemicals, including industrial by-products and pesticides, and seven chemical mixtures [chlordane, toxaphene, and the poly-chlorinated biphenyls (PCBs)] commonly found in municipal and industrial wastewater.[4] The method involves extracting a liter of sample with 450 mL or more of dichloromethane, which is then evaporated to a volume of 1 mL. Next, a 1-μL aliquot of the concentrate is analyzed by GC or GC-mass spectrometry (MS). A continuous extractor using 500 mL of solvent can also be used for the extraction, to limit emulsion formation.

The major advantages of this method are simplicity and the ability to extract many analytes at once into a solvent compatible for analysis by GC-MS, which in turn is capable of separating, identifying, and quantifying each analyte. The major disadvantages of the method are the large amounts of solvent used for the extraction, the manual labor involved, emulsions sometimes formed when using a separatory funnel, the time required for the extraction (up to 24 h for a continuous extractor), and the fact that the method actually does not concentrate the analytes significantly. True, 1 L of water is concentrated into 1 mL of dichloromethane, a 1000-fold enrichment, but only 1 μL is analyzed, 1/1000 of the extract. Thus, if an analyte is present at a concentration of 1 μg/L, only 1 ng is actually analyzed. A consequence of the need to evaporate the solvent is that this method is not applic-able to volatile chemicals. Another major disadvantage is the relatively large amount of water sample, 1 L, that is required, making this method difficult to automate. Finally, the method will require two preparations and analyses per sample if both basic and neutral/acidic components are present, effectively doubling the analysis time. However, despite these difficulties, method 625 does yield method detection limits (MDLs) of 1 to 5 μg/L (1 to 5 ng/mL) for each analyte, and fairly good precision and accuracy.

1.2.1.2 *EPA Method 551.1: Micro Liquid–Liquid Extraction*   Several new methods have been developed to address the disadvantages of solvent use, manual labor, and time required for sample preparation methods such as 625. One method includes EPA method 551.1, which is a micro extraction method for analysis of the chlorination disinfection by-products, including the trihalomethanes, and several chlorinated industrial solvents and pesticides.[5] The method involves extracting 50 mL of drinking water saturated with salt with either 3 mL of methyl *tert*-butyl ether or 5 mL of pentane (compared to 500 mL of solvent for method 625, thus the term *micro*), followed by analysis of 1 to 2 μL of the extractant by GC with an electron capture detector (ECD). If pentane is the solvent of choice and 2 μL of the concentrate is injected, the analytes have been concentrated tenfold, but only 1/2500 of the analytes are then analyzed. For example, if the original sample contained 1 μg/L of an analyte, only 20 pg would be analyzed. However, because the ECD is very sensitive and selective for halogenated compounds, the effective method detection limit is approximately 0.1 μg/L (0.1 ng/mL) or less. This method does require significantly less solvent and water sample than method 625, and no solvent concentration is needed, enabling analysis of volatile as well as nonvolatile

chemicals and decreasing the preparation time significantly, to about an hour per sample. The major difficulties are that the method again is not easily fully automated and there is little analyte enrichment.

1.2.1.3 Solvent Extraction–Flow Injection Analysis

Semiautomated solvent extraction (SE) approaches utilizing the technique of flow injection analysis (FIA) were first reported in 1978,[6,7] and the kinetics and mechanism of the extraction process have been described in detail.[8–10] SE-FIA involves the injection of an aqueous sample plug in a flowing aqueous stream which may or may not contain chemical reagents. The aqueous stream is "teed" to an organic solvent stream, resulting in a segmented flow of alternating aqueous and organic segments which flow through a sufficient length of tubing (the extraction coil). Following the extraction process, the segmented flow passes through a phase separator, which typically separates and directs the organic phase to a flow-cell type of detector for quantitation of the analyte by absorbance or fluorescence spectroscopy. The principal advantages of this approach are the reduction in volumes of sample and solvent, and the ability to achieve high sample throughput through a semiautomated approach. However, owing to the comparable volumes of aqueous and organic phases, even exhaustive extraction results in little, if any, analyte enrichment. The technique has been reviewed extensively by Kubáň.[11]

1.2.2 Liquid–Solid Extraction

Solid-phase extraction is actually based on HPLC media technology. It had been observed that very dilute organic analyte samples in water could be concentrated at the head of a reversed-phase (C18) coated silica gel chromatography column as several milliliters of the dilute water sample passed through the column. This led to the development of small cartridges containing either silica gel or alumina (to retain polar compounds), or reversed-phase adsorbent (to retain nonpolar analytes) for sample cleanup and concentration for HPLC. The success of this method led in turn to the development of cartridges and disks containing a variety of retentive adsorbents which could be used for concentrating analytes from large volumes of water.[12] More recently, the method has been transformed by the development of molecularly imprinted polymers for use as the stationary phase.[13] Molecularly imprinted polymers are polymers synthesized in the presence of a specific chemical. When the three-dimensional form of the polymer is complete and the chemical has been removed from the polymer, holes or receptor sites are left throughout the polymer which have bonding affinities for the chemical.

1.2.2.1 EPA Method 525.2: Solid-Phase Extraction

Method 525 was originally a duplicate of method 625, using liquid–liquid extraction to extract analytes from drinking water. The method has been modified to use either a cartridge or a disk containing octadecylsilane (C18) bonded to silica gel to retain the organic analytes as the water is passed through.[14] The analytes are then extracted from the adsorbent with solvent, usually 10 mL of a mixture of ethyl acetate and dichloromethane. The solvent is dried, concentrated to 1 mL, and 1 μL is injected for analysis. This method works well for relatively clean matrices such as drinking water, but less well for "dirty" samples since the silica adsorbents are easily plugged. Using a prefilter material ahead of the cartridge or disk is often mandatory for these samples. The filtration process takes about 1 h, and the entire preparation method about 2 h, and can be semiautomated. The sample is concentrated to the same extent and analyzed in the same manner as in method 625, yielding similar method detection limits. Although much less solvent is used, the method still requires 1 L of water sample and because the solvent is still evaporated, only non-volatile chemicals can be determined using this method.

1.2.3 Headspace Extraction

A third general method widely used for sample preparation is headspace analysis. Two types of headspace analysis are commonly used: static headspace analysis and purge-and-trap. *Static headspace analysis* is the simplest of all sampling methods, involving sampling, usually with a gastight syringe, and injection of a portion of the headspace gas above a sample in a sealed container.[15] Only a small portion of the total analyte present in the sample is thus analyzed. This also results in a requirement for very careful calibration if this method is to be used for quantification. The method has the disadvantages that only about 10% of the headspace can be injected for analysis, and it is generally useful only for fairly volatile analytes. However, static headspace sampling has the advantage that it can be done either manually or with a standard GC auto-sampler. Sophisticated automated samplers are also available that can heat the sample and take headspace samples under pressure, enhancing sample enrichment in the headspace. This type of instrumentation is used for *United States Pharmacopeia* (USP) method 467 and the *European Pharmacopeia* method for the analysis of residual solvents in pharmaceutical products. In contrast to static headspace sampling, the *purge-and-trap technique*, used for EPA method 624, is a dynamic sampling procedure, capable of exhaustive extraction and concentration of all the volatile components in water or solid samples.

1.2.3.1 EPA Method 624: Purgeables
Purge-and-trap method 624 was designed to analyze 31 volatile chemicals in municipal and industrial wastewater.[16] Other purge-and-trap methods extend the technique to different sample types and analytes. The main advantage of PT is the ability to effectively determine all the analytes in a 5- to 25-mL sample of water. The sample is sparged with a stream of helium gas, which carries the analytes to a solid adsorbent trap consisting of silica gel, Tenax, graphitized carbon, or layers of these materials. The trap is quickly heated and the analytes released are carried in a helium stream to a GC inlet for analysis. The high flow rate of the desorbing helium gas (10 to 30 cm^3/min) requires the use of a wide-bore analytical column and a split-column outlet flow (a jet separator), or a split inlet flow, or liquid-nitrogen cooling of a pre-column to trap the analytes in a tight sample plug which can then be released upon heating onto a narrow-bore GC analytical column at a reduced flow rate. Often, the concentrating effect of the PT method allows a simple split inlet technique to be used successfully.

Typically, method detection limits are 1 μg/L or less with a modern GC-MS and capillary column. Therefore, for a typical 5-mL sample, around 5 ng of each analyte is analyzed. The method does suffer, however, from the possibility of sample carryover from one sample to another, degradation of the trap over time, potential leaks in the plumbing, and the cost of the instrumentation. In addition, foaming of the sample due to the presence of detergents or natural products in soil can cause major contamination problems if the foam is allowed to enter the heated plumbing of the instrument. Foaming can sometimes be avoided by purging the headspace rather than by sparging the water sample with helium. The technique is fully automated, and typically, 30 to 50 samples can be run sequentially.

A variant of purge and trap is membrane extraction with a sorbent interface.[17] This technique utilizes a nonporous membrane, usually silicone, which allows selective transport of nonpolar compounds from a water sample across the membrane barrier. The extracted analyte is swept by a stream of helium and concentrated on a cold or sorbent trap and then released for analysis thermally. This technique is therefore similar to purge-and-trap in the ability to extract analytes exhaustively. Limits of detection below 0.5 ng/L have been reported for toluene and benzene.[18] The limitation for the method is the nonpolar nature of the silicone membrane. Thus, only nonpolar chemicals can be extracted.

1.2.3.2 *EPA Method 524.2: Purgeables*

This is an updated PT method for 84 volatile chemicals ranging from dichlorodifluoromethane to naphthalene in surface waters, groundwater, and drinking water.[19] The method detection limits are in the range of 0.1 μg/L or less. In fact, PT is a very sensitive technique for volatile compounds and is so sensitive that even when using a split inlet with a narrow-bore capillary column, the sample may still need to be diluted to bring the extracted components into a concentration range compatible with the chromatography column and detector. The same advantages and disadvantages apply to method 524.2 as to method 624.

1.2.3.3 *USP Method 467 and the* European Pharmacopeia *Method for Residual Solvents*

These methods were developed for the analysis of residual solvents present in pharmaceutical products after manufacture. The USP method, until recently, involved the analysis of 7 chemicals, but has been extended to cover 64 solvents and essentially duplicates the European method.[20] These methods involve the use of static headspace sampling of pharmaceuticals dissolved in either water, dimethylformamide–water or dimethyl sulfoxide–water. Five-milliliter solutions are heated to 80°C under pressure in a 20-mL headspace vial and 1 mL of the headspace withdrawn and injected into a GC for analysis. Static headspace analysis is not very sensitive for chemicals which are very soluble in water and chemicals with high boiling points. However, the solvents analyzed under method 467 are at relatively high concentrations, ranging from a maximum allowable concentration of 3000 ppm for the xylenes to 2 ppm for benzene. The elevated extraction temperature increases the concentration in the headspace sufficiently so that the samples can be analyzed using a GC with a flame ionization detector. The major disadvantages of the method are the low sensitivities for some analytes and the initial expense required for the headspace autosampler.

1.2.4 Solid-Phase Microextraction

SPME is a relatively new and important sample preparation method with many advantages and some very important shortcomings. As SPE was developed using HPLC column technology, SPME was developed using GC column technology. SPME is basically an inside-out GC column. A 100- to 250-μm fused silica or metal fiber is coated with 7 to 100 μm of a coating that functions to extract analytes from a water or headspace sample. The coating is either a silicone-based polymer, identical to the polymers used for GC columns, which absorbs the analytes, or a polymer with bonded porous carbon particles (carbon molecular sieve), which adsorbs volatile chemicals. Many hundreds of application notes and papers are available for this technique, which cover samples ranging from industrial discharge waters to arson analyses to biological samples to flavorings in foods.[2,3] SPME is a true solventless method, and that is a major advantage. The technique also allows one to extract a sample and inject the extract using one device. Several fiber polymer coatings are available in various thicknesses, ranging from the nonpolar dimethylsiloxane to relatively polar polyacrylate and the porous adsorbent Carboxen (porous carbon molecular sieve). The method can be carried out manually using a special syringe-type holder or automated completely for GC or HPLC. Finally, the method can and has been used as a field sampling device for atmospheric, water, agricultural, and forensic analyses. So why hasn't SPME replaced all other sampling techniques? This method, like all others, has not only advantages but some important limitations.

One limitation often overlooked is the limited volume of the polymer extractant. One centimeter of a 100-μm-thick poly(dimethylsiloxane) (PDMS) coating on a 100-μm fiber is calculated to have a volume of approximately 0.628 μL. A 7-μm coating on a 250-μm core would have a volume of only 0.056 μL. These volumes are not severe limitations for the application, but must be considered when designing an experiment. SPME theory is almost identical to SME theory, and in Chapters 3 and 4 we show clearly that the total amount of analyte that can be extracted from a water sample, *no matter how large the volume of the sample*, is limited by the distribution coefficient between the coating and water (the water/sorbent distribution ratio) and air and water (the Henry's law constant), as well as the limited capacity of the coating to dissolve analyte. These factors may result in competitive adsorption, especially for the porous adsorbent fibers. Basically, this means that a large amount of one analyte in the sample may prevent the adsorption of components present at lower concentrations. A second limitation results from the fact that the polymer is, in fact, a very viscous, gummy material. This means that prolonged periods (up to an hour or more) are needed for the analyte concentrations in the sample and polymer to come to equilibrium. In fact, in most cases the system does not come to equilibrium. This is not a major problem if very careful manual sampling or an autosampler are used for the procedure, since the method does not rely on exhaustive extraction, and reproducible results can be achieved if extraction conditions (extraction time, sample temperature, stirring rate, and salt concentrations) are reproduced exactly. Fiber coating thicknesses and types must be chosen carefully for a particular sample type, however. For example, if a sample contains not only relatively volatile but also relatively nonvolatile components, a problem could exist. If a thick polymer coating is used to extract the volatile components, prolonged fiber cleanup (using a heating block with helium sparge) of up to an hour may be needed to remove all the nonvolatile components from the fiber, since they will not be removed with the typical 5-min desorption time in the GC inlet normally used for SPME. Without complete thermal cleaning, carryover of nonvolatiles is likely. On the other hand, a 7-μm coating would extract and efficiently release the nonvolatiles upon injection, but might not retain the volatile analytes during the extraction.

Originally, problems with SPME involved mostly the fragility of the fibers and differences in extraction efficiencies between individual fibers. These problems have been largely overcome, especially with the advent of metal fiber cores and better production techniques. However, fiber lifetimes may still vary from only one use to approximately 100 uses. Another factor is that the fiber coatings are reported to be subject to degradation by high salt concentrations, required for maximizing extractions of very dilute volatile components with fiber–water partition coefficients less than 1000. The limited lifetime of the fibers must be taken into account when the cost per sample is considered, since each fiber costs between $85 and $170 (2007 prices).[21]

A modified SPME technique addresses the fragile fiber problem by replacing the coated fiber with a syringe needle trap. The inside of the needle is coated with an immobilized sorbent such as PDMS or even packed with a solid adsorbent such as Carboxen, with resulting sorbent volumes up to six times the extraction phase possible for SPME. Not only is the needle less fragile than a coated fiber, but dynamic sampling is possible, lending the term *solid-phase dynamic extraction* (SPDE) to this method.[22] SPDE can be used for direct (DI-SPDE) and headspace (HS-SPDE) microextractions with an autosampler.[23] In the analytical literature, this technique is also called *microextraction in a packed syringe* (MEPS).

1.2.5 Solvent Microextraction

Solvent microextraction, in its four main modes—single-drop microextraction (SDME), headspace single-drop microextraction (HS-SDME), hollow fiber–protected microextraction (HFME), and dispersive liquid–liquid microextraction (DLLME)—are the methods discussed here. The first two modes (SDME and HS-SDME) are also easily automated and can involve either static or dynamic sampling, which are explained fully later. SME is actually based on all of the methods discussed earlier in the chapter, but can best be compared directly to SPME, with which it shares the same basic operational theory. The technique can be traced to several articles: one by Liu and Dasgupta[24] which described a device for continuous monitoring of a stream of gas with a microliter volume extractant, and one by Jeannot and Cantwell describing a polymer rod device for extraction using an 8-μL drop at its tip.[1] Jeannot and Cantwell[25] and He and Lee[26] realized almost immediately, however, that replacing the polymer rod with a standard GC syringe would not only allow for extraction from a water sample, but the microdrop could then be withdrawn into the syringe and the extract injected directly into the GC for analysis. This technique, which Jeannot and Cantwell referred to as solvent microextraction, has been referred to variously as single-drop microextraction and liquid-phase microextraction[27,28] Thus, the solvent microdrop is used to extract and concentrate the analytes, while effectively cleaning up the sample and changing the solvent to one compatible with GC. The enrichment factor for this method can range up to 1000 or more, and the extraction times range from a few seconds to 1 h. The method is easily automated, allowing for precision, reduced labor and faster extraction times, using dynamic extraction, which will be described in detail in the next chapter. Soon, publications appeared extending the method to headspace samples[29] and even biological samples.[30] The method has even been extended to the use of derivatizing agents for analytes such as aldehydes and ketones in the extraction solvent, thus increasing the sensitivity and specificity of extraction.[31]

A major advantage of SME can be illustrated by comparing extractions using headspace extraction versus HS-SDME, which was introduced through a set of papers and presentations by Przyjazny, et al.,[29,32] Theis et al.,[33] Shen and Lee,[34] and Kokosa and Przyjazny[35] (See Section 7.3 and Figures 7.1 to 7.4 for the chromatograms for the following discussion.) The standard test method for gasoline diluent in used motor oils is ASTM method D 3525-93.[36] This method involves injecting a tetradecane solution of the oil directly into a GC for analysis. Obviously, this would present major contamination problems when using a GC with a capillary column. An alternative would be to analyze a headspace sample (Figure 7.1). The problem is that whereas volatile components of the gasoline are present at high concentrations in the headspace, higher-boiling components are present in decreasing amounts, resulting in a chromatogram very different from that resulting from the ASTM procedure. However, using tetradecane or hexadecane as the extracting solvent for HS-SDME results in a chromatogram (Figure 7.4) almost identical to the ASTM results.[35] This is because the higher-boiling components have increasingly larger oil/extracting solvent partition coefficients, resulting in higher extraction efficiencies for them.

In the hollow fiber–protected mode developed by Pedersen-Bjergaard and Rasmussen in 1999 and referred to by them as liquid–phase microextraction, the extracting system consists of a porous polypropylene hollow fiber, usually sealed at one end and containing between 4 and 20 µL of the extracting solvent.[30] HFME is actually a liquid–liquid membrane extraction which is very similar in principle to supported liquid membrane (SLM) methodology reported extensively in the literature by Jönsson and Mathiasson.[37–39] The porous polymer effectively prevents the biological matrix, including proteins, from contaminating the extractant. Two modes for HFME exist: a two-phase system developed by Shen and Lee[40] [which we refer to as HF(2)ME] in which the fiber (actually, a small-diameter polymer tube) is filled with an organic extraction solvent such as 1-octanol and a three-phase system [which we refer to as HF(3)ME]. The two-phase method, which is sometimes called microporous membrane liquid–liquid extraction, may be useful for water samples highly contaminated by solids such as silt.[41] The three-phase system, which is also called the SLM technique, has the polymeric hollow fiber saturated with an organic solvent, while the lumen of the fiber contains an aqueous phase (acceptor phase), usually acidic or basic. This method may be useful for extracting pharmaceuticals or metabolites from biological fluids.[42,43] The main disadvantage of HFME is that, at present, it is not easy to automate the method and the fiber extraction devices are constructed manually, resulting in reproducibility problems.[44,45] Despite these difficulties, however, the method has proved useful, especially for bioresearch and applications where drop stability can be a problem.[46] It may be noted that three-phase systems can also be accomplished without the use of porous hollow fibers.[47]

A recently developed mode, dispersive liquid–liquid microextraction, is actually based on the long-known technique of trituration used by synthetic chemists to purify chemicals. Thus, a contaminated chemical would be dissolved in a solvent such as ethanol or acetone and the solution rapidly pipetted into vigorously stirred water, which dispersed the water-insoluble chemical and allowed efficient extraction into the water of impurities. DLLME, first reported by Rezaee et al. in 2006,[48] involves dissolving 8.0 to 50 µL of a water-insoluble extraction solvent such as tetrachloroethene in 0.5 to 2.0 mL of a water-soluble solvent such as acetone, and rapidly injecting it into 5 mL of the water sample contained in a centrifuge tube.[49–51] The tube is then centrifuged and a portion of the extracting solvent is removed using a syringe and injected into a GC or concentrated, redissolved in acetonitrile, and injected into an HPLC. A major limitation of the technique has been that only solvents slightly soluble in water and denser than water, such as tetrachloroethene, carbon tetrachloride, carbon disulfide or chlorobenzene, could be used as extractants. This limitation has been partially circumvented in a modification developed by using a liquid extractant such as undecanol, which has a melting point close to room temperature and a density less than that of water.[52] By cooling the solution after centrifugation, the solidified drop can be removed from the vial, melted, and analyzed. The method appears not to have any major advantages over SDME or HS-SDME for extracting most samples for GC analysis. However, DLLME does have a major advantage over SDME when used for the preparation of HPLC samples for high-boiling nonpolar analytes such as polycyclic aromatic hydrocarbons (PAHs), PCBs, and pesticides, since the larger extraction volumes of water and solvent used result in the larger amounts of extracted analytes required by HPLC. The technique has also been used effectively with a complexing agent for the extraction of cadmium from water, followed by AA analysis.[53] One major disadvantage of this technique compared to SDME is the fact that several discrete steps must be taken, including centrifugation. This limits the method to semiautomation, since extraction and injection are not performed in one device.

SME theory shows that the amount of analyte extracted is almost directly proportional to the volume of the microdrop, but this has practical limitations for SDME and HS-SDME, since a drop larger than 3 µL in a stirred solution or headspace has a tendency to fall off the syringe needle. The practical drop size is therefore 1 to 2 µL, at least two to three times the volume of an SPME fiber coating.

Another major advantage of SME is that the extractant is renewed with each sampling, eliminating the problem of carryover possible with SPME, SPDE, and PT. Although SME is not a solventless technique, only microliters of solvent are actually used. Although one might think that traditional extraction solvents such as dichloromethane, chloroform, ethyl acetate, and ethyl ether would be the solvents of choice for SME, this is not the case. These solvents are too soluble in water to be used for SME in most cases. See chapter 4 for a more detailed discussion. As mentioned above, DLLME requires either a water-insoluble solvent denser than water or a high-melting liquid less dense than water. Volatile solvents such as dichloromethane cannot be used at room temperature or above for SDME, HS-SDME, or HFME, because they evaporate immediately in the headspace and/or dissolve in water. Thus, the most volatile solvents commonly used for SDME and HS-SDME extractions are toluene and the xylenes, which are used to extract analytes with higher boiling points, such as the PAHs. High-boiling solvents such as 1-octanol or tetradecane are used in turn for extracting volatile chemicals. One limitation of the method when using a solvent such as 1-octanol for extraction is that the GC injector must be set at a split from $10:1$ to $40:1$ to yield resolved peaks. On the other hand, this leads to sharp peaks with large signal-to-noise ratios. It has been observed that the split inlet, as with PT, tends to act somewhat as a jet separator, concentrating the analytes, and thus a $10:1$ split does not necessarily mean that 90% of the sample is lost. If a solvent lower boiling than the extracted analytes is used, traditional splitless injections are possible, with greater chromatographic sensitivity.

One might think that using an extractant such as tetradecane or even 1-octanol would pose difficulties in extracting more polar analytes, such as alcohols and acids. In fact, extraction efficiencies are directly dependent on the water/solvent distribution ratio and polar or hydrogen-bonding-capable analytes are extracted more efficiently with 1-octanol rather than tetradecane. However, even analytes containing polar functional groups have significant solubility in tetradecane at the very low concentrations typically encountered in SME. The water/1-octanol distribution coefficient, K_{ow}, is easily calculated (references and examples are given in Chapters 3 and 4) or available in the literature and distribution coefficients for other solvents, such as toluene or tetradecane, are also available in the literature or can be estimated fairly accurately from K_{ow}.[54] Before this technique is attempted, however, the reader is strongly advised to carry out the calculations illustrated in Chapter 4 to determine whether enough analyte will be extracted to meet the requirements needed for its quantitative determination.

One approach often taken to increase the amount of analyte extracted and thus analyzed is to increase the volume or mass of sample. This may not work for methods such as SME and SPME. The theory for SME clearly shows that the amount of analyte extracted depends on the size of the drop (~ 1 to $2\,\mu L$) (or fiber volume for SPME), the solvent, the salt concentration, the sample temperature, and most importantly, the water/solvent distribution coefficient (K_{ow}) and Henry's law constant (K_{aw}) for headspace extractions. For analytes with K_{ow} values below about 1000, the amount extracted does *not* increase linearly with sample volume and quickly reaches a maximum. See Chapters 4 and 5 for example calculations and plots. This means that for all practical purposes, the typical amount of water sample that should normally be extracted for chemicals up to the boiling point of naphthalene is 1 to 4 mL. Larger sample volumes (10 to 40 mL) often seen in the literature simply will not give better results. This is not the case for chemicals such as the PAHs, PCBs, and nonpolar pesticides, however, since theory shows that the extraction for chemicals with large K_{ow} values is essentially exhaustive, and larger sample volumes can be extracted effectively. However, it should be realized that the larger the volume extracted, the longer it will take the system to come to equilibrium. In addition, there is a practical limit to the actual amount of analyte that can be dissolved in 1 to 2 μL of the extracting solvent. This is a common limitation not only for SME, but also SPME, that is often forgotten. For example, in one comprehensive study, the experiment was designed to extract 1 μg of each of several PAH analytes into 1 μL of undecane.[55] Calculations for the study above might show that the PAHs should be extracted exhaustively, but this ignores the fact that as a practical matter, PAHs cannot dissolve completely in undecane at these concentrations.

One remaining major disadvantage of SME has recently been overcome. No new sampling method will be widely accepted unless it is sensitive, accurate, and precise and amenable to automation. SME has similar sensitivity, accuracy, and precision to the other methods listed in this chapter, but until recently it was not automated. This has been overcome by the development of computer programs allowing a commercial autosampler to carry out the SME extraction and injection of the sample into a GC or HPLC.[56] Similar programming should allow interfacing to commercial CE and AA systems. Automation of the extraction is important, not just because it removes the need for intense manual labor, but also because it allows the use of dynamic SME. As discussed above, SME, like SPME, is often not an exhaustive extraction process but an equilibrium process. Manual extraction, with practice, does give very reproducible results. However, to maximize the extraction of analytes, it is necessary for the extraction system to come to equilibrium, and this often can take 10 to 30 min or more. If, however, a dynamic extraction is used, equilibrium can be achieved in 10 min or less. For example, one variant of dynamic extraction involves depositing the drop at the tip of the needle withdrawing it into the needle, and repeating the process 10 to 30 times. This repeated extraction increases the flux of analyte through the surface of the drop into the interior and thus decreases equilibration time. Dynamic extraction is, however, practical only when using a computer-controlled autosampler, due to the need for accurate and precise timing and syringe plunger movement. The ability to conduct dynamic sampling, an option not available for SPME, decreases the time required to bring the extraction to equilibrium, thus increasing sample throughput.

1.3 SUMMARY

In this chapter we have presented a summary of some of the most commonly used sample preparation techniques. Each method has distinct advantages and disadvantages. For instance, liquid–liquid extraction is simple, straightforward, relatively sensitive, and technically simple. The method can be tedious and requires sizable amounts of sample as well as large amounts of hazardous and expensive solvents, which are its major limitations. Solid-phase extraction decreases the manual labor and solvent requirements dramatically and has good sensitivity, selectivity, accuracy, and precision. However, sample preparation is only semi-automated, requires 1 to 2 h per sample, and if analyte concentration through solvent evaporation is necessary, the method is limited to less volatile analytes.

Static headspace analysis is technically the simplest sample preparation method and fairly sensitive for volatile chemicals. However, it requires careful calibration to be useful for quantification. In addition, the method is of limited usefulness for the less volatile chemicals. Purge-and-trap, on the other hand, is one of the most sensitive sample preparation methods available for volatile chemicals, and has good accuracy and precision. The major disadvantages of PT are the possibilities of sample carryover and potential leaks in the system, as well as the expense of the instrumentation.

Solid-phase microextraction and solvent microextraction are very similar in both operational theory and practical method development, and potentially can replace or supplement the sample preparation methods discussed above. In many ways these two techniques are complementary. For instance, SPME may be the method of choice for sensitivity when analyzing very volatile chemicals if a Carboxen extraction fiber is used. On the other hand, SME may be the method of choice for sensitivity when analyzing nonpolar PAHs, PCBs, and pesticides. SME may also be the method of choice when analyzing samples containing a complex matrix or analytes with wide differences in boiling points, as discussed in Section 1.2.4.

In the remaining chapters we help the analyst to decide whether SME is an appropriate method to use for a sample preparation. In Chapter 2 we give a more detailed view of each SME mode and instrumentation requirement. Chapter 3 is a comprehensive examination of the theory for SME. Chapter 4 is a practical chapter, intended to introduce the analyst quickly to the basic example calculations resulting from SME theory and the suggested experimental conditions needed to develop an SME method. In Chapter 5 we then look in detail at each step of SME method development. In Chapter 6 we summarize the literature on SME, including recent developments in the field. Finally, in Chapter 7 we present a number of detailed, validated SME experimental procedures that the analyst can use as a starting point for a specific sample.

REFERENCES

1. Jeannot, M. A.; Cantwell, F. F., Solvent microextraction into a single drop. *Anal. Chem.* 1996, *68* (13), 2236–2240.
2. Pawliszyn, J., *Solid Phase Microextraction: Theory and Practice*, Wiley-VCH, New York, 1997.
3. Wercinski, S. A., *Solid Phase Microextraction: A Practical Guide*, CRC Press, Boca Raton, FL, 1999.
4. *Code of Federal Regulations* (CFR) 40, Part 136, revised as of July 1, 1995, Appendix A to Part 136: Methods for Organic Chemical Analysis of Municipal and Industrial Wastewater, method 625: Base/neutrals and acids.

5. *Code of Federal Regulations* (CFR) 40, Part 136, revised as of July 1, 1995, Appendix A to Part 136: Methods for Organic Chemical Analysis of Municipal and Industrial Wastewater; method 551.1: Determination of chlorination disinfection byproducts, chlorinated solvents and halogenated pesticides/herbicides in drinking water by liquid-liquid extraction and gas chromatography with electron-capture detection, revision 1.0.
6. Karlberg, B.; Thelander, S., Extraction based on the flow-injection principle: Description of the extraction system. *Anal. Chim. Acta* 1978, *98* (1), 1–7.
7. Bergamin F, H.; Medeiros, J. X.; Reis, B. F.; Zagatto, E. A. G., Solvent extraction in continuous flow injection analysis: determination of molybdenum in plant material. *Anal. Chim. Acta* 1978, *101* (1), 9–16.
8. Nord, L.; Bäckström, K.; Danielsson, L. G.; Ingman, F.; Karlberg, B., Extraction rate in liquid–liquid segmented flow injection analysis. *Anal. Chim. Acta* 1987, *194*, 221–233.
9. Lucy, C. A.; Cantwell, F. F., Kinetics of solvent extraction-flow injection analysis. *Anal. Chem.* 1989, *61* (2), 101–107.
10. Lucy, C. A.; Cantwell, F. F., Mechanism of extraction and band broadening in solvent extraction-flow injection analysis. *Anal. Chem.* 1989, *61* (2), 107–114.
11. Kubáň, V., Liquid–liquid extraction flow injection analysis. *Crit. Rev. Anal. Chem.* 1991, *22* (6), 477–557.
12. Thurman, E. M.; Mills, M. S., *Solid-Phase Extraction: Principles and Practice*, Wiley, New York, 1998.
13. Alexander, C.; Andersson, H. S.; Andersson, L. I.; Ansell, R. J.; Kirsch, N.; Nicholls, I. A.; O'Mahony, J.; Whitcombe, M. J., Molecular imprinting science and technology: a survey of the literature for the years up to and including 2003. *J. Mol. Recogn.* 2006, *19* (2), 106–180.
14. *Code of Federal Regulations* (CFR) 40, part 136: revised as of July 1, 1995, Appendix A to Part 136: Methods for Organic Chemical Analysis of Municipal and Industrial Wastewater, method 525.2: Determination of organic compounds in drinking water by liquid–solid extraction and capillary column gas chromatography/mass spectrometry, revision 2.0.
15. Russo, M. V.; Campanella, L.; Avino, P., Identification of halocarbons in the Tiber and Marta rivers by static headspace and liquid–liquid extraction analysis. *J. Sep. Sci.* 2003, *26* (5), 376–380.
16. *Code of Federal Regulations* (CFR) 40, Part 136, revised as of July 1, 1995, Appendix A to Part 136: Methods for Organic Chemical Analysis of Municipal and Industrial Wastewater, method 624: Purgeables.
17. Jakubowska, N.; Polkowska, Ż.; Namieśnik, J.; Przyjazny, A., Analytical applications of membrane extraction for biomedical and environmental liquid sample preparation. *Crit. Rev. Anal. Chem.* 2005, *35* (3), 217–235.
18. Creaser, C. S.; Lamarca, D. G.; Freitos dos Santos, L. M.; New, A. P.; James, P. A., A universal temperature controlled membrane interface for the analysis of volatile and semi-volatile organic compounds. *Analyst* 2003, *128* (9), 1150–1156.
19. *Code of Federal Regulations* (CFR) 40, Part 136, revised as of July 1, 1995, Appendix A to Part 136, Methods for Organic Chemical Analysis of Municipal and Industrial Wastewater, method 524.2: Measurement of purgeable organic compounds in water by capillary column gas chromatography/mass spectrometry, revision 4.0.
20. Chapter 467, Residual solvents, *Pharmacop Forum.* 2007, *33* (3), 1.
21. *Supelco Catalog*, 2007.
22. Jochmann, M. A.; Kmiecik, M. P.; Schmidt, T. C., Solid-phase dynamic extraction for the enrichment of polar volatile organic compounds from water. *J. Chromatogr. A* 2006, *1115* (1–2), 208–216.

23. Ridgway, K.; Lalljie, S. P. D.; Smith, R. M., Comparison of in-tube sorptive extraction techniques of non-polar volatile organic compounds by gas chromatography with mass spectroscopic detection. *J. Chromatogr. A* 2006, *1124* (1–2), 181–186.

24. Liu, S.; Dasgupta, P. K., Liquid droplet: a renewable gas sampling interface. *Anal. Chem.* 1995, *67* (13), 2042–2049.

25. Jeannot, M. A.; Cantwell, F. F., Mass transfer characteristics of solvent extraction into a single drop at the tip of a syringe needle. *Anal. Chem.* 1997, *69* (2), 235–239.

26. He, Y.; Lee, H. K., Liquid-phase microextraction in a single drop of organic solvent by using a conventional microsyringe. *Anal. Chem.* 1997, *69* (22), 4634–4640.

27. Basheer, C.; Lee, H. K., Analysis of endocrine disrupting alkylphenols, chlorophenols and bisphenol-A using hollow fiber–protected liquid-phase microextraction coupled with injection port-derivatization gas chromatography–mass spectrometry. *J. Chromatogr. A* 2004, *1057* (1–2), 163–169.

28. Psillakis, H. K.; Kalogerakis, N., Developments in single-drop microextraction. *Trends Anal. Chem.* 2002, *21* (1), 53–63.

29. Przyjazny, A.; Austin, J. F.; Essenmacher, A. T., *Headspace liquid-phase microextraction: a novel preconcentration technique for volatile organic pollutants. Proc. 6th Polish Conference on Analytical Chemistry,* Gliwice, Poland, 2000, *2*, 135–136.

30. Pedersen-Bjergaard, S.; Rasmussen, K. E., Liquid–liquid–liquid microextraction for sample preparation of biological fluids prior to capillary electrophoresis. *Anal Chem.* 1999, *71* (14), 2650–2656.

31. Deng, C.; Yao, N.; Li, N.; Zhang, X., Headspace single-drop microextraction with in-drop derivatization for aldehyde analysis. *J. Sep. Sci.* 2005, *28* (17), 2301–2305.

32. Przyjazny, A.; Kokosa, J. M., Analytical characteristics of the determination of benzene, toluene, ethylbenzene and xylenes in water by headspace solvent microextraction. *J. Chromatogr. A* 2002, *977* (2), 143–153.

33. Theis, A. L.; Waldack, A. J.; Hansen, S. M.; Jeannot, M. A., Headspace solvent microextraction. *Anal. Chem.* 2001, *73* (23), 5651–5654.

34. Shen, G.; Lee, H. K., Headspace liquid-phase microextraction of chlorobenzenes in soil with gas chromatography–electron capture detection. *Anal. Chem.* 2003, *75*(1), 98–103.

35. Kokosa, J. M.; Przyjazny, A., Headspace microdrop analysis: an alternative test method for gasoline diluent and benzene, toluene, ethylbenzene and xylenes in used engine oils. *J. Chromatogr. A* 2003, *983* (1–2), 205–214.

36. Standard Test Method for Gasoline Diluent in Used Gasoline Engine Oils by Gas Chromatography, ASTM Test Method D3525-93, American Society for Testing and Materials, West Conshohockin, PA.

37. Jönsson, J. Å.; Mathiasson, L., Supported liquid membrane techniques for sample preparation and enrichment in environmental and biological analysis. *Trends Anal. Chem.* 1992, *11* (3), 106–114.

38. Jönsson, J. Å.; Mathiasson, L., Liquid membrane extraction in analytical sample preparation: I. Principles. *Trends Anal. Chem.* 1999, *18* (5), 318–325.

39. Jönsson, J. Å.; Liquid membrane techniques. In J. Pawliszyn, ed., *Sampling and Sample Preparation for Field and Laboratory*, Elsevier, New York, 2002, pp. 503–530.

40. Shen, G.; Lee, H. K., Hollow fiber-protected liquid-phase microextraction of triazine herbicides. *Anal. Chem.* 2002, *74* (3), 648–654.

41. Fontanals, N.; Barri, T.; Bergström, S.; Jönsson, J. Å., Determination of polybrominated diphenyl ethers at trace levels in environmental waters using hollow–fiber microporous membrane liquid–liquid extraction and gas chromatography–mass spectrometry. *J. Chromatogr. A* 2006, *1133* (1–2), 41–48.

42. Hou, L.; Lee, H. K., Dynamic three-phase microextraction as a sample preparation technique prior to capillary electrophoresis. *Anal. Chem.* 2003, *75* (11), 2784–2789.
43. Kuuranne, T.; Kotiaho, T.; Petersen-Bjergaard, S.; Rasmussen, K. E.; Leinonen, A.; Westwood, S.; Kostiainen, R., Feasibility of a liquid-phase microextraction sample clean–up and liquid chromatographic/mass spectrometric screening method for selected anabolic steroid glucuronides in biological samples. *J. Mass Spectrom.* 2003, *38* (1), 16–26.
44. Hou, L.; Shen, G.; Lee, H. K., Automated hollow fiber–protected dynamic liquid–phase microextraction of pesticides for gas chromatography–mass spectrometric analysis. *J. Chromatogr. A* 2003, *985* (1–2), 107–116.
45. Ouyang, G.; Pawliszyn, J., Kinetic calibration for automated hollow fiber–protected liquid-phase microextraction. *Anal. Chem.* 2006, *78* (16), 5783–5788.
46. Pawliszyn, J.; Pedersen-Bjergaard, S., Analytical microextraction: current status and future trends. *J. Chromatogr. Sci.* 2006, *44* (6), 291–307.
47. Cantwell, F. F.; Losier, M., Liquid–liquid extraction. In J. Pawliszyn, ed., *Sampling and Sample Preparation for Field and Laboratory,* Elsevier, New York, 2002, pp. 297–340.
48. Rezaee, M.; Assadi, Y.; Milani Hosseini, M. R.; Aghaee, E.; Ahmadi, F.; Berijani, S., Determination of organic compounds in water using dispersive liquid–liquid micro-extraction. *J. Chromatogr. A* 2006, *1116* (1–2), 1–9
49. Berijani, S.; Assadi, Y.; Anbia, M.; Milani Hosseini, M. R.; Aghaee, E., Dispersive liquid–liquid microextraction combined with gas chromatography–flame photometric detection. Very simple, rapid and sensitive method for the determination of organophosphorus pesticides in water. *J. Chromatogr. A* 2006, *1123* (1), 1–9.
50. Kozani, R. R.; Assadi, Y.; Shemirani, F.; Milani Hosseini, M. R.; Jamali, M. R., Part-per–trillion determination of chlorobenzenes in water using dispersive liquid–liquid microextraction combined gas chromatography–electron capture detection. *Talanta* 2007, *72* (2), 387–393.
51. Farajzadeh, M. A.; Bahram, M.; Jönsson, J. Å., Dispersive liquid–liquid microextraction followed by high performance liquid chromatography–diode array detection as an efficient and sensitive technique for determination of antioxidants. *Anal. Chim. Acta* 2007, *591* (1), 69–79.
52. Zanjani, M. R. K.; Yamini, Y.; Shariati, S.; Jönsson, J. Å., A new liquid-phase micro-extraction method based on solidification of floating organic drop. *Anal. Chim. Acta* 2007, *585* (2), 286–293.
53. Zeini Jahromi, E.; Bidari, A.; Assadi, Y.; Milani Hosseini, M. R.; Jamali, M. R., Dispersive liquid–liquid microextraction combined with graphite furnace atomic absorption spectrometry: Ultra trace determination of cadmium in water samples. *Anal. Chim. Acta* 2007, *585* (2), 305–311.
54. Schwarzenbach, R. P.; Gschwend, P. K.; Imboden, D. M., *Environmental Organic Chemistry,* 2nd edition, Wiley-Interscience, Hoboken, NJ., 2002, pp. 213–244.
55. Ouyang, G.; Zhao, W.; Pawliszyn, J., Automation and optimization of liquid-phase microextraction by gas chromatography. *J. Chromatogr. A* 2007, *1138* (1–2), 47–54.
56. Kokosa, J. M., Automation of liquid phase microextraction, U.S. patent 7,178,414 B1, Feb. 20, 2007.

CHAPTER 2

BASIC MODES OF OPERATION FOR SOLVENT MICROEXTRACTION

2.1 BASIC PRINCIPLES OF SME

2.1.1 Introduction

Solvent microextraction (SME) is based on classic solvent extraction, with the exception that the extraction solvent volumes are limited to a range of 1 μL to several hundred microliters, rather than 1 L to several liters. The most commonly used techniques consist of extraction of 1 to 25 mL of aqueous sample with a single microliter solvent drop at the tip of a syringe or with 3 to 20 μL of solvent contained in a microporous hollow tube. Extraction can consist of direct immersion of the extracting solvent in the sample or exposure of the solvent to the headspace of an aqueous, liquid, or solid sample. There are several variants of these two general techniques, some more practical or popular than others, each with advantages and disadvantages. A full listing of SME modes is available in the Appendix. The most commonly used extraction modes (Table 2.1) are covered in some detail in the following sections of this chapter. Chapter 3 covers the theory necessary to understand SME, and in Chapter 4 we present example calculations for commonly used extraction modes. SME method development and specific literature applications are covered in Chapters 5 and 6, respectively. In Chapter 7 we present a detailed description of selected experiments for commonly used SME methods. See these

Solvent Microextraction: Theory and Practice, By John M. Kokosa, Andrzej Przyjazny, and Michael A. Jeannot
Copyright © 2009 John Wiley & Sons, Inc.

TABLE 2.1 Commonly Used Solvent Microextraction Modes

	Definition	Section
Direct immersion modes		
SDME	Single-drop microextraction	2.2.1.1
DLLME	Dispersive liquid–liquid ME	2.2.1.2
HF(2)ME	Hollow fiber–protected two-phase solvent ME	2.2.1.3
HF(3)ME	Hollow fiber–protected three-phase solvent ME	2.2.1.4
Headspace mode		
HS-SDME	Headspace single-drop ME	2.2.2.1
HS-HF(2)ME	Headspace hollow fiber–protected two-phase SME	2.2.2.2
Dynamic modes		
In-syringe	Repeated withdrawal of sample into syringe	2.2.3
In-needle	Repeated withdrawal of solvent into needle	2.2.3

chapters for more detailed lists of references. We provide only a general list in this chapter.

2.1.2 Comparison of Classical Solvent Extraction and SME

Classic solvent extraction utilizes a separatory funnel ranging from 10 mL to 6 L in size, allowing for the extraction of water samples ranging from 5 mL to several liters in volume. Some environmental procedures also use a Soxhlet or continuous extractor, which can also extract water or solid samples. More recent environmental procedures reduce the sample volume to 40 to 50 mL and extraction solvent volume to 1 to 2 mL. Typical classical extraction solvents include ethyl ether, ethyl acetate, methylene chloride, chloroform, hexane, and toluene. The extraction solvents are normally reduced in volume to 1 to 2 mL before analysis, necessitating low-boiling extractants. Typically, only 1 to 10 μL of the concentrate is analyzed, however, resulting in little overall increase in analyte concentration. Often, a salt is added to water samples to decrease analyte (and extraction solvent) solubility in the water and increase extraction efficiency. Analyte derivatization (methylation, acetylation, and silanization) is also used to decrease analyte polarity or increase analyte volatility. The greatest advantages of classical extraction techniques are that they are relatively simple and quite reproducible. The obvious disadvantages are the use of large amounts of sometimes toxic extraction solvents and the necessity to use relatively large (10 mL to several liters) volumes of sample.

In principle, SME is nearly identical to classical solvent extraction and has been used to carry out very similar extractions of a wide range of organic and inorganic analytes. However, as the name implies, SME uses very little extraction solvent and greatly reduced sample amounts. Since all or a large portion of the extracting solvent is injected into the analytical instrument, analyte enrichment can range from 10- to 1000-fold, necessitating sample volumes of only 1 to 25 mL. In many cases a sample size of 1 mL is sufficient, but as little as 10 μL of sample can be used in one variant of SME. Extraction equipment requirements are also often very simple. Extractions can be carried out manually using a standard GC syringe containing the extraction solvent and a 2- to 40-mL sample vial with a septum closure. Typically, the sample is stirred with a magnetic stir bar. Both direct immersion and headspace extractions can be accomplished in this way. Aqueous samples that contain very complex or dirty matrices (biological or silt-laden samples), however, are best extracted using either headspace extraction or by direct immersion with a protective sleeve to protect the solvent from contamination, or dislodgement of the solvent drop from the syringe needle by particulate or protein matter. The protective sleeve is usually a 2- to 3-cm hollow tube of microporous polypropylene. The extracting solvent can vary widely for this mode and can include acidic or basic aqueous solutions to extract basic or acidic analytes, respectively. Finally, SME is very suitable for the use of derivatization techniques to decrease analyte polarity or increase analyte volatility. Derivatization can be carried out in the sample, in extracting solvent, and in the injection port of a GC using standard derivatizing reagents.

There are several experimental parameters, detailed in the following chapters, which are crucial to a successful SME procedure. These include sample volume, extraction solvent volume, solvent type, agitation, temperature, incubation and extraction time, and static versus dynamic extraction. Since most of these parameters need to be controlled carefully, computer-controlled automation greatly simplifies the technique, although very good accuracy and precision can be achieved by manual extraction.

2.2 EXTRACTION MODES

It can be said that a good idea breeds imitation, but a really good idea breeds innovation. That is certainly true for SME. Since Jeannot and Cantwell, and Lee separately and simultaneously recognized that a standard GC syringe could be used for SME, the variations for the method have exploded. Each variation, or mode, has been given separate acronyms by the developing authors, but unfortunately, more than one designation is often used for the same mode. We try to simplify and codify each mode as much as possible by a unique designation. In this chapter we discuss only three of the most common, and easiest to use, SME modes: single-drop microextraction (SDME), hollow fiber–protected microextraction (HFME), and dispersive liquid–liquid microextraction. A complete list of applications involving all the various SME modes and their variants is given in Chapters 5 and 6.

There are several ways that SME extraction modes can be classified, including the number of phases involved in the extraction (see the Appendix), but in general all mode classifications fall into two main categories: direct immersion sampling (DI) and headspace sampling (HS). The most commonly used solvent microextraction DI and HS mode involves the extraction of a sample with a single microdrop of solvent suspended at the tip of a syringe. This mode has been referred to variously as single-drop microextraction or liquid-phase microextraction (LPME). Since the acronym LPME has been used for other modes as well, the term *SDME* will be used here. The prefix HS- will be added when headspace sampling is used. The second most widely used mode involves the use of a microtube of porous polymer, usually polypropylene, to contain the extracting solvent. The walls of a microtube, usually referred to as a hollow fiber (HF), are saturated with a water-immiscible solvent and the lumen filled with the same solvent or with an aqueous solution as the final extractant. If the fiber pores and interior contain only the water-immiscible solvent, the extraction is essentially two-phase. If the lumen is filled with an aqueous final extractant, the extraction is three-phase. The modes using the hollow fiber have been referred to variously as liquid-phase micro-extraction (LPME), supported liquid membrane extractions (SLM or SLMME), hollow fiber–supported liquid membrane extraction (HF-SLM), solvent bar micro-extraction (SBE), or even liquid–liquid–liquid microextraction (LLLME). To avoid this confusion, the terms HF(2)ME and HF(3)ME will be used for two- and three-phase modes respectively, with the prefix HS- added for headspace extraction. A third simple and promising mode involves dispersal of a small amount of water-insoluble solvent (dissolved in a small amount of water-soluble solvent) in 5 to 10 mL of aqueous sample. The solvent is separated, by centrifugation or by solidifi-cation after cooling, and removed for analysis. This mode will be referred to as dispersive liquid–liquid microextraction (DLLME). In addition, the SDME and HF modes can be carried out under either static or dynamic extraction conditions. In dynamic extraction, the solvent is agitated separately during extraction. An example would be to move the sample drop in and out of the syringe barrel during extraction for continuous exposure of a new solvent surface to the sample.

2.2.1   Direct-Immersion Modes

Each of the following SME modes involves extraction of analytes from aqueous samples using a small amount of solvent, which can be an exposed drop at the tip of a syringe needle, contained within a microporous membrane, or even dispersed within the sample. The hollow fiber mode can also use acidic or basic water or water-soluble solutions as the final extraction medium to extract basic or acidic analytes. The hollow-fiber mode has also been used with microelectrodes, to de-crease extraction times for biological analytes and drugs—essentially a miniature electrophoresis extraction. Derivatization reagents can also be added to the sample or to the extracting solvent to facilitate extraction.

2.2.1.1   *Single-Drop Microextraction*   Much of the development work for SDME was conducted by Jeannot and Cantwell[1] at the University of Alberta, Canada, and by Lee's group at the National University of Singapore.[2] Refer to Figures 1.2, 4.1, and 4.2 for typical manual and autosampler equipment setups. Direct-immersion single-drop microextraction is most useful for extracting rela-tively nonpolar and semivolatile analytes from water samples that contain little or no particulate matter. A static extraction can be performed manually, although a computer-controlled autosampler is preferred when running many samples. Dynamic extraction is best accomplished with a computer-controlled autosampler.

The technique involves filling a 5- to 10-μL GC syringe (a No. 2 needle point style works best) with 1 to 3 μL of water-insoluble solvent, inserting the needle through a polytetrafluoroethylene (PTFE)-coated septum into a 2- to 40-mL vial and exposing the solvent as a drop of liquid. Magnetic stirring (PTFE-coated stir bar) during the extraction shortens the required extraction time dramatically. Stirring rates up to about 600 rpm can be used. Higher stirring rates can dislodge the drop. If an orbital agitator is used, the agitator must be turned off during extraction or the drop will be displaced from the syringe needle. Dynamic extraction (discussed later) allows orbital agitation while the solvent is contained within the syringe. Addition of sodium chloride or sodium sulfate to the sample solution often increases extraction efficiency. This is especially useful for extraction of smaller and more polar molecules. In a few cases, addition of salt can have a negative effect. If salt is added, use a constant concentration of the salt. A concentration of 300 mg NaCl/mL water sample is useful. Saturated solutions run the risk of having undissolved particles present that can dislodge the drop. Saturated solutions of sodium sulfate can result in the hydrate of the salt solidifying on sitting.

A recent modification of the method developed by Wu's group at the National Sun Yat-Sen University, Taiwan, referred to as drop-to-drop extraction, is useful when dealing with very small amounts of sample solution, such as biological samples (Figure 2.1).[3] Approximately 10 μL of sample in a conical-bottomed microvial is extracted with 0.5 to 1 μL of solvent. The limitation of the method is the limit of detection for the analytes, but it does provide a quick and simple method for extracting drugs from blood, serum, and urine.

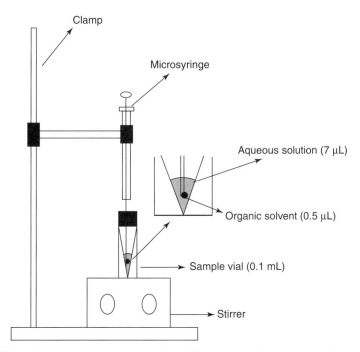

FIGURE 2.1. Drop-to-drop microextraction. (Reprinted with permission from Ref. 3; copyright © 2006 American Chemical Society.)

There are some practical matters concerning the extracting solvent, which are often overlooked by those using this technique. Some of these are expanded on in Section 5.6. First, the solvent must have low solubility in water, or it will dissolve. Thus, ethyl acetate, ethyl ether, and chloroform, while excellent extraction solvents when using a separatory funnel, are not practical solvents for microextraction in most cases because they are too soluble in water. In addition, these solvents are so low-boiling that a 1- to 4-µL drop will evaporate in less than a minute. Even some higher-boiling solvents will form a gaseous bubble within the drop at elevated temperatures, leading to irreproducible extraction results. See Chapter 5 for a complete list of SME solvents and their physical–chemical properties.

Second, in the single-drop method, when the drop is extended out onto the tip of the needle, the solvent tends to creep up and around the needle, and when the drop is withdrawn, not all of the liquid can be withdrawn back into the needle. The Hamilton No. 2 needle point works best at retaining the drop, since the curved bevel of the needle point provides a larger surface area for the drop to cling to than do flat-cut needles. On the downside, the Hamilton needle can core injection port septa quickly if the fragile tip is bent. About 95% of the drop is recovered with the curved bevel tip, and about 90% with the flat bevel tip. A standard GC autosampler tip has very little surface area, and the solvent drop is easily dislodged during extraction. Therefore, the use of a standard autosampler syringe for SDME is strongly discouraged.

In any case, when the drop is withdrawn into the syringe after extraction, some of the solvent is lost, by a combination of wicking up the needle, evaporation, and water solubility. Thus, if a 2-µL drop is used for an extraction and 2 µL is then withdrawn into the syringe for analysis, you are actually injecting a fraction of a microliter of sample water, which may be contaminated with soil, protein, salt, and other nonextractable impurities harmful to the GC injector and column. The solution to this problem is to use a drop size 0.2 to 0.5 µL larger than the amount of solvent drawn back into the syringe, leaving a small amount of solvent at the tip of the needle which acts as a protective barrier against contamination by the sample solution. In some cases, addition of salt to the water solution will not only enhance extraction but also decrease the solubility of the extracting solvent.

2.2.1.2 Dispersive Liquid–Liquid Microextraction

Developed by Assadi's group at the Iran University of Science and Technology, DLLME is one of the simplest of all extraction modes (Figure 2.2).[4] However, due to the need for several manual manipulations, the technique cannot at present be completely automated. Between 10 and 50 µL of water-insoluble solvent is dissolved in 0.5 to 2 mL of methanol, acetone, or other water-soluble solvent, and added quickly with a syringe to 5 to 10 mL of water sample. If a solvent heavier than water (such as carbon tetrachloride or carbon disulfide) is used, a conical-bottomed vial or centrifuge tube is used and the sample is centrifuged to separate the extraction solvent, which is removed with a syringe and analyzed.

A modification of the technique developed by Yamini's group at the Tarbiat Modarres University, Iran, uses a solvent less dense than water, such as 1-undecanol

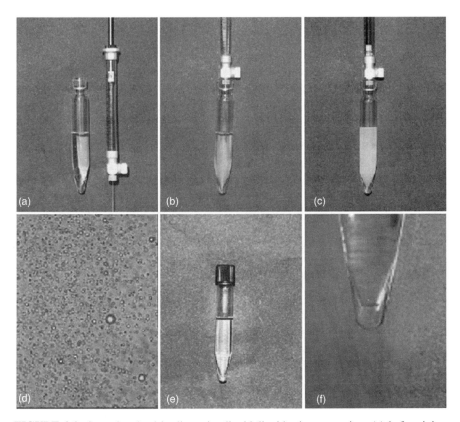

FIGURE 2.2. Steps involved in dispersive liquid–liquid microextraction: (a) before injection of mixture of disperser solvent (acetone) and extraction solvent (C_2Cl_4) into sample solution; (b) beginning of injection; (c) end of injection; (d) optical microscopic photography, magnification 1000 (shows fine droplets of C_2Cl_4 in cloudy state); (e) after centrifuging; (f) enlarged view of sedimented phase (5.0 $\pm$ 0.2 μL). (Reprinted with permission from Ref. 4; copyright © 2006 Elsevier.)

or hexadecane, which will solidify when the sample is cooled in an ice bath.[5] The solidified extractant is removed from the sample with tweezers, allowed to warm and re-liquefy, and then analyzed. This extraction mode is very effective for extraction of analytes, such as the PAHs and PCBs, which have large organic solvent–water partition constants (K_{ow}).

2.2.1.3 Hollow Fiber–Protected Two-Phase Solvent Microextraction

This technique, first utilized by Lee's group at the National University of Singapore, involves the use of a small-diameter microporous polypropylene tube (a hollow fiber) to contain the organic extracting solution (Figure 2.3).[6] This two-phase technique is often referred to in the literature as liquid-phase microextraction (LPME), which is confusing, since this is the same designation as that often used for SDME as well as for the corresponding three-phase hollow fiber–protected mode

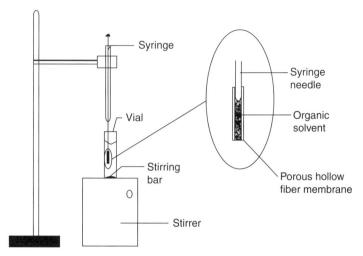

FIGURE 2.3. Hollow fiber–protected two-phase microextraction system. (Reprinted with permission from Ref. 6; copyright © 2002 American Chemical Society.)

(see Section 2.2.1.4). Thus, the hollow fiber–protected two-phase microextraction mode is referred to here as HF(2)ME. The HF(2)ME mode has the advantages that the solvent cannot be dislodged and lost, as with SDME, and very rapid stirring (rates up to 1600 rpm) can be used. Also, larger extractant volumes (4 to 20 μL) are used, with resulting increased extraction efficiencies. The membrane also protects the solvent from particulate matter or soluble polymeric material (humus or proteins) present in the sample. The major disadvantages to the method are that longer extraction times are often required (20 to 60 min vs. 5 to 15 min with SDME) and that unless a large-volume injector is used for the analysis, not all of the extract is analyzed. In addition, the method is not easily fully automated, and fibers must be sized and prepared individually from available bulk commercial materials. The most commonly used fiber is Accurel Q 3/2 polypropylene microporous hollow tubing (Membrana GmbH, Wuppertal, Germany), which has an inner diameter of 600 μm, a wall thickness of 200 μm, a pore size of 0.2 μm, and a wall volume porosity of about 70%.

HF(2)ME is usually conducted with sample amounts in the range 3 to 20 mL. Actually the fibers are, of course, small-diameter polypropylene microporous tubing. To conduct an extraction, tubing of the appropriate inner diameter (usually, 0.6 mm) must be cut to an appropriate length (3 cm is usual for a 5-mL water sample, although lengths of up to 50 cm have been used with larger sample amounts) and cleaned with acetone, often using sonication. Normally, one end is sealed to provide an enclosed membrane through which the analytes must migrate into the solvent, although the ends can be left open in some variants of the mode. The fiber end can be sealed by crimping mechanically, heat-sealing with a flame or soldering iron, tying with wire, or plugging with wire or other material. The open end of the fiber is attached to an autosampler-style syringe needle either before or after the fiber is immersed for a few seconds in the extracting solvent, to saturate the pores of the polymer. The attached syringe contains the same extraction solvent. The fiber is then inserted and held in place through a hole in the septum and the cap attached to the sample vial. In some cases the fiber is attached to the septum without also attaching the syringe. Extracting solvent is then injected into the fiber with a syringe. The internal capacity of a fiber is approximately 2.8 μL/cm of length. Extraction times are much longer for this mode than SDME techniques, since the analytes must pass through the fiber into the bulk extracting solvent. After extraction, the solvent is removed with the syringe for analysis. Unless a large-volume injector is used, however, only a fraction of the extract is analyzed. If the sample will be analyzed using HPLC, the solvent may have to be evaporated and changed to one compatible with high-performance liquid chromatography.

HF(2)ME has been adapted for use with an autosampler (Figure 2.4) by Pawliszyn's group at the University of Waterloo, Canada.[7] The open end of the fiber was attached to the end of a micropipette, which was in turn held in place in a hole in the vial septum. The autosampler syringe could then be guided into the fiber through the pipette tip for addition and withdrawal of the organic solvent. However, the fiber

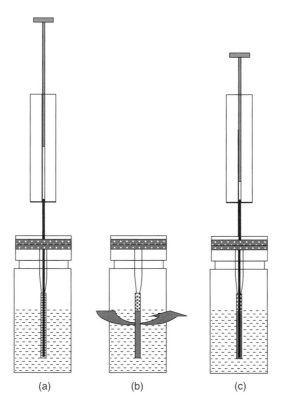

(a)   (b)   (c)

FIGURE 2.4. Automation of the HF(2)ME procedure: (A) filling of extraction solvent; (B) agitation; (C) withdrawing the solvent to syringe and injection to GC. (Reprinted with permission from Ref. 7; copyright © 2006 American Chemical Society.)

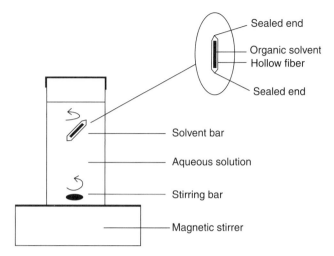

FIGURE 2.5. Solvent bar microextraction. (Reprinted with permission from Ref. 8; copyright © 2004 American Chemical Society.)

and pipette tip devices must still be prepared manually, and this is labor intensive for large numbers of samples.

Finally, if the fiber is filled with solvent and sealed at both ends, it can be placed directly into a stirred solution for extraction (Figure 2.5). The fiber is then retrieved, the polymer punctured by a syringe needle, and the solvent removed for analysis. Lee has coined the term *solvent bar microextraction* for this mode.[8] This technique can also be used as a three-phase system, as in HF(3)ME discussed below.

2.2.1.4 Hollow Fiber–Protected Three-Phase Solvent Microextraction

As the name HF(3)ME implies, this mode involves three liquid phases: the aqueous sample, a water-immiscible organic phase that fills the pores of the hollow fiber polymer, and an aqueous acceptor phase in the lumen of the fiber (Figure 2.6). This technique was introduced and has been explored thoroughly by Pedersen-Bjergaard and Rasmussen's research group at the University of Oslo, Norway.[9] Other groups that contributed heavily to early development of this method include Andrews, de Jager, and Kramer at Ohio University,[10,11] Lee's group at the National University of Singapore,[12] and Psillakis and Kalogerakis[13] at the Technical University of Crete. Since the final acceptor solution is aqueous, this technique is used to extract water-soluble analytes from water samples. Thus, to extract compounds with basic (amine) groups, the pH of the sample solution is made basic to suppress ionization of the amine and allow it to dissolve into the organic solvent within the walls of the fiber. The amine is then drawn irreversibly into the interior acceptor solution, which is acidic, which causes the amine to ionize and become insoluble in the organic layer. Acidic analytes are extracted by reversing the pH values of the sample and acceptor solutions. A modification of this method utilizes ion-pairing reagents to complex with a polar analyte, allowing it to pass through the organic barrier into the acceptor solution.[14]

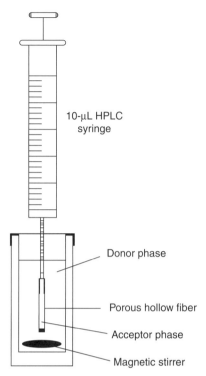

10-µL HPLC
syringe

Donor phase

Porous hollow fiber

Acceptor phase

Magnetic stirrer

FIGURE 2.6. Hollow fiber-protected three-phase microextraction. (Reprinted with permission from Ref. 12; copyright © 2001 Elsevier.)

The HF(3)ME procedure is carried out in a fashion similar to HF(2)-SME. Usually, one end is sealed (an early method was to curve the fiber into a U-shape so that both ends exited the sample) and the pores impregnated with a water-immiscible solvent, such as 1-octanol. A syringe containing the aqueous acceptor solution is attached to the open end of the fiber and the aqueous phase introduced. The fiber is placed in the stirred sample for extraction. The acceptor solution is then withdrawn by syringe and analyzed by capillary electrophoresis, HPLC, or atomic absorption spectroscopy.

As with HF(2)ME, the major disadvantages of the method are long extraction times (20 to 60 min or more), lack of full automation, and the manual labor required to prepare the fibers individually. Despite these disadvantages, this technique, along with the HF(2)ME method, is a powerful tool for the extraction of pharmaceuticals, metabolites, and drugs of abuse from biological media and environmental contaminants from dirty matrices. It is therefore finding wide application.

The term *electromembrane extraction* (EME) has been coined by Pedersen-Bjergaard and Rasmussen group.[15] This technique is a modification of HF(3)ME and involves inserting a small electrode inside the hollow fiber (Figure 3.8). When a potential is applied across the membrane with a second electrode in the sample solution, extraction times may be reduced to as little as 5 min. While requiring additional experimental apparatus, including an appropriate power supply, this technique may prove very useful for the rapid analysis of biological samples.

2.2.2 Headspace Modes

At the present time, headspace solvent microsampling is conducted almost universally using the single-drop method, which we refer to as HS-SDME. However, the hollow-fiber modes have also been used in a few cases for headspace sampling.

2.2.2.1 Headspace Single-Drop Microextraction   HS-SDME is accomplished by exposing a microdrop of solvent to the headspace of a sample. The sample can be aqueous, solid, a nonvolatile liquid such as motor oil, or a gas sample. Contributors to the early development of this technique include Kokosa and Przyjazny's group at Kettering University, Michigan,[16,17] Jeannot's group at St. Cloud State University, Minnesota,[18] Lee's group at the National University of Singapore,[19] Yamini's group at Tarbiat Modarres University, Iran,[20] and Xu's group at the Chinese Academy of Sciences.[21]

The equipment needs for this mode are essentially the same as for direct-immersion SDME, with the exception that the drop is exposed to the headspace of the sample. Normally, stirring rates are set as high as possible, without splashing, and elevated temperatures are more commonly used. See Figures 1.2, 4.1, and 4.2 for typical extraction setups.

The concentration of an analyte in the headspace over a water solution depends on the Henry's law constant (K_H), which is a ratio of the analyte concentration in the headspace to the concentration in the sample. This is dependent, in turn, on the boiling point and water solubility of the analyte. The analyte partitions between the headspace and the extracting solvent. In Chapters 3 and 4 we detail the theory and practical aspects of headspace sampling. The rate of partition between the headspace and the extracting solvent is very rapid in most cases, so that the partition constant between the water and the headspace often determines the effectiveness of this technique. Headspace sampling has several advantages over SDME. First, contaminants that are not volatile, such as silt, salt, and very polar chemicals, will not contaminate the extracting solvent. Second, headspace extractions usually require much less time than direct immersion modes. A HS-SDME extraction usually requires between 3 and 15 min, compared to 5 to 60 min for direct-immersion SDME extractions.

Headspace extractions for volatile nonpolar analytes such as benzene or chloroform can be carried out readily at room temperature. Higher-boiling or more polar analytes such as PCBs and PAHs will require elevated temperatures. It should be remembered, however, that if elevated temperatures are used, the drop temperature will also increase during the extraction. This will affect negatively the partition between the drop and the headspace. Thus, a compromise extraction temperature must be found by experimentation leading to the best extraction efficiency.

As with direct-immersion SDME, the addition of salt to an aqueous sample will increase extraction efficiency and decrease evaporation of the drop that would result from its solubility in the sample solution (see Section 3.2.6 for an example).

Solids such as pharmaceutical tablets, polymers, food, or soil can be analyzed in their natural form, ground into a fine powder, or dispersed in water. Very volatile components can be lost during a grinding process, however. Soil is often treated with a small amount of acetone or methanol prior to the addition of water, to aid in release of volatile components. Liquids other than water can also be analyzed for volatile components. As an example, motor oil can be analyzed for the presence of gasoline contamination by HS-SDME (see Chapter 7).

2.2.2.2 Hollow Fiber–Protected Headspace Extraction Headspace extractions can also be carried out using the HF mode. The obvious advantages to this technique compared to the single-drop mode are that the solvent cannot be dislodged from the needle and that larger volumes of solvent can be used. However, there are also two major disadvantages to using the HF mode compared to using HS-SDME. First, extraction times are longer than with the single-drop mode, since the analyte has to pass through the solvent-saturated polymer into the bulk extraction solvent. Thus, extraction times range from 10 to 30 min for HS-HF(2)ME versus 3 to 10 min for HS-SDME. In addition, the fibers, which typically range from 1.3 to 5 cm in length, require a rather large headspace volume compared to the single-drop mode. As detailed in the following chapters, headspace volume must be minimized to maximize extraction efficiency.

2.2.3 Static vs. Dynamic Extraction Modes

A major factor in the extraction efficiency and extraction time is the rate of absorption of a chemical by the extracting solvent. This is not a problem with dispersive liquid–liquid microextraction, since the surface area of the microdrops present in the dispersed state is very large, resulting in efficient extraction within a few seconds. The surface area of a single drop in the SDME modes is quite small, however, and chemicals tend to migrate from the surface to the interior of the drop slowly, especially when using a viscous solvent such as 1-octanol. Although the solvent volume and the surface area are larger for the HF mode, migration through the polymer membrane is quite slow. The migration time is shortened for less viscous solvents such as toluene or when higher extraction temperatures are used (see Chapter 3). Stirring the solution and increasing the temperature also aids migration for both SDME and HF modes.

The surface of the drop in the SDME modes and the polymer solvent layer in HF modes may be renewed constantly if, rather than using a static extraction, the solvent boundary is renewed by using dynamic extraction. Two types of dynamic extraction are possible. In the first, developed by Shen and Lee,[19] the in-syringe dynamic method, the water or headspace is withdrawn into the syringe needle or barrel and pushed out repeatedly. This technique can be used for both SDME and HF modes. The second approach, developed by Ouyang et al.[22] and Kokosa et al.[23] which can be used for the SDME mode, consists of pulling 90% of the drop back into the syringe needle and then pushing it back out repeatedly, called the exposed or in-needle dynamic method. The in-syringe method is more effective when used with sample solutions containing no salt or major matrix contaminants. However, when used with the

SDME mode a small amount of water may contaminate the syringe contents and be injected during the analysis. The in-needle method may be the more useful method when the sample matrix contains impurities that could harm the analytical instrument injector or column. It should be pointed out that neither technique is practical, however, unless a computer-controlled autosampler is used, since reproducible syringe plunger movements and timing are crucial to obtaining reproducible analyses.

2.3 SOLVENTS

We end with a short discussion of extraction solvents. Choosing the right extraction solvent is crucial for an SME extraction. The choice is determined in part by the medium and matrix the analytes are in, the chemical nature of the analytes, the mode of extraction, and the instrument that will be used for the analysis. The most widely used extraction solvent for SME is 1-octanol. This solvent has both polar (alcohol) and nonpolar (the alkyl chain) moieties and is a nearly universal extraction solvent. It was originally used in pharmaceutics to mimic drug partitioning between cellular components in the body. Other common extraction solvents include toluene, the xylenes, and alkanes, such as octane and tetradecane. Table 2.2 lists a few of the more common and useful SME solvents that can be used as starting points for many SME applications. The log K_{ow} (sometimes referred to as log P) value is the log of the 1-octanol/water partition constant for a solvent and is a measure of the polarity of a solvent. The larger the log K_{ow} value, the less water soluble and more hydrophobic the solvent. A full list of SME solvents and their physical properties is provided in Chapter 5. A detailed methodology for choosing a correct solvent is also covered in Chapter 5. Chapter 7 has specific recommendations for several commonly extracted analytes using SME.

2.3.1 General Rules for Choosing a Solvent

In general, you must first consider the type of instrumentation needed for the analysis. For example, if you are using an electron capture detector with a gas chromatograph, you will need to avoid most oxygenated and all halogenated solvents, since their detector signals can overwhelm signals from extracts. Even alkanes such as tetradecane can contain oxidation products, which interfere with analyses of very

TABLE 2.2 Common Water-Immiscible SME Extraction Solvents

Solvent	Boiling Point (°C)	Water Solubility (μL/mL)	Log K_{ow}
Tetrachloroethene	121	0.092	3.40
Toluene	110	0.55	2.73
o-Xylene	144	0.19	3.12
Decane	174	1.2×10^{-5}	5.98
1-Octanol	195	0.65	3.00
1-Undecanol	245 (m.p. = 19)	0.023	4.72
Tetradecane	254	2.9×10^{-6}	7.20

low concentrations of analytes. In general, for analyses at very low concentrations, all solvents should be of the best quality, vacuum distilled, and stored refrigerated to prevent autoxidation. The solvent must also be compatible with the chromatography system. As an example, toluene extraction solvent may not be directly compatible with the HPLC mobile phase and would therefore have to be evaporated and exchanged with the HPLC solvent, and benzyl alcohol may not give sharp, resolved peaks when used with a nonpolar GC column.

For GC analyses, the extraction solvent must not overlap with analyte peaks and therefore must be lower or higher boiling than the chemicals extracted. If a low-boiling solvent such as toluene is chosen, the sample may be analyzed by GC using a splitless injection. When a higher-boiling solvent such as 1-octanol or tetradecane is used, a split injection with a split ratio of 10:1 to 50:1 must be used. In general, a higher-boiling solvent is used with volatile components in the range from chloromethane to naphthalene. A lower-boiling solvent is used with components with semivolatile components such as the PAHs.

As discussed in Section 2.2.1.1, except for HF(3)ME and headspace procedures, the extraction solvent cannot be soluble in water. In fact, the direct-immersion single-drop method cannot use a solvent with even slight solubility in water. Calculations in Chapter 4 show that a 5-μL drop of chloroform would dissolve completely in 1 mL of sample water. Even in 30% by weight salt solution, 1 μL of chloroform would dissolve completely in 1 mL of the sample. When using HS-SDME, chloroform would evaporate in the headspace even when extracting a solid. This is one major reason why solvents with boiling points lower than toluene are rarely used for SME. The problem is exacerbated when extractions are carried out at higher temperatures. The solubilization is usually slow, however, but may have skewed experimental results in many cases when sample water was drawn back into the syringe and injected. Conservatively, it may be best to expect solvent loss from evaporation, solubility, and from wicking up on the outside of the needle for single-drop modes and use a drop 0.2 to 0.5 μL larger than the solvent amount drawn back into the syringe and injected for analysis. Volatile extraction solvents have been used successfully, however, when the solvent is used in larger amounts, as with membrane-assisted solvent extraction (see Chapters 5 and 6), or when the solvent is not exposed directly to the bulk of the sample, as in the in-syringe dynamic extraction procedure devised by Lee. A more promising but rarely explored possibility is to use the volatile solvent as a co-solvent with a water-insoluble solvent. This is a fruitful area for future research.

A further problem often seen with SDME modes, especially with more volatile extraction solvents, is the formation of a bubble within the microdrop during the extraction. This can lead to varied amounts of solvent extracted back into the syringe following extraction. The problem exists even with higher-boiling solvents, especially at higher extraction temperatures. One way to minimize the problem has been to avoid even a tiny amount of air within the syringe needle when filling the syringe. This can be accomplished by filling and flushing the syringe repeatedly, up to 10 times. It may also prove useful to pull up slightly more solvent into the syringe than will be used and then push out a fraction of a microliter of solvent before the needle is placed into the vial septum.

With these solvent restrictions in mind, solvents are generally chosen with the adage "like dissolves like," which really means that the solvent should be able to bond well with the analyte. For instance, a solvent with hydrogen-bonding capability, such as 1-octanol, would be a good extraction solvent candidate for an analyte capable of hydrogen bonding. Toluene or *o*-xylene, on the other hand, which bond primarily by London dispersion (van der Waals) forces, would be good candidates for extraction of PAHs. As an example, tetradecane or hexadecane can be used for the HS-SDME extraction of water disinfection by-products—halogenated hydrocarbons from tap water—because they are higher boiling than the extracted analytes and do not readily evaporate, are good solvents for the halogenated compounds, and are very insoluble in water. Water containing the sugar-based β-cyclodextrin has been used successfully for the HF(3)ME extraction of PAHs. This water–cyclodextrin solvent mixture bonds strongly with the PAHs and is directly compatible with the methanol used as HPLC elution solvent in the reversed-phase separation. Water containing extraction solvent is also compatible with capillary electrophoresis instrumentation and atomic absorption instrumentation. Many more examples are presented in Chapter 6.

2.3.2 Internal and Surrogate Standards

Most standard methods recommend the use of internal and surrogate standards to increase reliability and accuracy of data. Surrogates are chemicals similar to those being extracted but not expected to be present in the unspiked sample. They are generally used to ensure that the instrument or method is working properly. Internal standards are used to account for variations in extractions and for quantification. Sometimes, surrogates and internal standards are interchangeable. In SME the standards can be spiked into the sample before extraction, but one should always add a standard to the extraction solvent as well when using the SDME mode. In the single-drop mode, this is an alarm bell that tells you immediately whether the drop has been lost during the extraction and the data should be discarded. The in-drop standard is also used to account for inevitable small differences in the percentage of the drop that is drawn back into the syringe. This has often been ignored in experiments. This is even true in HF(2) and HF(3) experiments, although to a lesser extent. Sources of readily available standards in most solvents are the naturally occurring impurities present, even after careful distillation. Often, these are minor amounts of oxidized solvent, such as an alcohol or aldehyde in an alkane such as tetradecane, or chemical analogs such as mesitylene in xylene. A more complete discussion of standards and quantification is presented in Chapter 5.

REFERENCES

1. Jeannot, M. A.; Cantwell, F. F., Mass transfer characteristics of solvent extraction into a single drop at the tip of a syringe needle. *Anal. Chem.* 1997, *69* (2), 235–239.
2. He, Y.; Lee, H. K., Liquid-phase microextraction in a single drop of organic solvent by using a conventional microsyringe. *Anal. Chem.* 1997, *69* (22), 4634–4640.

3. Wu, H. F.; Yen, J. H.; Chin, C. C., Combining drop-to-drop solvent microextraction with gas chromatography/mass spectrometry using electronic ionization and self-ion/molecule reaction method to determine methoxyacetophenone isomers in one drop of water. *Anal. Chem.* 2006, *78* (5), 1707–1712.

4. Rezaee, M.; Assadi, Y.; Milani Hosseini, M.-R.; Aghaee, E.; Ahmadi, F.; Berijani, S., Determination of organic compounds in water using dispersive liquid–liquid microextraction. *J. Chromatogr. A* 2006, *1116* (1–2), 1–9.

5. Zanjani, M. R. K.; Yamini, Y.; Shariati, S.; Jönsson, J. Å., A new liquid-phase microextraction method based on solidification of floating organic drop. *Anal. Chim. Acta* 2007, *585* (2), 286–293.

6. Shen, G.; Lee, H. K., Hollow fiber–protected liquid-phase microextraction of triazine herbicides. *Anal. Chem.* 2002, *74* (3), 648–654.

7. Ouyang, G.; Pawliszyn, J., Kinetic calibration for automated hollow fiber–protected liquid-phase microextraction. *Anal. Chem.* 2006, *78* (16), 5783–5788.

8. Jiang, X.; Lee, H. K., Solvent bar microextraction. *Anal. Chem.* 2004, *76* (18), 5591–5596.

9. Pedersen-Bjergaard, S.; Rasmussen, K. E., Liquid–liquid–liquid microextraction for sample preparation of biological fluids prior to capillary electrophoresis. *Anal. Chem.* 1999, *71* (14), 2650–2656.

10. Kramer, K. E.; Andrews, A. R. J., Screening method for 11-nor-delta9-tetrahydrocannabinol-9-carboxylic acid in urine using hollow fiber membrane solvent microextraction with in-tube derivatization. *J. Chromatogr. B* 2001, *760* (1), 27–36.

11. de Jager, L.; Andrews, A. R. J., Development of a screening method for cocaine and cocaine metabolites in saliva using hollow fiber membrane solvent microextraction. *Anal. Chim. Acta* 2002, *458* (2), 311–320.

12. Zhu, L.; Zhu, L.; Lee, H. K., Liquid–liquid–liquid microextraction of nitrophenols with a hollow fiber membrane prior to capillary liquid chromatography. *J. Chromatogr. A* 2001, *924* (1–2), 407–414.

13. Psillakis, E.; Kalogerakis, N., Hollow-fibre liquid-phase microextraction of phthalate esters from water. *J. Chromatogr. A* 2003, *999* (1–2), 145–153.

14. Ho, T. S.; Pedersen-Bjergaard, S.; Rasmussen, K. E., Experiences with carrier-mediated transport in liquid-phase microextraction. *J. Chromatogr. Sci.* 2006, *44* (6), 308–316.

15. Gjelstad, A.; Andersen, T. M.; Rasmussen, K. E.; Pedersen-Bjergaard, S., Microextraction across supported liquid membranes forced by pH gradients and electrical fields. *J. Chromatogr. A* 2007, *1157* (1–2), 38–45.

16. Przyjazny, A.; Kokosa, J. M., Analytical characteristics of the determination of benzene, toluene, ethylbenzene and xylenes in water by headspace solvent microextraction. *J. Chromatogr. A* 2002, *977* (2), 143–153.

17. Kokosa, J. M.; Przyjazny, A., Headspace microdrop analysis: an alternative test method for gasoline diluent and benzene, toluene, ethylbenzene and xylenes in used engine oils. *J. Chromatogr. A* 2003, *983* (1–2), 205–214.

18. Theis, A. L.; Waldack, A. J.; Hansen, S. M.; Jeannot, M. A., Headspace solvent microextraction. *Anal. Chem.* 2001, *73* (23), 5651–5654.

19. Shen, G.; Lee, H. K., Headspace liquid-phase microextraction of chlorobenzenes in soil with gas chromatography–electron capture detection. *Anal. Chem.* 2003, *75* (1), 98–103.

20. Shariati-Feizabadi, S.; Yamini, Y.; Bahramifar, N., Headspace solvent microextraction and gas chromatographic determination of some polycyclic aromatic hydrocarbons in water samples. *Anal. Chim. Acta* 2003, *489* (1), 21–31.

21. Zhao, R. S.; Lao, W. J.; Xu, X. B., Headspace liquid-phase microextraction of trihalo-methanes in drinking water and their gas chromatographic determination. *Talanta* 2004, *62* (4), 751–756.
22. Ouyang, G.; Zhao, W.; Pawliszyn, J., Automation and optimization of liquid-phase microextraction by gas chromatography. *J. Chromatogr. A* 2007, *1138* (1–2), 47–54.
23. Kokosa, J. M.: Przyjazny, A.; Jones, R., A comparison of direct and headspace micro-extraction for the determination of selected volatile compounds. Paper 1680-4, presented at PittCon 2007, March 6, 2007, Chicago,

CHAPTER 3

THEORY OF SOLVENT MICROEXTRACTION

3.1 INTRODUCTION

This chapter is focused on providing a sound physical basis for understanding solvent extraction phenomena on the microscale. General concepts that apply to all types of extractions are dealt with briefly, with a more in-depth focus on aspects unique to SME. Thermodynamic considerations will be used to gain insight into equilibrium aspects of SME, and a detailed kinetic treatment is employed in order to understand and control rate-determining features of the microextraction process. This is important since SME is often performed under nonequilibrium conditions. Kinetic aspects of different SME approaches are discussed. Experimental verification of theoretical considerations is described wherever possible. Finally, implications for analytical calibration methods are presented. Every effort has been made to fully explain the significance of the equations and ideas with illustrative examples and interpretations.

3.2 THERMODYNAMICS

3.2.1 Phase Distribution: Fundamental Considerations

Solvent microextraction, in common with other sample cleanup and preconcentration techniques, is based on the distribution of analyte molecules between the

Solvent Microextraction: Theory and Practice, By John M. Kokosa, Andrzej Przyjazny, and Michael A. Jeannot
Copyright © 2009 John Wiley & Sons, Inc.

sample phase (normally, aqueous solution) and the extracting phase (normally, an organic solvent). We begin with a brief review of the fundamentals of phase distribution of simple nonreactive analyte molecules in liquid–liquid extraction.

An equilibrium distribution isotherm is a plot of equilibrium concentration of the analyte in the organic phase versus its concentration in the aqueous (water) phase. Hypothetical isotherms are shown in Figure 3.1. Such isotherms will typically exhibit linear behavior at low concentrations, but may deviate from linearity at higher concentrations. The equilibrium distribution coefficient, K, is defined as

$$K = \frac{C_o}{C_w} \tag{3.1}$$

where C_o and C_w represent concentrations of analyte in the organic and water phases, respectively, at any point on the distribution isotherm. The assumption that K is constant is typically valid only in dilute solutions. At higher concentrations, solute–solute interactions become important, and deviations from linear distribution behavior may become apparent, as illustrated in Figure 3.1. This assumption of a constant equilibrium distribution coefficient is generally valid in SME, where we are working with low concentrations of analyte in the sample solution (ppb or ppm) and still relatively low concentrations in the organic extracting phase (ppm or parts per thousand). However, in cases of higher analyte concentrations in either or both phases, this issue may be worthy of investigation, as it may result in nonlinear calibration curves. In the discussion that follows, we assume that K is constant.

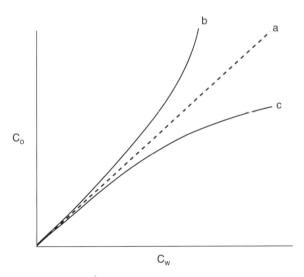

FIGURE 3.1. Hypothetical equilibrium distribution isotherms between water (w) and organic solvent (o). Curve a represents ideal behavior, and curves b and c represent nonlinear isotherms due primarily to solute–solute interactions. The equilibrium distribution coefficient K is simply C_o/C_w at any point on the isotherm, and is generally constant at low concentrations for all isotherms.

By combining the definition of the equilibrium distribution constant [equation (3.1)] with the condition for mass balance given an initial analyte concentration C_w^0 with water and organic volumes V_w and V_o respectively,

$$C_w^0 \cdot V_w = C_w V_w + C_o V_o \tag{3.2}$$

we obtain an expression for the equilibrium concentration in the organic phase as a function of the initial analyte concentration in the aqueous (sample) phase:[1]

$$C_o = \frac{K C_w^0}{1 + K V_o / V_w} \tag{3.3}$$

Although these equations apply in general to most types of extraction processes, it is important to recognize a unique feature of SME compared to conventional solvent extraction. Because the phase ratio (V_o/V_w) is typically very small for SME (10^{-3} or less), the function $K V_o/V_w$ is often small, and the denominator in equation (3.3) may be close to unity under circumstances where solutes with reasonably small distribution constants (10^2 or smaller) are employed. For example, if a 1- μL drop is used to extract a solute with a distribution constant of 50 from a 2-mL water sample, $K V_o/V_w$ is 0.025 and $C_o \approx K C_w^0$ within 2% error. This is equivalent to saying that the number of moles of analyte in the organic phase is small compared to the number of moles in the aqueous phase, and a negligible amount of analyte is removed from the sample. Any equilibria involving the analyte in the aqueous phase are therefore not perturbed. This has important consequences for speciation studies, as discussed in Section 3.2.5, but is not necessary for routine analysis.

The other extreme in equation (3.3) is the case of exhaustive extraction where K is so large that the second term in the denominator is much greater than unity and C_o is independent of K and is simply $C_w^0 V_w/V_o$. For example, a solute with a distribution constant of 5000 being extracted with 5 μL of solvent from a 0.5-mL water sample would result in $K V_o/V_w = 50$ and $C_o \approx C_w^0 V_w/V_o$ (exhaustive extraction) within 2% error.

However, most often the situation is intermediate between these two extremes, where a significant amount (but far less than 100%) of analyte is removed from aqueous solution. We shall see in Section 3.4 that linear calibration curves can be obtained in all these cases.

Equilibria in headspace (three-phase) SME can be described simply by incorporating two equilibrium distribution coefficients and a mass balance equation involving all three phases:[2]

$$K_{aw} = \frac{C_a}{C_w} \tag{3.4}$$

$$K_{oa} = \frac{C_o}{C_a} \tag{3.5}$$

$$C_w^0 V_w = C_w V_w + C_a V_a + C_o V_o \tag{3.6}$$

where K_{aw} and K_{oa} are the air–water and organic–air distribution coefficients respectively, and C_a and V_a refer to the equilibrium air (headspace) concentration and air volume. The overall (organic–water) distribution coefficient (K_{ow}) may be expressed as

$$K_{ow} = \frac{C_o}{C_w} = K_{aw}K_{oa} \qquad (3.7)$$

Combining the equations above gives

$$C_o = \frac{K_{ow}C_w^0}{1 + K_{aw}V_a/V_w + K_{ow}V_o/V_w} \qquad (3.8)$$

Note that if V_a is negligibly small, equation (3.8) reduces to the two-phase situation [equation (3.3)]. Like the two-phase system, the equilibrium situation may be nonperturbing ($C_o = K_{ow}C_w^0$) in the case where the organic and headspace volumes are small and/or the distribution coefficients are not too large. Exhaustive extraction ($C_o = C_w^0 V_w/V_o$) would require very small K_{aw} and/or V_a, along with very large K_{ow} and reasonably large V_o. Most realistic situations will result in an appreciable number of moles of analyte in all three phases at equilibrium. Detailed example calculations for various chemical systems are presented in Chapter 4.

3.2.2   Solvation and Solvent Selection

Because of the very small solvent volume and potential for drop dissolution or evaporation in SME, solvents that are very nonpolar (e.g., hydrocarbons) and of low volatility are normally required. Thus, SME is most easily applied to an analysis of compounds of low to moderate polarity. However, polar compounds can still partition effectively into these solvents, although improved partitioning may require either derivatization prior to (or during) SME, or hollow fiber approaches which allow for the use of more polar extracting solvents. Also, in theory at least, it should be possible to presaturate the aqueous sample solution and headspace with a more polar or volatile solvent to improve the extraction of polar substances. As with any extraction technique, acid–base speciation and the charge state of the analyte is obviously of great importance, and this topic is discussed in Section 3.2.5.

Typical nonpolar solvents such as aliphatic or aromatic hydrocarbons work well for most analytes via interaction through dispersion forces and/or the hydrophobic effect and are capable of extracting a wide range of analytes. Various functional groups, such as alcohol, ketone, and halide, may be incorporated in the solvent to enhance dipole–dipole or even hydrogen-bonding interactions with analyte molecules, or mixed solvents may be employed. Numerical parameters such as the dielectric constant, dipole moment, polarity index, and the Hildebrand solubility parameter may be used to predict appropriate solvation properties.[3] Detailed solvent properties and useful applications are presented in Chapter 5 (see Tables 5.2 and 5.3).

One of the major advantages of SME is the ability to easily investigate different solvents tuned for their ability to interact effectively with the analyte. Issues such as water solubility and vapor pressure must also be examined carefully. These topics are addressed in more detail in Section 3.2.6 and Chapter 4.

3.2.3 Octanol–Water Partition Coefficients and Henry's Law Constants

1-Octanol is a nearly ideal extracting solvent, owing to its low water solubility, low vapor pressure, long hydrocarbon chain that can easily accommodate analyte molecules of low polarity, and hydroxyl group, which may be important for stabilizing any polar functional groups on the molecule. Furthermore, there is a wide body of empirical and semiempirical data and estimation methods for the partitioning of various solutes between water and 1-octanol.[4] Thus, octanol–water partition coefficients are frequently used as a measure of the "extractability" of organic solutes from water and can be used as estimates of the water–organic distribution coefficients in the equations in this chapter.

In headspace SME (or simple two-phase SME where an air space is present in the flask and volatile or semivolatile analytes are being studied), it is important to understand the partitioning of analyte molecules into the gas phase. The importance of the headspace as a "reservoir" for analyte molecules can be estimated by examining the magnitude of the second term in the denominator of equation (3.8), as discussed previously.

Henry's law constants in the literature (K_H) are typically defined for the partitioning of solute from water into air.[5] Various units are employed; for example:

$$K_H = \frac{p\,(\text{bar})}{C\,(\text{mol/L})} \tag{3.9}$$

Dividing by RT according to the ideal gas law, one obtains a dimensionless partition coefficient based on molar concentrations. This coefficient was used in equation (3.4).

$$K_{aw} = \frac{K_H}{RT} \tag{3.10}$$

As was the case for water–organic partition coefficients discussed earlier, it is generally assumed that K_H and K_{aw} are constant and independent of concentration, an assumption that is generally true for dilute solutions of an organic solute in water.

3.2.4 Temperature and Salt Effects

Water–organic solvent partitioning of most compounds is only weakly dependent on temperature. The effect of temperature on the distribution constant K from basic thermodynamic considerations is

$$\ln K = -\frac{\Delta H}{RT} + \text{const} \qquad (3.11)$$

where ΔH represents the enthalpy of transfer of solute from the water phase to the organic phase. The practical effect of T on K is small over temperature ranges normally accessible in microextraction systems because ΔH is relatively small for liquid–liquid partitioning.[6] On the other hand, the effect of temperature on water–air (and air–solvent) partitioning is substantial. Here, too, equation (3.11) can be used to describe the effect of temperature, with ΔH approximately equal to the (relatively large) heat of vaporization of the analyte.[7] Thus, there is a stronger temperature dependence for liquid–air partitioning.

In headspace SME, a higher temperature results in a larger K_{aw} according to equation (3.11), since ΔH is positive for the transfer of analyte from water to headspace. On the other hand, the transfer from headspace to organic solvent would be favored by a lower temperature because ΔH is negative. Thus, by independent control of the temperature of the sample (water) phase and extracting (organic) phase, it is possible in principle to enhance the overall distribution constant.

The effect of salt on the distribution of analyte among phases may be explained by the lowering of analyte solubility (and equilibrium aqueous concentrations) in the aqueous phase. Since salt is obviously not present in the headspace or organic phases to any significant extent, the effect of salt is to increase K in two-phase systems [equation (3.1)], and to increase K_{aw} (and therefore K_{ow}) in three-phase systems [equations (3.4) and (3.7)]. The effect of salt concentration ([salt]) on the distribution constant is exponential and can be deduced from the empirical equation of Setschenow as follows:

$$K_{\text{salt}} = K \times 10^{K^S [\text{salt}]} \qquad (3.12)$$

where K_{salt} is the distribution coefficient in the presence of salt and K^S is the empirically deduced Setschenow constant.[8] Detailed example calculations of the effect of salt concentration are presented in Chapter 4.

3.2.5 Solute Equilibria and Speciation: pH and Back-Extraction

The distribution equilibria discussed so far assume that the molecule partitioning between phases is not involved in any homogeneous chemical equilibria, such as acid–base dissociation or self-association. Now we consider the case of analyte species with acid–base functional groups and the effect of pH on the microextraction process. When considering such compounds, it is useful to define the distribution ratio, D, which is the ratio of the sum of the concentrations of all species in the organic phase (formal concentration, F_o) to the sum of the concentrations of all species in the water phase (F_w):

$$D = \frac{F_o}{F_w} \tag{3.13}$$

Consider, for example, the extraction of a neutral weak acid such as benzoic acid from water into octanol. The protonated form (HA) is neutral, and this species will have a reasonably large K_{ow}, whereas the deprotonated form (A^-) is ionic and will have a negligibly small K_{ow}. Thus, the only species present in the organic phase is HA. The distribution ratio will clearly depend on the speciation in the water phase, as demonstrated by the following relationships:

$$D = \frac{[HA]_o}{[HA]_w + [A^-]_w} \tag{3.14}$$

$$K_a = \frac{[H^+]_w [A^-]_w}{[HA]_w} \quad \text{acid dissociation constant} \tag{3.15}$$

$$K_{ow} = \frac{[HA]_o}{[HA]_w} \tag{3.16}$$

Solving equation (3.15) for $[A^-]_w$ and substituting in equation (3.14) gives

$$D = \frac{[HA]_o}{[HA]_w + K_a[HA]_w/[H^+]_w} \tag{3.17}$$

which rearranges to

$$D = \frac{[HA]_o}{[HA]_w} \frac{[H^+]_w}{[H^+]_w + K_a} \tag{3.18}$$

Substituting for K_{ow} from (3.16) gives

$$D = K_{ow}\left(\frac{[H^+]_w}{[H^+]_w + K_a}\right) \tag{3.19}$$

The term in parentheses can be recognized from acid–base chemistry as α_{HA}, the fraction of benzoic acid present in its protonated (neutral) form, and we may simply write

$$D = K_{ow}\alpha_{HA} \tag{3.20}$$

This fraction depends, of course, on pH and approaches unity when $[H^+]$ becomes much larger than K_a (pH $\ll$ pK_a), and approaches zero when $[H^+]$ becomes much smaller than K_a (pH $\gg$ pK_a). Thus, the distribution ratio for benzoic acid may vary from zero at high pH to K_{ow} for HA at low pH.

In a similar way, it is possible to show that for a neutral weak base such as an amine, the reverse situation would apply. (At pH well above the pK_a, value, the amine would be deprotonated and neutral, and thus D would approach K_{ow} for the neutral amine.) In light of these considerations, it is always important to control the pH of the water phase when extracting ionizable solutes.

This also has important consequences for three-phase liquid–liquid–liquid extraction where the analyte is transported from the donor (water) phase to an organic solvent and then back-extracted into a second receiver (water) phase. Returning to the benzoic acid example, it would be necessary to employ pH $\ll$ pK_a in the donor phase to protonate the molecule, making it neutral and extractable. The receiver phase should be at a higher pH $\gg$ pK_a, such that the molecule becomes ionized and "trapped" in the receiver phase. In this case, the distribution ratio for the second step is the reciprocal of equation (3.19) or (3.20), and D becomes very large at high pH. The overall donor–receiver distribution ratio ($F_{\text{receiver}}/F_{\text{donor}}$) is the product of these two D's. It is independent of K_{ow} and can be shown to be approximately equal to $K_a/[H^+]_{\text{receiver}}$.

A further consequence of selective partitioning of species is the potential for the determination of specific chemical species in a complex mixture. For example, in a two-phase system, under non-perturbing SME conditions (Section 3.2.1), it is possible to quantitate only the neutral (extractable) form of a weak acid/base molecule without altering its concentration in the aqueous phase and without perturbing the acid-base equilibrium position. This idea can also be extended to the analysis of the free (unbound) concentrations of small molecules in the presence of binding proteins.[9]

3.2.6   Dissolution and Evaporation of Solvent

The small volumes of solvent employed in SME techniques imply that special consideration must be given to solvent losses both in the aqueous sample phase and the headspace phase. Solubility data and the ideal gas law may be used to estimate potential worst-case (equilibrium) drop losses through dissolution and evaporation. The potential volume loss of the solvent may be estimated from the following equation, where the first and second terms in parentheses represent equilibrium losses in the water and headspace phases, respectively:

$$V_{\text{loss}} = \frac{\text{MM}}{\rho}\left(S_w V_w + \frac{P_{\text{vap}}V_a}{RT}\right) \tag{3.21}$$

Here S_w represents the molar solubility of the solvent in water (sample) solution, P_{vap} is the vapor pressure of the solvent at temperature T, MM is the solvent molar mass, and ρ is its density. As an example, we can see why the commonly used solvent dichloromethane is generally not suitable for SME: Given 1 mL of water and 1 mL of air phase, equation (3.21) predicts an equilibrium loss of about 12 μL to dissolution and about 1 μL to evaporation, for a total loss of about 13 μL, much larger than a typical microdrop. (See Table 5.3 for solvent data.) Solvents with lower water solubility and lower volatility are therefore preferred.

Losses through dissolution in the sample solution can therefore be minimized by working with solvents of low water solubility (or altering the composition of the sample solution to lower the solubility by salt addition, for example). Minimizing the volume of the water phase will also result in smaller drop losses.

Theoretically, of course, headspace losses can be eliminated by working with a closed system containing no air space. Alternatively, solvents with low vapor pressure (high boiling point) are necessary. Since P_{vap} in equation (3.21) increases exponentially with temperature according to the Clausius–Clapeyron equation, low temperatures can also help prevent headspace losses.

It should be pointed out that small solvent losses are not necessarily problematic in terms of analytical applications, provided, of course, that the loss is significantly less than 100%! Care and consideration must be taken in ensuring reproducible parameters in equation (3.19) (such as solubility) among all solutions under study for a particular calibration by paying attention to consistent ionic strength, temperature, and so on. The equilibrium distribution coefficients themselves may also change with significant solvent dissolution, but again, consistency in solution composition will maintain the constancy of the distribution coefficient. If it is desired to eliminate losses from dissolution and evaporation altogether, the hollow fiber approach is helpful in retaining the solvent, and it may be conceivable to presaturate the system with the solvent prior to the extraction process. If losses do occur, care must be taken to account for this when retracting the solvent into the syringe, being careful not to uptake any aqueous sample. But in general, suitable choice of extracting solvent makes this a nonissue for single-drop and other techniques.

3.2.7 Interfacial Adsorption

The equilibrium distributions considered in this chapter have assumed that there is no significant adsorption of analyte molecules at the water–air or water–organic interfaces. This assumption is generally valid unless the analyte molecules are amphiphilic and highly surface active in nature. In such cases, the interface itself may be considered a unique two-dimensional phase. Equilibrium distribution coefficients and adsorption isotherms may be defined involving adsorption on this "phase." However, water–air and water–organic interfacial adsorption is rarely important for typical analyte molecules[10] and is not considered further except for a brief consideration of the mass transfer effect of adsorbed species at the interface in Section 3.3.3. On the other hand, adsorption at various interfaces and surfaces, including the vial septum, glass vial, stir bar, or dispersed matrix components is always at least a potential source of analyte loss.

3.3 KINETICS

3.3.1 Diffusive Mass Transfer and Fick's Laws

The rate of transport of analyte molecules among phases is governed at the molecular level by random molecular motion or diffusion. We begin to develop kinetic models for various modes of SME by reviewing the mathematical laws that govern diffusion processes in stagnant (unstirred) systems and the variables that influence the rate of diffusion. In Section 3.3.2 we will see how convection (stirring or agitation) combined with molecular diffusion is used to enhance the rate of mass transfer in SME.

It is well known that molecules in solution (at least gas- and liquid-phase solutions) undergo translational motion, and the direction and speed of this motion is constantly changing due to collisions and interactions with solvent molecules. Because of the random nature of this motion, the mean velocity of an ensemble of molecules is zero. How, then, can this random molecular motion result in mass transfer of analyte molecules? The answer lies in the fact that there is statistical variation in the velocity distribution of molecules. In other words, although the mean velocity for a large ensemble of molecules may be zero, there is a variance associated with this mean, and a given molecule will probably experience a net translational motion in a given period of time. Thus, one can see how an ensemble of molecules originally concentrated in a point, along a line, or in a plane will spread out or diffuse over time. In general, then, because of random molecular motion, concentrations of molecules will tend to equalize throughout the solvent over time and move from regions of high concentration to regions of low concentration.

There are two well-known mathematical laws that describe this process. *Fick's first law of diffusion* describes the flux of molecules (moles, n per area A per time t) as a function of the concentration gradient (dC/dx):

$$-\frac{1}{A}\frac{dn}{dt} = D\frac{dC}{dx} \qquad (3.22)$$

The proportionality constant between the concentration gradient and the flux is the diffusion coefficient, D, with units of cm^2/s (with A in cm^2, dC in mol/cm^3, and dx in cm). The negative sign signifies diffusion from high concentration to low concentration.

Diffusion coefficients of molecules in the gas phase are relatively large (on the order of $0.1\ cm^2/s$), owing to the mostly unhindered translational motion in the gas phase. On the other hand, diffusion coefficients for solute molecules in liquid solution are much smaller (on the order of $10^{-5}\ cm^2/s$). Because of this, understanding and controlling diffusion rates in the liquid phase is much more important. Liquid diffusion coefficients may be measured experimentally (e.g., via the Taylor dispersion method[11]) or estimated by one of a number of theoretical or empirical models (e.g., via the Stokes–Einstein equation or the Wilke–Chang model, to name just two[12]). In general, most models feature a direct relationship of D with temperature and an inverse relationship with solvent viscosity and the size of the diffusing molecule. Thus, we see immediately three experimental parameters that

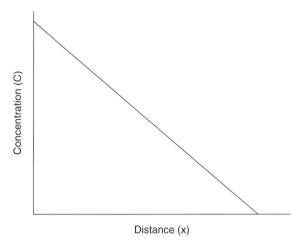

FIGURE 3.2. Steady-state molecular diffusion with a linear concentration gradient. The direction of flux is to the right according to Fick's first law, but there is no change in concentration at any point over time according to Fick's second law.

control the rate of diffusion, although the exact dependence of diffusion rates on these parameters may not be exactly known.

Fick's second law of diffusion is a differential equation for the rate of change of concentration with time that may be solved exactly for a number of simple cases:

$$\frac{dC}{dt} = D\frac{d^2C}{dx^2} \tag{3.23}$$

Fick's second law tells us that for linear concentration gradients (where d^2C/dx^2 is zero), there is no net change in concentration, although we do have a net transport of molecules according to Fick's first law. In other words, with a linear concentration gradient, we have a steady-state situation. On the other hand, nonlinear concentration gradients do result in a buildup or depletion of concentration (unsteady state) according to Fick's second law. We refer to steady-state and unsteady-state molecular diffusion in Section 3.3.2. Figures 3.2 and 3.3 illustrate the difference between these two types of concentration gradients.

3.3.2 Convective–Diffusive Mass Transfer

Mechanical agitation (convection) of one or all of the liquid phases in SME is generally recommended to facilitate mass transfer of analyte molecules among the various phases. In a most simplistic sense, convection may be thought of as a means of reducing the distance over which slow diffusion processes must occur. We develop specific models to account for the combined contributions of convection and diffusion to the overall mass transfer rate in SME.

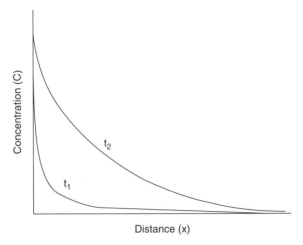

FIGURE 3.3. Unsteady-state molecular diffusion with a nonlinear concentration gradient. The direction of flux is also to the right, but because of the positive second derivative of the gradient, the concentration at any point increases over time from t_1 to t_2.

The two most common methods of liquid-phase agitation are magnetic stirring and mechanical vibration. In both cases, bulk flow of the liquid phase is employed to achieve mixing within the liquid phase or to renew the interfacial area of contact between the phases. An exact description of hydrodynamic behavior of the liquid phase and its dependence on experimental variables such as stirring rate and stir bar geometry is generally unattainable. However, simple but useful models of convective–diffusive phenomena may be constructed, and the various modes of agitation may be treated similarly in terms of mass transfer modeling.

Our basic postulates of mass transfer theory will be as follows:[1]

1. The rate of mass transfer to or from an interface within any phase will be proportional to the difference in bulk solution concentration and the concentration at the interface. The proportionality constant is called the *mass transfer coefficient*.

2. We will assume that there is negligible adsorption of analyte molecules at the interface itself. In other words, there will be no accumulation or depletion of analyte at the interface.

3. We will assume that there is no resistance to mass transfer at the interface itself. In other words, distribution equilibrium will prevail at all times in each phase immediately adjacent to the interface. With these basic assumptions, we may formulate general mass transfer models for two- and three-phase systems and adapt these models to specific circumstances associated with different modes of SME.

3.3.3 Two-Phase Kinetics

3.3.3.1 *General Mass Transfer Model* Our first two postulates may be combined and stated mathematically as

$$\frac{1}{A_i}\frac{dn}{dt} = \beta_w(C_w - C_{w,i}) = \beta_o(C_{o,i} - C_o) \tag{3.24}$$

where A_i is the water–organic interfacial area, n represents moles of analyte transferred per time t; β_w and β_o are the water and organic phase mass transfer coefficients; C_w and $C_{w,i}$ are the water phase concentrations in bulk water solution and at the water side of the interface, respectively; and $C_{o,i}$ and C_o are the organic phase concentrations at the organic side of the interface and in bulk organic solution, respectively. The equal fluxes of molecules to and from the interface satisfy the second postulate above.

The third postulate may be stated simply as

$$K = \frac{C_{o,i}}{C_{w,i}} \tag{3.25}$$

where K is the equilibrium distribution coefficient.

The rate of change of concentration in the organic phase as a function of time may be derived from equations (3.24) and (3.25) by solving equation (3.25) for $C_{o,i}$, substituting in equation (3.24) and solving the second equality for $C_{w,i}$, and finally, substituting this in the first equality in equation (3.24). The resulting differential equation is

$$\frac{dC_o}{dt} = \frac{A_i}{V_o}\frac{\beta_w\beta_o}{\beta_w + K\beta_o}(KC_w - C_o) \tag{3.26}$$

The term $\beta_w\beta_o/(\beta_w + K\beta_o)$ is referred to as the overall mass transfer coefficient, $\overline{\beta}$:

$$\overline{\beta} = \frac{\beta_w\beta_o}{\beta_w + K\beta_o} \tag{3.27}$$

or

$$\frac{1}{\overline{\beta}} = \frac{K}{\beta_w} + \frac{1}{\beta_o} \tag{3.28}$$

The overall mass transfer coefficient may be thought of as the proportionality constant between the flux of analyte molecules at the interface and the difference between the hypothetical organic concentration if the organic phase were at equilibrium with the current water-phase concentration and the actual organic phase concentration. When the system is actually at equilibrium, $KC_w = C_o$ and the rate equation (3.26) goes to zero.

Writing the reciprocal of the overall mass transfer coefficient as in equation (3.28) is useful because it allows us to think of the mass transfer process as the sum of two "resistances" to mass transfer: a water-phase resistance (K/β_w) and an organic-phase resistance ($1/\beta_o$). This gives us some insight into the relative importance of the mass transfer process in both phases. Since K is normally much larger than unity for a typical SME application, the resistance to mass transfer in the water phase becomes more important as a rate-determining step in the overall process. Thus, it is normally more important to increase the mass transfer coefficient in the water phase (via enhanced agitation of this phase) than the organic phase.

If the condition for mass balance [equation (3.2)] is solved for C_w and substituted in equation (3.26), a first-order linear differential equation results. The general solution is[13]

$$C_o = \frac{KC_w^0}{KV_o/V_w + 1} + \frac{\text{const}}{e^{kt}} \tag{3.29}$$

where k is the rate constant for the extraction process:

$$k = \frac{A_i}{V_o}\overline{\beta}\left(K\frac{V_o}{V_w} + 1\right) \tag{3.30}$$

Finally, the constant term may be evaluated by substituting initial conditions of $C_o = 0$ at $t = 0$ and recognizing that the first term on the right-hand side of equation (3.29) is the equilibrium concentration in the organic phase, which we now call C_o^∞. Equation (3.29) becomes simply

$$C_o = C_o^\infty(1 - e^{-kt}) \tag{3.31}$$

Thus, the concentration in the organic phase is zero at time zero and approaches a plateau value of C_o^∞ at long times. Concentration versus time data typically fit equation (3.31) quite well in various SME systems.[14]

3.3.3.2 Experimental Variables
The parameters that determine the rate constant in equation (3.30) are now apparent. This equation shows that there are several fundamental parameters that can be adjusted to enhance the rate of extraction.

1. Increasing the interfacial area has a direct effect on k. Later we shall see how different modes of SME take advantage of this effect.
2. The effect of water- and organic-phase volumes is much more complex. Consider first the case of negligible removal of solute as discussed previously, where KV_o/V_w is much smaller than unity. In this case, the term in parentheses in equation (3.30) is unity, and k is independent of V_w. This makes sense since we are not really changing the composition of the water phase and its

concentration remains essentially unchanged. Also in this case, the rate constant is related inversely to V_o. Again, this makes sense since a smaller volume of organic phase will equilibrate more quickly with the aqueous phase under these circumstances. Going to the other extreme (exhaustive extraction where KV_o/V_w is much greater than unity), the rate becomes independent of V_o but is inversely related to V_w. In the most typical scenario in between these extremes, we can see that the rate is enhanced by making both V_o and V_w as small as possible. Note that this is only a rate effect; the actual amount extracted at equilibrium actually increases with both V_o and V_w according to a reworking of equation (3.3) in terms of moles extracted, n_o^{∞}:

$$n_o^{\infty} = \frac{KC_w^0}{1/V_o + K/V_w} \tag{3.32}$$

Because of the opposing effects of volumes on the rate of extraction and the actual amount extracted at equilibrium, optimization of water and organic volume is less critical in SME method development.

3. Increasing the overall mass transfer coefficient has a direct effect on k. This is generally achieved by increasing the rate of convection through stirring or mechanical vibration. Factors that enhance the diffusion coefficient (e.g., temperature) can also be used to enhance the mass transfer coefficient. As stated previously, the impact of water versus organic mass transfer resistances depends on the magnitude of K.

4. The rate of extraction is semi-independent of K, as can be seen by examining equations (3.27) and (3.30), unless $KV_o/V_w \ll 1$, in which case the rate constant may approach an inverse relationship with K (a smaller K results in a faster rate). On the other hand, the amount extracted at equilibrium obviously increases with increasing K.

3.3.3.3 Model for Drop Techniques

For drop-based SME where the extracting solvent is present in the form of a drop suspended in a stirred or otherwise agitated aqueous solution, there is evidence for induced convection within the organic drop as a result of momentum transfer across the interface.[1] Thus, we have convective–diffusive mass transfer in both phases. There are at least two specific convective–diffusive models that can be considered in describing the observed rate of mass transfer; film theory and penetration theory.[15]

Film theory assumes complete mixing of both the water and organic phases up to a distance, δ_w and δ_o, respectively, to the interface. Within these "films" of thickness δ_w and δ_o, the solution is assumed to be stagnant, with mass transfer via molecular diffusion. A higher degree of stirring or agitation results in smaller film thicknesses in this model. Steady-state (linear) concentration gradients are assumed on both sides of the interface. Combining Fick's first law [equation (3.22)] with equation (3.24) in the case of linear concentration gradients implies a simple definition for the mass transfer coefficients:

$$\beta_w = \frac{D_w}{\delta_w} \qquad \beta_o = \frac{D_o}{\delta_o} \qquad\qquad (3.33)$$

Here D_w and D_o represent the diffusion coefficients in the respective phases. Figure 3.4 illustrates some hypothetical concentration gradients adjacent to the interface using film theory. An initial concentration of unity (arbitrary units) in the aqueous (water) phase and zero in the organic phase is assumed. The interfacial concentrations are calculated by simultaneously solving equations (3.24), (3.25), and (3.33) with nominal diffusion coefficients, film thicknesses, and distribution coefficient, as indicated. This figure is useful because it demonstrates graphically the idea of resistance to mass transfer. We concluded earlier that for compounds with a reasonably large K, the main resistance to mass transfer is in the aqueous phase. In Figure 3.4, when the organic film thickness doubles (dashed lines), there is little effect on the gradients (and hence the rate of transfer). On the other hand, when the water-phase film thickness doubles (dotted line), the concentration gradients are almost halved, cutting the rate of transfer by approximately one-half.

Film theory is useful because of its simplicity and elegance in explaining transport phenomena in convective modes of SME. We know that the true hydrodynamic situation must be more complex than this simple model, but nevertheless, there is at least some experimental evidence for direct dependence of the mass transfer coefficient on the diffusion coefficient, as is required by film theory.[16]

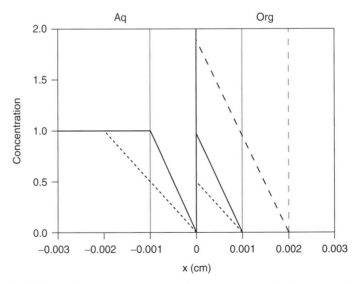

FIGURE 3.4. Hypothetical concentration gradients consistent with film theory. Solid lines are for $\delta_w = \delta_o = 10^{-3}$ cm. Dashed lines are for $\delta_w = 10^{-3}$cm, $\delta_o = 2 \times 10^{-3}$cm. Dotted lines are for $\delta_w = 2 \times 10^{-3}$cm, $\delta_o = 10^{-3}$cm. $K = 40$, $D_w = D_o = 5 \times 10^{-6}$ cm²/s. (From Ref. 13.)

Penetration theory is an alternative model that can be used to describe the mass transfer process in drop-based SME. In this model there are no stagnant films. Rather, the contact or exposure time (t_e) associated with the bulk flow of an element of the water phase and induced flow within an element of the organic drop before separation is the parameter of interest. Non-steady-state diffusion results, as illustrated in Figure 3.5. The flux or "penetration" of analyte molecules decreases during t_e, and the average flux during this time is used to define the mass transfer coefficient for either phase:

$$\beta = 2\sqrt{\frac{D}{\pi t_e}} \tag{3.34}$$

In this model, faster stirring or enhanced agitation results in a smaller t_e (more rapid renewal of the interfacial region), producing a larger mass transfer coefficient. Also, a square-root dependence of β on D is predicted.

There are numerous other more complex models for mass transfer in two-phase systems in the chemical engineering literature, and most of these models exhibit a power dependence of β on D ranging from ½ (the limit from penetration theory) to 1 (the limit from film theory). Also, it should be noted that the presence of surface-active material which adsorbs at the interface can be problematic in SME. Such material may inhibit momentum transfer across the interface and reduce the degree of convection within the organic drop.[17] In the film model, this has the effect of increasing the film thicknesses. In the penetration model, this can be interpreted as an increase in the exposure time, and therefore slower surface renewal.

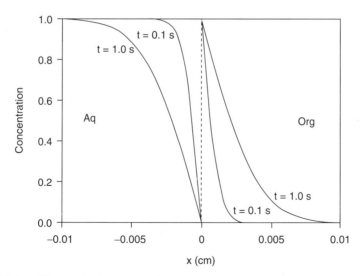

FIGURE 3.5. Hypothetical concentration gradients consistent with penetration theory. (From Ref. 13.)

3.3.3.4 Model for Dynamic Techniques Dynamic SME (or dynamic LPME) was introduced by He and Lee in 1997.[18] In this technique, the organic liquid phase is contained within the microsyringe, and microliter volumes of the aqueous sample phase are repeatedly drawn into and expelled out of the syringe. This technique provides an alternative to magnetic stirring as a means of inducing convection within the liquid phases, and is very amenable to automation. One of the advantages of this technique is said to be the formation of a thin film of organic solvent along the inside walls of the syringe, providing a large interfacial area of contact with the aqueous phase. The aqueous plug is renewed with each cycle (mixed with the bulk solution), and the organic film is mixed with the rest of the organic plug and renewed also. A diagrammatic representation of this dynamic technique is shown in Figure 3.6.

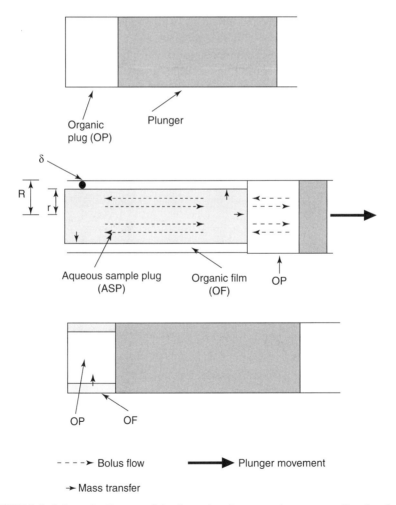

FIGURE 3.6. Schematic diagram of the dynamic microextraction process. (Reprinted with permission from Ref. 18; copyright © 1997 American Chemical Society.)

Understanding the mass transfer kinetics in dynamic systems is complicated by the following factors: First, the interfacial area is continually changing, owing to the formation and removal of the organic film with each cycle. Second, there are really two different mechanisms of transfer: aqueous plug to organic film, and aqueous plug to organic plug. (The former is considered to be the main contributor to mass transfer because of the much larger interfacial area.) Third, the extraction process must be defined as a stagewise process with each cycle representing an extraction stage.

Lee's model assumes instantaneous equilibrium between the circulating aqueous sample plug and the thin organic film (within the time frame of a single cycle) and neglects any transfer directly to the organic plug. It is also assumed that complete mixing of the phases occurs after each cycle, and the overall composition of the phases does not change significantly with each cycle. Under these conditions, a linear relationship exists between the amount of analyte extracted and both the volume of aqueous solution drawn into the syringe and the number of sampling cycles.[18] Experimental results support this hypothesis over a limited volume and sample cycling range.[18] Dynamic microextraction may be seen as a fast (equilibrium-based) technique between the aqueous plug within the syringe and the organic film. However, the actual amount extracted in each cycle is very small. Over the course of a small number of cycles, there is no significant change in the overall aqueous or organic solvent composition, and homogeneous equilibria are not perturbed, as discussed earlier. However, with many more cycles, the entire system would begin to approach equilibrium and the amount of analyte extracted would become more significant, increasing sensitivity. Automated approaches based on this technique will undoubtedly expand its utility.

3.3.3.5 Model for Hollow-Fiber SME

Hollow fiber SME, in which the solvent is held within the pores and inside a hollow, porous fiber (normally attached to the end of a GC syringe needle), is analogous in mass transfer behavior to solid-phase microextraction (SPME). Pawliszyn has developed an equation that describes the mass of analyte extracted in nanograms (n) as a function of time in seconds (t) for short exposure times for SPME:[19]

$$n = \frac{2\pi D_w L C_w t}{\ln[(b + \delta)/b]} \tag{3.35}$$

Here D_w represents the diffusion coefficient of the analyte in the sample (water) phase (cm^2/s), L is the length of the fiber (cm), C_w is the analyte concentration in the sample (water) phase (ng/mL), b is the outer radius of the fiber (cm), and δ is the diffusion film thickness of the water layer (cm, as discussed previously). This equation implies a linear extraction-time profile which is commonly observed for short exposure times. Furthermore, the amount extracted depends directly on the diffusion coefficient, length of fiber, and concentration of analyte, all of which can be understood intuitively. Increasing the rate of stirring will reduce δ, which in turn will increase n.

Note that, in general, equation (3.31) is still applicable for this and other modes of SME over a wide range of extraction times. Although the use of hollow fibers adds a level of complexity and perhaps inconvenience to the SME process, it does provide kinetic advantages in terms both of increased interfacial area (A_i) compared to the single drop method, and the ability to generate greater agitation in the water phase (reducing δ_w in the film model) due to the protection provided by the hollow fiber. Dynamic hollow fiber techniques have also been reported[20] in which the aqueous solution is drawn repeatedly into and out of the hollow fiber providing mass transfer analogous to the dynamic case discussed above (Section 3.3.3.4) combined with transfer through the outside of the fiber analogous to Section 3.3.3.5. Dynamic hollow fiber SME therefore provides an optimally large interfacial area (transfer on both the inside and outside of the fiber) and agitation mechanism (solution stirring outside and dynamic plug flow mixing inside the fiber). Nevertheless, hollow fibers involve significant manual manipulation, which may restrict their general utility.

3.3.3.6 Model for Dispersive SME   Dispersive SME methods involve the dispersion of organic solvent as a "cloudy mixture" of tiny nanoliter-scale droplets suspended within the aqueous phase. The extremely large interfacial area and larger V_o/V_w ratio associated with dispersive SME means that equilibrium can be reached rapidly, and dispersive SME can therefore be considered an equilibrium technique. To illustrate this, consider the root-mean-square distance (d) traveled by a molecule with diffusion coefficient (D) in a short time period (t). This distance is given by the *Einstein–Smoluchowski relation:*[21]

$$d = \sqrt{2Dt} \qquad (3.36)$$

For a nominal diffusion coefficient of 5×10^{-6} cm^2/s, and a short time of 10 s, this distance is 0.01 cm, or 100 μm. Diffusion across aqueous (or organic) distances larger than this is normally not necessary in dispersive SME, and therefore equilibrium can be achieved quite quickly. A rigorous treatment of diffusion from aqueous solution into stationary small organic droplets is given elsewhere,[22,23] but such mass transfer modeling is of less practical importance in dispersive SME.

3.3.4 Three-Phase Kinetics

3.3.4.1 General Mass Transfer Model for Headspace SME   Here we consider a general model for three-phase kinetics applicable to headspace SME.[24] We begin with the same three postulates as for the two-phase system (see Section 3.3.3). Furthermore, we invoke a steady-state approximation for the intermediate (headspace or air) phase.[24,25] This approximation is implicit in the single dn/dt applicable to both transfer processes, neglecting the moles of analyte in the headspace in the mass balance [equation (3.41) below]. In other words, this model is applicable to semivolatile compounds with relatively low Henry's law constants. As explained elsewhere,[2] such compounds are most amenable to preconcentration by headspace SME versus direct headspace analysis.

Below are the starting equations for headspace SME, which place our postulates in mathematical form:

$$\frac{1}{A_{aw}}\frac{dn}{dt} = \beta_w(C_w - C_{w,i}) = \beta_{a1}(C_{a1,i} - C_a) \tag{3.37}$$

$$\frac{1}{A_{oa}}\frac{dn}{dt} = \beta_{a2}(C_a - C_{a2,i}) = \beta_o(C_{o,i} - C_o) \tag{3.38}$$

$$K_{aw} = \frac{C_{a1,i}}{C_{w,i}} \tag{3.39}$$

$$K_{oa} = \frac{C_{o,i}}{C_{a2,i}} \tag{3.40}$$

$$C_w V_w + C_o V_o = C_w^0 V_w \tag{3.41}$$

where A_{aw} and A_{oa} are the air–water and organic–air interfacial areas, respectively, and the subscripts $a1$ and $a2$ refer to the first and second air interfacial regions (adjacent to the water and organic phases, respectively). We begin to solve this system of equations by solving equation (3.40) for $C_{o,i}$, substituting in equation (3.38), and solving the second equality for $C_{a2,i}$. The first equality in equation (3.38) gives

$$\frac{dn}{dt} = A_{oa}\overline{\beta}_o(K_{oa}C_a - C_o) \tag{3.42}$$

where

$$\overline{\beta}_o = \frac{\beta_{a2}\beta_o}{\beta_{a2} + K_{oa}\beta_o} \tag{3.43}$$

Similarly, equation (3.37) becomes

$$\frac{dn}{dt} = A_{aw}\overline{\beta}_a(K_{aw}C_w - C_a) \tag{3.44}$$

where

$$\overline{\beta}_a = \frac{\beta_w\beta_{a1}}{\beta_w + K_{aw}\beta_{a1}} \tag{3.45}$$

Combining equations (3.42) and (3.44) gives an expression for the concentration in the headspace (air) phase:

$$C_a = \frac{A_{oa}\overline{\beta}_o C_o + A_{aw}\overline{\beta}_a K_{aw} C_w}{A_{oa}\overline{\beta}_o K_{oa} + A_{aw}\overline{\beta}_a} \tag{3.46}$$

Substituting this expression into equation (3.42), rearranging, and simplifying gives the following differential equation:

$$\frac{dC_o}{dt} = \frac{A_{oa}A_{aw}\overline{\beta}_o\overline{\beta}_a}{V_o(A_{oa}\overline{\beta}_oK_{oa} + A_{aw}\overline{\beta}_a)}(K_{ow}C_w - C_o) \tag{3.47}$$

Solving the mass balance equation (3.41) for C_w and substituting in equation (3.47) reduces the differential equation to a single variable that can be solved analogous to the two–phase case, resulting in the same first-order kinetic behavior as was observed for two-phase systems [equation (3.31)]. This "simple" kinetic behavior and steady-state approximation have been experimentally verified.[24,26] In the case of three-phase headspace SME, the resulting rate constant used in equation (3.31) is given by

$$k = \frac{A_{oa}A_{aw}\overline{\beta}_o\overline{\beta}_a}{V_o(A_{oa}\overline{\beta}_oK_{oa} + A_{aw}\overline{\beta}_a)}\left(K_{ow}\frac{V_o}{V_w} + 1\right) \tag{3.48}$$

To better understand the rate-limiting features of headspace SME, it is useful to examine equation (3.48) and the definitions of the overall mass transfer coefficients [equations (3.43) and (3.45)] more closely. In the denominator of equation (3.43) it would seem reasonable to assume that $\beta_{a2} \gg K_{oa}\beta_o$, since diffusion coefficients in the gas phase are so much larger than in the condensed phase as discussed earlier, and even though K_{oa} is normally large, it is probably not large enough to compensate for the difference in the mass transfer coefficients. Furthermore, there is probably some convection in the air phase, with little or none in the organic phase,[26] further exacerbating the difference between β_{a2} and β_o. So the resistance to mass transfer for the second (air–organic) step lies mainly with diffusion into the organic drop, and equation (3.43) may be approximated as

$$\overline{\beta}_o \approx \beta_o \tag{3.49}$$

Similar arguments can be made for the water–air mass transfer process [equation (3.45)], resulting in resistance to mass transfer within the water phase:

$$\overline{\beta}_a \approx \frac{\beta_w}{K_{aw}} \tag{3.50}$$

Substituting these approximate equations into the rate constant expression [equation (3.48)] gives

$$k = \frac{A_{oa}A_{aw}\beta_o\beta_w}{V_o[A_{oa}\beta_oK_{oa} + A_{aw}(\beta_w/K_{aw})]K_{aw}}\left(K_{ow}\frac{V_o}{V_w} + 1\right) \tag{3.51}$$

The magnitudes of the two terms in parentheses in the denominator are probably similar for a typical headspace SME analyte. A_{oa} and β_o are going to be approximately an order of magnitude each smaller than their water-phase counterparts given the small organic drop size versus the larger water–air interfacial area, and given the lack of convection within the organic drop. However, the difference in distribution coefficients (K_{oa} vs. $1/K_{aw}$) offsets this effect since a typical K_{oa} is on the order of 10^3, and K_{aw} is on the order of 10^{-1} for compounds such as the BTEX series (benzene, toluene, ethylbenzene, and xylene). This suggests that under these typical circumstances, mass transfer in both the water and organic phases limits the overall rate of extraction, consistent with experimental studies of BTEX compounds.[26]

On the other hand, for compounds of lower volatility where K_{aw} is extremely small, the second term in the denominator in equation (3.51) may dominate, and equation (3.51) reduces to

$$k \approx \frac{A_{oa}\beta_o}{V_o} \left(K_{ow} \frac{V_o}{V_w} + 1 \right) \qquad (3.52)$$

In this case, the rate-limiting step is diffusion into the organic drop, and the establishment of equilibrium between water and headspace phases is a fast process. This can also be understood intuitively in the sense that very little transfer from the water to headspace phase is necessary to reach equilibrium for compounds of low volatility. This observation has also been reported[27] in a study of headspace SME analysis of aldehydes with K_{aw} on the order of 10^{-4} or 10^{-5}.

Thus, in the case of analytes with low volatility, there is little benefit to agitating the aqueous solution. On the other hand, mass transfer into the organic drop is always a contributor to the kinetics of headspace SME. Headspace SME techniques employing a "dynamic" organic phase take advantage of this fact by inducing convection within the organic phase, enhancing the rate of extraction according to equation (3.52). Hollow fiber techniques may also enhance the rate via an increased A_{oa} as shown in equation (3.52).

It has commonly been reported that enhanced rates of extraction are possible with headspace SME (or headspace SPME) compared to direct two-phase extraction.[2,19] This observation is consistent with the model above. Consider the two-phase rate constant from equation (3.30), simplified assuming a large distribution constant (K_{ow} in this case) from equation (3.27):

$$k_{\text{2-phase}} \approx \frac{A_{ow}\beta_w}{V_o K_{ow}} \left(K_{ow} \frac{V_o}{V_w} + 1 \right) \qquad (3.53)$$

Comparison of the three-phase rate constant from equation (3.52) with this two-phase rate constant reveals that the three-phase (headspace) k is higher by a factor of $K_{ow}\beta_o/\beta_w$. (A_{oa} and A_{ow} are approximately the same.) As long as K_{ow} is large, this will override any smaller differences in the mass transfer coefficients and result in a larger rate constant for the headspace SME system. Intuitively, this is due to the ideas presented earlier: rapid equilibration of the water–headspace system for analytes with low volatility, and the very fast mass transfer rates in the headspace phase (versus the water phase in a two-phase system) to the organic drop surface.

3.3.4.2 Mass Transfer Model for Liquid–Liquid–Liquid SME

The adaptation of the general three-phase model [equations (3.37) through (3.41)] to liquid–liquid–liquid (aqueous–organic–aqueous) systems is possibly complicated by the fact that the mass balance condition may need to include the analyte present in the intermediate (in this case, organic) phase since the distribution constant from the donor to organic phase is normally large in these systems, unlike the case with headspace SME. An assumption of fast acid–base chemistry and the presence of sufficient concentration of strong acid or base near the interface in the receiver phase would also be required. (Equilibria among the donor, organic, and receiver phases may be described as in Section 3.2.5.) Ma and Cantwell have developed a rigorous kinetic treatment of such systems based on analogy with consecutive reversible homogeneous first-order chemical reactions.[28] Rate control was presumed to be on the donor side of the donor–organic interface and the organic side of the organic–receiver interface, implying the importance of generating convection within both the donor and organic phases for maximum rate of extraction.

In a second paper, Ma and Cantwell invoke the steady-state approximation for the organic "intermediate" for the study of the kinetics of preconcentration into a small microdrop of receiver phase.[29] Their model incorporates a lag time related to the time required to achieve steady state. The enrichment factor (E_f = concentration in receiver / initial concentration in donor) is given by

$$E_f \approx \frac{V_{\text{donor}}}{V_{\text{receiver}}} \left[1 - e^{-k(t - t_{\text{lag}})} \right] \tag{3.54}$$

An important conclusion from this study is that although going to a smaller V_{receiver} results in a smaller interfacial area between the organic and receiver phases (hence somewhat reducing k), the E_f is always increased at any time by going to a smaller volume of receiver phase according to equation (3.54).

3.3.4.3 Carrier-Mediated and Electrokinetic Transport in Liquid–Liquid–Liquid SME

The analysis of polar compounds by liquid–liquid–liquid SME is limited by the low distribution constant of the analyte between the donor and organic phases. Carrier-mediated transport has been applied successfully to solve this type of analytical problem.[30] Figure 3.7 shows a schematic model for this approach.

Basic analyte molecules (A) with relatively high pK_a are ionized (AH$^+$) in the donor solution, which is maintained at a moderate pH (e.g., 7). The donor solution also contains an organic carboxylic acid carrier (**R-COOH**) with a much lower pK_a which is deprotonated (**R-COO$^-$**) at this pH. Thus, an ion pair (**AH$^+$R-COO$^-$**)

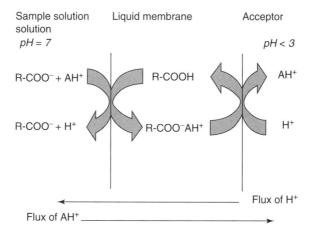

A = analyte, R-COOH = carboxylic acid

FIGURE 3.7. Schematic diagram of carrier-mediated liquid–liquid–liquid SME. (Reprinted with permission from Ref. 30; copyright © 2003 Elsevier.)

forms which is readily extracted into the organic layer. On the receiver side, the pH is much lower (well below the pK_a value of the carrier, which becomes neutralized), leaving the protonated analyte (AH^+) to remain in the receiver phase. The carrier acts as a proton and analyte "shuttle," effectively carrying H^+ back to the donor phase and becoming ionized once again, now capable of transporting another AH^+ ion across the membrane.

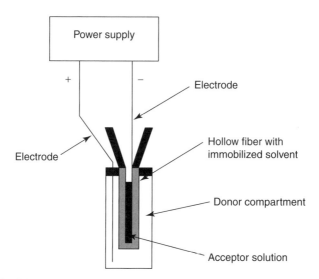

FIGURE 3.8. Schematic diagram of the electromembrane isolation technique. (Reprinted with permission from Ref. 31; copyright © 2006 Elsevier.)

Bjergaard and Rasmussen have also reported on electrokinetic transport as a means of driving basic hydrophobic substances through liquid membranes in liquid–liquid–liquid SME, which they termed *electromembrane isolation* (EMI).[31] This technique is illustrated in Figure 3.8.

Basic hydrophobic analyte molecules are protonated (AH^+) in all three phases in EMI. The positive electrode is placed in the donor side, and the negative electrode is placed in the receiver side, thus driving AH^+ from the donor phase through the organic layer (2-nitrophenyl octyl ether) in the hollow fiber and into the receiver phase. In common with the carrier-mediated process, the goal is primarily sample cleanup for drug analysis in complex biological systems such as plasma or urine.

3.4 CALIBRATION METHODS

We conclude our chapter on SME theory with a consideration of issues related to analytical calibrations based on our theoretical models. Quantitative analysis by SME ultimately involves an empirical correlation of a measured property (e.g., chromatographic peak area of the analyte in the extracting phase) as a function of the original concentration of the analyte in the aqueous or donor phase. (Since SME is not typically an exhaustive extraction technique, preparation of standards in the extracting phase and calibration based on this is not normally useful.) SME calibration is therefore based on equilibria among the phases, or analysis of mass transfer–controlled systems that are not at equilibrium but are carefully controlled in terms of kinetic factors. Equations (3.3) and (3.8) show that the equilibrium concentration in the extracting phase is directly proportional to the original sample concentration for two- and three-phase systems, respectively, provided that the distribution constants and volumes are constant. This also implies that temperature and salt concentrations must remain constant, as discussed in Section 3.2.4. Furthermore, other matrix effects on the distribution constant must be absent (or handled using a standard-addition calibration approach). So assuming a linear chromatographic response of peak area versus concentration, peak areas should be proportional to the original analyte concentrations.

In cases where the time required to reach equilibrium is too long, linear calibrations can also be obtained under nonequilibrium conditions under certain circumstances. Equation (3.31) (applicable to both two- and three-phase systems) shows that the concentration in the extracting phase at any time is proportional to the equilibrium concentration provided that the rate constant, k, is indeed constant and the extraction time is fixed. Thus, all the factors that go into the rate constant [equations (3.30) and (3.48)] must be held constant throughout the analysis of all standards and unknown samples. These factors include interfacial areas (drop shapes, etc.), phase volumes, mass transfer coefficients (diffusion coefficients: i.e., temperature and solvent viscosity; degree of agitation or stirring) and distribution constants (temperature and salt concentration). Many of these factors are best controlled and most reproducible with automated approaches.

In most cases, the use of an internal standard is required for maximum precision. Peak area ratios of analyte to internal standard are used as the measured property and correlated with initial sample concentration. The internal standard may be present and soluble only in the extracting phase, or it may be present in the sample (aqueous phase) and coextract along with the analyte. The latter approach has the advantage of partially correcting for variations in the kinetic or equilibrium aspects of the extraction process in addition to chromatographic variations. Having an internal standard present in the extracting drop is also useful as a way to monitor drop losses during the extraction process, particularly when using an autosampler where there is no visual check of the drop size prior to analysis.

Pawliszyn and co-workers have also presented an alternative calibration method based on preloading the extracting phase with a known amount of analyte (or isotopically labeled analyte).[32,33] The rate constant for the desorption of the preloaded analyte is assumed to be the same as the rate constant for the absorption of analyte from the sample solution. This method allows for calibration without adding anything to the aqueous sample solution or preparing any aqueous standards and is capable of dealing with complex sample matrixes since the same sample matrix is used for the desorption and absorption processes. This method does require knowledge of the equilibrium distribution constants to determine the aqueous sample concentrations.

3.5 SUMMARY

Solvent microextraction is a simple and elegant approach to sample cleanup and preconcentration. The intent of this chapter was to give the reader important insights into the physical and chemical principles of SME, to provide a transition into more practical considerations (Chapter 4) and to establish a sound basis for rational analytical method development (Chapter 5). The equilibrium considerations have revealed that SME (under equilibrium conditions) is generally intermediate between nonperturbing and exhaustive extraction. Equilibrium partitioning can be predicted from octanol–water partition coefficients, Henry's law constants, standard thermodynamic temperature and ionic strength effects, and chemical speciation. Special consideration must be given to solvent water solubility and volatility in SME.

Since the attainment of equilibrium in SME techniques may require up to an hour or more, analyses are commonly carried out under nonequilibrium conditions. This necessitates a detailed understanding of the experimental variables that control the rate of extraction. The discussion in this chapter sheds light on the effects of variables such as interfacial areas, phase volumes, stirring conditions in water and/or organic phases, temperature, and distribution coefficients on the rate of extraction. These variables must be optimized and carefully controlled for high sensitivity, good reproducibility, and linear calibration curves.

REFERENCES

1. Jeannot, M. A.; Cantwell, F. F., Solvent microextraction into a single drop. *Anal. Chem.* 1996, *68* (13), 2236–2240.
2. Przyjazny, A.; Kokosa, J. M., Analytical characteristics of the determination of benzene, toluene, ethylbenzene and xylenes in water by headspace solvent microextraction. *J. Chromatogr. A* 2002, *977* (2), 143–153.
3. Schwarzenbach R. P.; Gschwend, P. M.; Imboden, D. M., *Environmental Organic Chemistry,* 2nd ed., Wiley-Interscience, Hoboken, NJ, 2003, Chap. 3.
4. Schwarzenbach R. P.; Gschwend, P. M.; Imboden, D. M., *Environmental Organic Chemistry,* 2nd ed., Wiley-Interscience, Hoboken, NJ, 2003, pp. 223–235.
5. Schwarzenbach R. P.; Gschwend, P. M.; Imboden, D. M., *Environmental Organic Chemistry,* 2nd ed., Wiley-Interscience, Hoboken, NJ, 2003, pp. 197–208.
6. Schwarzenbach R. P.; Gschwend, P. M.; Imboden, D. M., *Environmental Organic Chemistry,* 2nd ed., Wiley-Interscience, Hoboken, NJ, 2003, pp. 215–216.
7. Schwarzenbach R. P.; Gschwend, P. M.; Imboden, D. M., *Environmental Organic Chemistry,* 2nd ed., Wiley-Interscience, Hoboken, NJ, 2003, p. 199.
8. Schwarzenbach R. P.; Gschwend, P. M.; Imboden, D. M., *Environmental Organic Chemistry,* 2nd ed., Wiley-Interscience, Hoboken, NJ, 2003, pp. 199–203, 215–216.
9. Jeannot, M. A.; Cantwell, F. F., Solvent microextraction as a speciation tool: determination of free progesterone in a protein solution. *Anal. Chem.* 1997, *69* (15), 2935–2940.
10. Cantwell, F. F.; Losier, M., Liquid–liquid extraction. In J. Pawliszyn, ed., *Sampling and Sample Preparation for Field and Laboratory,* Elsevier, New York, 2002, p. 298.
11. Cussler, E. L., *Diffusion: Mass Transfer in Fluid Systems,* 2nd ed., Cambridge University Press, New York, 1997, pp. 133–134.
12. Cussler, E. L., *Diffusion: Mass Transfer in Fluid Systems,* 2nd ed., Cambridge University Press, New York, 1997, pp. 111–121.
13. Jeannot, M. A., Ph.D. dissertation, University of Alberta, 1997.
14. Psillakis, E.; Kalogerakis, N., Developments in single-drop microextraction. *Trends Anal. Chem.* 2002, *21* (1), 53–63.
15. Cussler, E. L., *Diffusion: Mass Transfer in Fluid Systems,* 2nd ed., Cambridge University Press, New York, 1997, pp. 332–338.
16. Jeannot, M. A.; Cantwell, F. F., Mass transfer characteristics of solvent extraction into a single drop at the tip of a syringe needle. *Anal. Chem.* 1997, *69* (2), 235–239.
17. Cantwell, F. F.; Losier, M., Liquid–liquid extraction. In J. Pawliszyn, ed., *Sampling and Sample Preparation for Field and Laboratory,* Elsevier, New York, 2002, p. 305.
18. He, Y.; Lee, H. K., Liquid-phase microextraction in a single drop of organic solvent by using a conventional microsyringe. *Anal. Chem.* 1997, *69* (22), 4634–4640.
19. Pawliszyn, J., Theory of solid-phase microextraction. *J. Chromatogr. Sci.* 2000, *38* (7), 270–278.
20. Zhao, L.; Lee, H. K., Liquid-phase microextraction combined with hollow fiber as a sample preparation technique prior to gas chromatography/mass spectrometry. *Anal. Chem.* 2002, *74* (11), 2486–2492.
21. Schwarzenbach R. P.; Gschwend, P. M.; Imboden, D. M., *Environmental Organic Chemistry,* 2nd ed., Wiley-Interscience, Hoboken, NJ, 2003, p. 788.
22. Crank, J., *The Mathematics of Diffusion*, 2nd ed., Oxford University Press, New York, 1975, pp. 89–96.
23. Cantwell, F. F.; Losier, M., Liquid-liquid extraction. In J. Pawliszyn, ed., *Sampling and Sample Preparation for Field and Laboratory,* Elsevier, New York, 2002, pp. 315–318.

24. Schnobrich, C. R.; Jeannot, M. A., Steady-state kinetic model for headspace solvent microextraction. *J. Chromatogr. A* 2008, *1215* (1–2), 30–36.

25. Ritchie, C. D., *Physical Organic Chemistry: The Fundamental Concepts,* 2nd ed., Marcel Dekker, New York, 1990, pp. 3–11.

26. Theis, A. L.; Waldack, A. J.; Hansen, S. M.; Jeannot, M. A., Headspace solvent microextraction. *Anal. Chem.* 2001, *73* (23), 5651–5654.

27. Fiamegos, Y. C.; Stalikas, C. D., Theoretical analysis and experimental evaluation of headspace in-drop derivatisation single-drop microextraction using aldehydes as model analytes. *Anal. Chim. Acta* 2007, *599* (1), 76–83.

28. Ma, M.; Cantwell, F. F., Solvent microextraction with simultaneous back-extraction for sample cleanup and preconcentration: quantitative extraction. *Anal. Chem.* 1998, *70* (18), 3912–3919.

29. Ma, M.; Cantwell, F. F., Solvent microextraction with simultaneous back-extraction for sample cleanup and preconcentration: preconcentration into a single microdrop. *Anal. Chem.* 1999, *71* (2), 388–393.

30. Ho, T. S.; Halvorsen, T. G.; Pedersen-Bjergaard, S.; Rasmussen, K. E., Liquid-phase microextraction of hydrophilic drugs by carrier-mediated transport. *J. Chromatogr. A* 2003, *998* (1–2), 61–72.

31. Pedersen-Bjergaard, S.; Rasmussen, K. E., Electrokinetic migration across artificial liquid membranes: new concept for rapid sample preparation of biological fluids. *J. Chromatogr. A* 2006, *1109* (2), 183–190.

32. Ouyang, G.; Zhao, W.; Pawliszyn. J., Kinetic calibration for automated headspace liquid-phase microextraction. *Anal. Chem.* 2005, *77* (24), 8122–8128.

33. Ouyang, G.; Pawliszyn, J., Kinetic calibration for automated hollow fiber–protected liquid-phase microextraction. *Anal. Chem.* 2006, *78* (16), 5783–5788.

CHAPTER 4

PRACTICAL CONSIDERATIONS FOR USING SOLVENT MICROEXTRACTION

4.1 INTRODUCTION

Solvent microextraction (SME) is a relatively new technique and is therefore developing rapidly. Although experimental conditions do need to be optimized for the specific analytes and matrices present in the sample, the literature has evolved enough so that recommended starting points for the development of a method are available.

In this chapter we present practical approaches for choosing the experimental conditions needed for developing an SME sample preparation method. This includes a suggested approach to deciding the general mode to use (direct immersion, SDME; headspace, HS-SDME; hollow fiber protected; HFME; dynamic, DY-SME; dispersive, DLLME) depending on available instrumentation and the class of chemicals to be extracted and analyzed. Experimental methodology can be as simple as using a standard GC syringe and a magnetic stirrer for extraction (Figure 4.1) or may involve the use of a computer-controlled autosampler system for the extraction (Figure 4.2). Chapters 2 and 3 provide insight into the background, general methodology, and theory of SME. Many of the points covered briefly in this chapter are discussed in detail in Chapter 5.

Solvent Microextraction: Theory and Practice, By John M. Kokosa, Andrzej Przyjazny, and Michael A. Jeannot
Copyright © 2009 John Wiley & Sons, Inc.

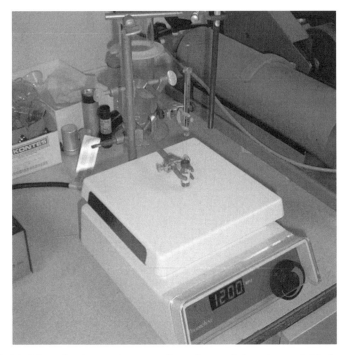

FIGURE 4.1. Manual solvent microextraction using a stir hot plate and 10-µL syringe.

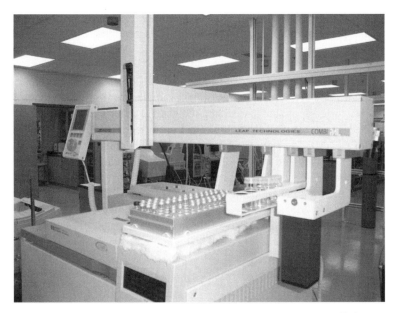

FIGURE 4.2. Automated solvent microextraction using a computer-controlled autosampler and stirring magnetic mixer.

4.2 GENERAL RECOMMENDATIONS

A summary of the most crucial factors necessary for the development and optimization of an SME method is given below.

1. Use headspace SME for volatile nonpolar chemicals.
2. Use direct-immersion SME (SDME) or HF(2)ME for nonvolatile, nonpolar, or mildly polar chemicals.
3. Use dispersive liquid–liquid solvent microextraction (DLLME) for nonpolar, semivolatile analytes such as the PAHs, PCBs, or pesticides in water at very low concentrations or when HPLC is used for the final determination.
4. Use HF(3)ME or LLLME for ionizable chemicals, such as amines, phenols, and carboxylic acids.
5. Use an extraction solvent with a boiling point higher than the analytes to be extracted, such as tetradecane or 1-octanol, to extract volatile chemicals, with a GC inlet split of 10:1 to 40:1 for SDME, HS-SDME, and HF(2)ME.
6. Use an extraction solvent with a boiling point lower than the analytes to be extracted, such as toluene, o-xylene, decane, or 1-octanol, for low-volatility chemicals, with splitless GC injection for SDME, HS-SDME, and HF(2)ME.
7. Use an extraction solvent insoluble in water with a specific gravity greater than 1, such as carbon disulfide, carbon tetrachloride, tetrachloroethene, or chlorobenzene, for DLLME or with a density lower than water and a melting point just below room temperature, such as 1-undecanol or hexadecane for the floating drop modification of DLLME.
8. Use a drop size ranging from 1.3 to 2.5 µL, with 1 to 2 µL injected into the GC for SDME and HS-SDME. This will protect the instrument from water and salt contamination and minimize variations in solvent injected caused by solvent creep up the needle.
9. If you use added salt to enhance extraction, use salt concentrations near but below, saturation (ca. 300 mg NaCl/mL water sample is adequate) and weigh the salt to the nearest milligram. Remember, adding salt will increase the total volume of the sample (by about 15% for nearly saturated NaCl solutions).
10. Use 1-to-4 mL samples where possible for SDME, HS-SDME, and HF(2)ME. Volumes larger than 4 mL are useful only for chemicals with very large (>1000) K_{ow} values when present at very low [parts per trillion (ppt)] levels. Headspace volumes should always be minimized for both direct and headspace extractions.
11. Always stir or agitate a sample before and during sampling. If an autosampler is available, use a dynamic extraction technique to increase extraction efficiency and decrease extraction time.
12. Use a relatively nonvolatile internal standard such as decafluorobiphenyl or p-bromofluorobenzene dissolved in the extracting solvent at a concentration

of about 5 to 50 µg/mL (depending on the sensitivity of the detector used). This allows correcting for small variations in the amount of solvent withdrawn into the syringe and monitoring possible drop loss.

13. Use a Hamilton 701N point (22°) syringe for manual SDME and HS-SDME modes. At the present time, this syringe is the most efficient in forming the microdrop and withdrawing the liquid back into the syringe. It also works well with an autosampler. Its only major drawback is that the GC septum will need to be replaced after 50 to 100 injections. If this is a serious issue, use a No.2 flat-bevel needle syringe.

14. Use a specialized technique such as hollow fiber–protected SME (HFME) for biological samples, which may need to be extracted using a two- or three-phase system (organic solvent in the lumen of the fiber and 10 to 20 µL of the organic or aqueous extractant inside the wall). Other specialized techniques involve using a complexing or derivatizing agent in the extracting solvent to increase extraction selectivity.

15. Use temperature control, if possible, for the sample. A temperature of 25 to 30°C is the best starting point. Elevated sample temperatures are counterproductive for most volatile chemicals, unless very short (1 min or less) extraction times are used or a cooled syringe needle and an autosampler are available, enabling dynamic extractions and a temperature-controlled microdrop.

16. Solvent purity is critical. Even the purest solvents can contaminate rapidly from the atmosphere. The level of purity required will depend on the concentrations of analytes in the sample and the type of detector used for the analysis

17. If at all possible, use a computer-controlled autosampler system for the method involving a large number of extractions. This will not only permit multiple unattended analyses, but also allow the use of dynamic extraction, which shortens extraction time.

4.3 GENERAL QUESTIONS TO CONSIDER BEFORE PERFORMING AN ANALYSIS

As is the case when developing any analytical method, a series of questions concerning the sample and the analytical instrumentation available need to be answered.

4.3.1 What Are the Properties of the Chemicals to Be Extracted?

1. What is the boiling point of the analyte? Is it relatively volatile or nonvolatile?
2. Is the analyte polar or nonpolar? What is its solubility in water? Is it capable of hydrogen bonding?

3. What is the log K_{ow} value for the chemical? The octanol–water partition coefficient may be found in the literature or calculated using a commercial program or by hand.[1,2]

4. What is the dimensionless Henry's law constant (K_{aw}) for the chemical? This value is also available in the literature, can be calculated using a commercial program or a free program available through the EPA[2] or can be calculated by hand.[3]

5. Do the analytes present in the sample contain halogens, nitrogen, oxygen, or sulfur? Are they alkanes, alkenes, or aromatic hydrocarbons? These data, along with the expected concentration levels of the analytes, will determine the type of instrumentation used for separation and detection.

4.3.2 What Type of Sample Matrix Will Be Analyzed?

1. Is the analyte in relatively clean water or water that contains a complex matrix?

2. Is the analyte in a solid matrix (such as residual solvents in pharmaceuticals or environmental contaminants in soil)?

3. Is the analyte in the atmosphere?

4. Is the analyte in a complex matrix such as engine oil?

4.3.3 What Analytical Instrumentation Is Available?

1. What analytical instrumentation will be used?

2. Is an autosampler available?

3. Is a syringe needle cooler available?

4. What detector are you using?

5. What are the sensitivity and flow rate capacity of the detector?

6. What chromatographic columns are available?

7. What type of stirrer or agitator are you using? Is temperature control possible?

8. How much time does the chromatographic method allow to carry out an extraction?

4.3.4 What Is the Concentration of the Analyte?

1. Is the analyte relatively concentrated, dilute, or present in trace amounts?

2. Is the analyte present at concentrations comparable to the concentrations of other components of the matrix?

3. Do you need to enhance extraction efficiency by using a technique such as salting out, dynamic sampling, or a chemical aid, such as the formation of a chemical derivative or formation of a salt with pH change?

4.4   CHOOSING THE SME MODE

4.4.1   Direct-Immersion Single-Drop Microextraction

If the analyte to be extracted has a relatively high molecular weight and melting point/boiling point or fair water solubility, you may want to consider SDME.[4,5] If the matrix is contaminated with salts, insoluble organic material (such as humus), or soluble proteins, you have to be careful that these contaminants do not enter the sampling syringe. Analytes that might be appropriate for SDME would include the higher-molecular-weight PAHs, such as benzo[a]pyrene, PCBs, and nonpolar pesticides. These analytes have relatively low vapor pressures (high boiling points) and large octanol–water partition coefficients (K_{ow} value above 2000). They can often be easily extracted into a relatively low-boiling liquid such as toluene or the xylenes, although a higher-boiling solvent, such as 1-octanol or tetradecane, can be used if lower boiling than the analytes. More polar analytes, such as ketones, alcohols, and amines, which can hydrogen bond, are more efficiently extracted into solvents such as 1-octanol, which have dipole–dipole and hydrogen-bonding capacity. As discussed in detail later, SDME extractions are slower than headspace SDME, and a dynamic extraction method using an autosampler will decrease extraction time.

4.4.2   Headspace Extraction

Headspace SDME is the preferred choice for chemicals that have relatively high vapor pressures and thus lower boiling points.[6,7] Additionally, this is the method of choice if the matrix is contaminated with nonvolatile materials such as salt, humus, or proteins. Examples of the types of analytes that are readily extracted include the halogenated alkanes, alkenes, aromatics, ketones and aldehydes, esters, alcohols, and lower-boiling carboxylic acids. The chlorination disinfection by-products and gasoline in oil and water are specific examples. The method is successful for chemicals such as naphthalene, which has a boiling point of 218°C. Typical extracting solvents used include 1-octanol, dodecane, tetradecane, and hexadecane. Extraction from water into the extracting microdrop is rapid for nonpolar volatile chemicals, lending this technique to successful static drop manual and automated extractions. However, dynamic extraction methods will decrease extraction time and increase extraction efficiency. The technique is less effective for aqueous solutions of analytes such as acetone with high solubilities in water.

Figures 4.3 and 4.4 illustrate the very rapid and efficient extraction of EPA method 524.2 volatiles from water by HS-SDME as compared to SDME. Figure 4.3 is a GC-MS chromatogram of a SDME extract of a solution of 1 mL of water saturated with salt and containing 4 µg/L of each analyte. Extraction time was 10 min; stirring rate was 500 rpm at 30°C, and the extracting solvent was 1 µL of tetradecane containing 5 ppm decafluorobiphenyl as an internal standard. For comparison, Figure 4.4 is a GC-MS of a HS-SDME extract under identical conditions, except that the extraction time was reduced to 3 min.

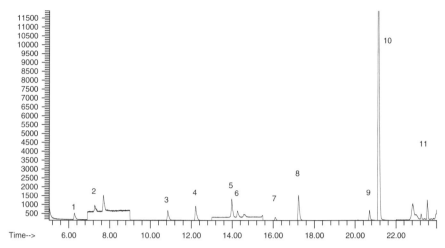

FIGURE 4.3. Direct immersion SDME of EPA method 524.2 volatiles; extraction time: 10 min; extraction temperature: 30°C; GC-MS SIM analysis; inlet split: 10:1; sample: 1 mL of water containing 4 µg/L of each analyte and 300 mg/mL NaCl in a 2-mL vial with a micro Teflon-coated stir bar; stirring rate: 500 rpm; extracting solvent: tetradecane (1.3 µL); volume injected: 1 µL. Peak identification: (1) chloroform, (2) benzene, (3) toluene, (4) tetra-chloroethylene, (5) chlorobenzene, (6) ethylbenzene, (7) bromoform, (8) bromobenzene, (9) *o*-dichlorobenzene, (10) decafluorobiphenyl (I.S.), (11) naphthalene.

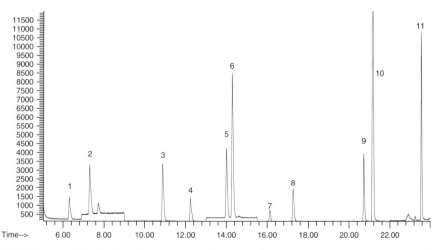

FIGURE 4.4. HS-SDME of EPA method 524.2 volatiles; extraction time: 3 min; GC-MS SIM analysis; sample: 1 mL of water containing 4 µg/L of each analyte. Conditions and peak identities as in Figure 4.3.

4.4.3 Dynamic Extraction

Dynamic SME may shorten the extraction times required for direct-immersion SME (SDME) and HS-SDME. This has been found to be especially useful for very dilute solutions. The technique is also useful when using elevated extraction temperatures for HS-SDME, since potentially the temperature of the microdrop will be lower than the temperature of the headspace, thus enhancing extraction into the drop by decreasing the air–organic solvent partition coefficient. Originally this technique was carried out manually,[8,9] then with a crude mechanical device[10] and a syringe pump,[11,12] but its potential can be fully realized only if a programmable auto-sampler is used.[13] Reproducibility depends on control of the syringe plunger speed and extraction time. Essentially, the technique is the same as a single or static extraction, except that the drop is exposed to the sample for only a short period, perhaps 12 s, and then withdrawn into the syringe. The drop is again exposed to the sample and the process repeated. Typically, 30 cycles are needed for SDME and 10 cycles for HS-SDME to maximize analyte extraction. Dynamic SDME can be especially helpful for very dilute solutions.

As seen in Figures 4.3 and 4.4, static SDME requires much longer extraction times than static HS-SDME. When the same conditions were applied to 1 mL of saturated salt solution containing 400 ng/L of each component, similar extraction efficiencies were obtained for 10-min extractions using static HS-SDME and dynamic SDME, as shown in Figures 4.5 and 4.6.

The extraction times required for dynamic SME are decreased because the surface of the drop is renewed continually. Withdrawal of the solvent into the syringe has the added benefit that the liquid is cooled by the syringe barrel rather than being heated continually by the headspace vapor, thus increasing the potential for

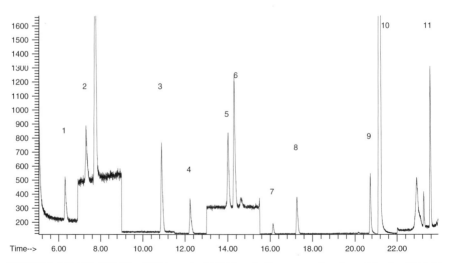

FIGURE 4.5. HS-SDME of EPA method 524.2 volatiles; extraction time: 10 min; GC-MS SIM analysis; sample: 1 mL of water containing 400 ng/L of each analyte. Conditions and peak identities as in Figure 4.3.

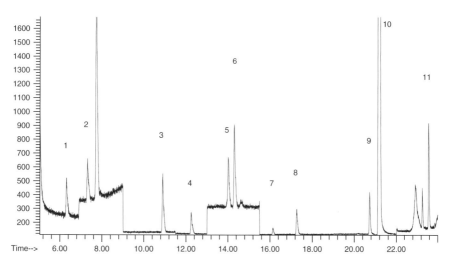

FIGURE 4.6. Dynamic in-needle direct-immersion SDME (30 cycles) of EPA method 524.2 volatiles; extraction time: 10 min; GC-MS SIM analysis; sample: 1 mL of water containing 400 ng/L of each analyte. Conditions and peak identities as in Figure 4.3.

enhanced analyte extraction into the drop. This advantage would be even more enhanced if a syringe needle cooling device were used. A computer-controlled thermoelectric cooled attachment is now commercially available for this purpose.[14]

4.4.4 Hollow Fiber–Protected Microextraction

When it is necessary to use direct extraction, but the sample matrix is contaminated by insoluble inorganic and organic material, or soluble high-molecular-weight organic material such as proteins, the HFME technique may prove useful.[15] The extracting solvent (ca. 10 to 20 µL) is contained in a small-diameter porous polymer tube (1 to 2 mm × 2 cm, polypropylene) with the bottom end pressure sealed, which is then attached to a holder or syringe needle and immersed in the stirred solution. The analytes equilibrate between the solution, the solvent-saturated polymer, and the extracting solvent, which can be the same as the solvent with which the fiber is saturated or even an aqueous solution with a low or high pH. The syringe is then used to withdraw 1 to 2 µL of the solvent, which is injected for analysis. Thus, only about 10% of the extracting solvent is analyzed if GC is used for the separation and final determination. Due to the larger extraction volume, more analyte is extracted and, by using a larger volume of water sample, it is possible to analyze nearly the same amount of analyte as extracted by SDME, but without the possibility of contamination with water, salt, or insoluble matrix components. For example, extraction of 1 mL of water containing 10 ng of benzene with 2 µL of 1-octanol would extract 2.28 ng of benzene. Extraction of 4 mL of the same solution with 20 µL of 1-octanol in a hollow fiber tube and injecting 2 µL of the extract would analyze 1.71 ng of extracted benzene. This method is especially useful for biological samples such as serum and urine, but has no major advantage over SDME or HS-SDME for most environmental water samples. One major difficulty with the technique is that while it can be performed with an autosampler, the fibers at present must be prepared manually, leading to potential problems with reproducibility and increasing the manual labor required.[16]

4.4.5   Dispersive Liquid–Liquid Microextraction

DLLME may prove useful for aqueous samples containing very low concentrations of high-boiling nonpolar chemicals, such as the PAHs, PCBs, and pesticides in water.[17–19] The method consists of dissolving a small amount (8 to 50 μL) of a water-insoluble extracting solvent with a specific gravity greater than 1.0 in a minimum amount (0.5 to 2.0 mL) of a water-soluble solvent and injecting the solution rapidly into 5 mL of the water sample in a centrifuge tube. The micro-dispersed insoluble extracting solvent is centrifuged to the bottom of the tube, removed with a syringe, and injected into a GC. The solvent can also be evaporated and replaced by methanol or acetonitrile and the solution then analyzed by HPLC.[20] This, in fact, may be the best SME method for preparing HPLC samples for non-polar analytes, since a relatively large amount of extracting solvent and sample ensures enough analyte availability for HPLC analysis. As with HFME, this technique requires manual collection of the extracting solvent, which is then analyzed using a standard manual or autosampler injection method.

4.5   EXTRACTION SOLVENT

Chloroform, methylene chloride, ether and ethyl acetate are traditional macro-extraction solvents. However, they perform poorly in SDME and HS-SDME. For a headspace extraction with 0.5 to 1.0 mL of headspace, these solvents evaporate in less than a minute. In SDME they start dissolving and evaporating immediately, even at room temperature, and you begin to see the formation of a vapor bubble in the drop. One researcher has reported the successful use of mixed solvents—small amounts of chloroform or other volatile solvent mixed with toluene or other higher-boiling solvents.[21] The mixtures had enhanced and more specific extraction capacity than those of pure solvents. Much more work must be done in this area, however. Several solvents are commonly used in SME. Useful SDME solvents include toluene, xylene, 1-octanol, and decane. HS-SDME solvents commonly used include 1-octanol, dodecane, tetradecane, and hexadecane. Alkane hydrocarbons, such as tetradecane, are of limited utility for macroextractions, since most even slightly polar organic chemicals have only limited solubilities in alkanes. However, the concentrations of analytes in most samples are very low, and they dissolve readily in 1 μL of tetradecane. It should be remembered that for most analytes, SME is not an exhaustive extraction technique and does not have to be so to work well. Many other solvents can be used, however, depending on the nature of analytes. In fact, one research group has successfully employed ionic liquids for extractions,[22,23] and water-based extractants[24] may be potential solvents for headspace extractions when needle cooling is possible.[25,26]

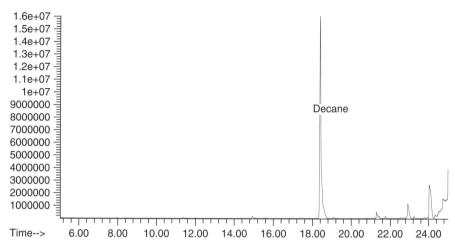

FIGURE 4.7. GC-MS chromatogram of 0.1 μL of neat tetradecane (99.9% purity); 50:1 inlet split. Peak at 18.5 min is decane impurity.

The solvent chosen is usually lower boiling than the analytes to be extracted when extracting high-boiling chemicals by SDME. The solvent selected for HS-SDME is typically higher boiling than the extracted analytes. There are also practical matters to consider. Even the best commercial solvents may need to be purified before they can be used for SME. Upon sitting, solvents will, of course, absorb organic chemicals from the atmosphere. The chromatograms below illustrate this point. The top chromatogram (Figure 4.7) is 99.9% commercial tetradecane as received.[27] The second chromatogram (Figure 4.8) is the same solvent after vacuum fractional distillation. The large peak at 18.5 min is decane. Because the

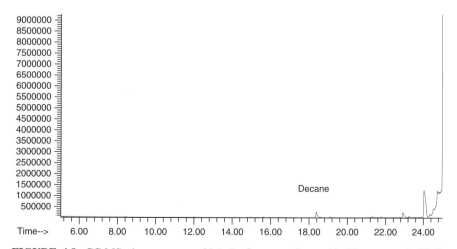

FIGURE 4.8. GC-MS chromatogram of 0.1 μL of neat tetradecane (doubly vacuum distilled); 50:1 inlet split. Peak at 18.5 min is decane impurity.

TABLE 4.1 Common Solvents Used in Solvent Microextraction

SDME Solvent	b.p./m.p. °C	HS-SDME Solvent	b.p./m.p. °C
Toluene	110/−93	Water	100/0
Octane	125/−57	Undecane	196/−26
o-Xylene	143/−26	1-Octanol	196/−15
Decane	174/−30	Dodecane	215/−10
Undecane	196/−26	Tetradecane	252/5.5
1-Octanol	196/−15	Hexadecane	287/18
Tetradecane	252/5.5	Diethyl phthalate	298/−3

amount of extracting solvent drawn back into the syringe tends to vary slightly, even when using an autosampler, precision is improved if a relatively nonvolatile internal standard such as decafluorobiphenyl (b.p. = 206°C, m.p. = 68 to 70°C) or 4-bromofluorobenzene (BFB; b.p. = 150°C, m.p. = −16°C) is added to the extracting solvent at concentrations of about 5 to 50 µg/mL as an internal standard. The decane impurity in tetradecane could also be used as an internal standard for this solvent, in some cases, since its concentration remains unchanged for a batch of distilled solvent.

Table 4.1 lists seven solvents appropriate for use with SDME and with HS-SDME. The boiling points and melting points are also listed. Since most SME applications involve extractions from water, these solvents (with the exception of water) have limited water solubility. All can be purchased at 99% or greater purity and all show good chromatographic behavior. Water can, in fact, be used for GC, with caution, depending on the detector being used. It may be necessary to distill these solvents under vacuum before use, however. One must exercise caution when using very high-boiling solvents such as hexadecane and diethyl phthalate, and to a lesser extent tetradecane and 1-octanol, as extracting solvents. These chemicals tend to condense out in the cooler points of the GC injector pneumatics and can cause problems with the flow valves and vapor traps, especially when splitless or low-ratio split injections are made. To avoid this problem, increase the flow rate through the injector split valve following the injection or after the analysis.

Good chromatographic practice when using 1-octanol, tetradecane, and especially hexadecane solvents is to inject one or two solvent blanks prior to injecting any sample extracts, since detector response stability seems to depend on a passivation or coating of the surfaces of the injector with the solvent. It takes one or two dry runs, in fact, to bleed all of this coating out of the GC system. Despite this, however, there seems to be no carryover problem with the extractants.

4.6 SAMPLE VOLUMES

The general approach taken by many chemists is that if a small sample works, why not use a larger sample to maximize method sensitivity? The first question that one should ask, however, is what is the minimum amount of analyte that can be detected by the analytical method? If the sample preparation technique cannot yield enough analyte in the extract to be detected at the minimum detectable level required, another technique should be chosen. Use the formulas presented below to do a rough calculation, as will be illustrated later, to see whether you need a 1- or a 40-mL sample. If the analyte is relatively volatile and has a K_{ow} value of less than 1000, increasing the sample volume beyond 2 to 4 mL will not give a significant increase in the amount of analyte extracted by SME, while increasing the extraction time needed for maximum extraction (i.e., to reach equilibrium). If the analyte has a K_{ow} value greater than 1000, larger volumes will yield more extracted chemical, although the yield is not directly proportional to the sample volume, and again, extraction times will be increased. Thus, the sample volume recommended for volatile chemicals up to the boiling point of naphthalene (b.p. $= 218°C$) is 1 to 2 mL. For analytes with boiling points higher than naphthalene, up to a 20-mL sample can be used. If SDME is employed, fill the sample vial, leaving little or no headspace. If HS-SDME is chosen, use a minimum headspace volume, just enough for the syringe needle and drop. This amounts to approximately about 0.3 to 0.5 mL of headspace for a 1.8 to 2.0-mL vial and about 5.0 mL of headspace for a 20-mL vial. Remember that you must account for the very minimal volume taken up by the stir bar and any added salt (up to a 15% increase in volume for a saturated NaCl solution).

4.7 SYRINGE AND MICRODROP

At the present time the best syringe for SME is a Hamilton 701 with a No.2 point style (26s gauge, 22° curved bevel tip). The 22° point bevel has a maximum surface area for the drop to cling to, and approximately 95 to 98% of the drop can be drawn back into the syringe. This type of syringe can be used successfully for 50 to 100 injections before the GC septum has to be replaced. However, to minimize septum degradation when using an autosampler, it is recommended that a syringe with a No.4 point (12° straight bevel) be used. This syringe gives nearly as good a pull-up solvent recovery as the standard No.2 tip, with less likelihood for septum damage. Theory shows that the larger the drop, the greater the amount of analyte extracted. This has practical limitations, however. The microdrop may not be stable for volumes greater than 3 μL. Most researchers use volumes of 1.0 to 2.0 μL. A 1.3-μL drop is usually sufficient for HS-SDME, and for SDME a 1.5-μL drop can be used. In both cases, 1.0 μL is drawn back into the syringe after sampling and injected into the GC. The additional 0.3 to 0.5 μL is sacrificed, to minimize variances in amounts of liquid withdrawn into the syringe for HS-SDME and to avoid drawing any water into the syringe when using SDME. Much larger extraction volumes—up to 20 μL —are used with hollow fiber–protected SME (HFME). Only 1 to 2 μL of the extracting solvent is withdrawn into the syringe and injected into the GC. Larger volumes could, of course, be used for HPLC, IC, or other separation techniques. As discussed above, DLLME uses larger amounts of extracting solvent, up to 50 μL. However, large volumes (up to 2.0 mL) of dispersing solvent are also used, and this results in some of the extracting solvent remaining in the water solution, along with some analyte, due to co-solvent effects.

4.8 CHROMATOGRAPHY AND DETECTOR REQUIREMENTS

Typically, SME is used with gas chromatography with a FID, ECD, or MS detector. Other detectors can also be used. A few researchers have also used HPLC,[28–32] CE,[33–35] and AAS[36–38] for the analysis of the extract. The following discussion will focus on gas chromatography, the most commonly used separation and detection method. The choice of detector will influence the sensitivity and selectivity for analytes. Both ECD and MS with selective ion monitoring are very sensitive and specific for chemical types. The MS detectors, however, are usually limited to column flows of 0.5 to 2.0 mL/min, restricting them to low-flow, smaller-diameter columns with diameters of 0.18 to 0.25 mm. These columns have very good resolution, but relatively low capacities, which limits the amount of analyte and solvent that can be injected onto them. Thus, most researchers use a split injection technique, with the split ratio ranging from 10:1 to 40:1 to avoid overloading of the column and achieve sharp, well-resolved peaks with high signal-to-noise ratios. When a solvent lower boiling than the analytes is employed, a splitless injection can be used, with increased sensitivity. If a detector can accept higher flow rates e.g., (FID), a larger diameter column (0.32 to 0.53 mm) can be used, which again increases column capacity, although with some loss in resolution. However, it may then be possible to use a smaller split ratio or even a splitless injection. A thin-film column (0.25 μm) can be used for high-boiling extracted analytes such as the PAHs, but a thick-film column (0.5 to 1.5 μm) is recommended for volatile analytes to help improve resolution. The chromatograms below illustrate some of these recommendations. The first chromatogram (Figure 4.9) is a headspace SDME with tetradecane and a splitless injection. The resolution is very poor for the lower-boiling components and the chromatogram is nearly useless. The second chromatogram (Figure 4.10) is for the same type of sample and extraction, but with a 10:1 split injection. All the peaks are sharp and well resolved. The third example (Figure 4.11) is a splitless injection with toluene as the extractant, which gave nice, sharp peaks for components higher boiling than toluene.

4.9 ADDITIONAL EXTRACTION PARAMETERS

Important extraction parameters that are not covered in detail here include temperature, stirring or agitation, equilibration time, ionic strength, and chemical aids such as derivative formation. For more details, see Chapter 5.

4.9.1 Sample Agitation

Many of the papers published on SME have included studies to maximize extraction by optimizing temperature, stirring rate, and salt concentrations. SME theory

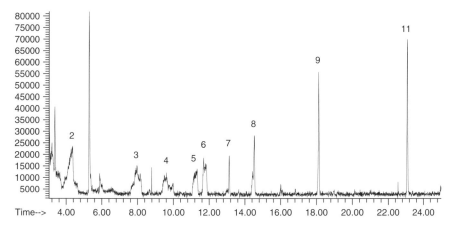

FIGURE 4.9. GC-MS chromatogram of EPA method 524.2 volatiles extracted using HS-SDME. Splitless injection. Extracting solvent: tetradecane (1.3 μL). Conditions and peak identities as in Figure 4.3.

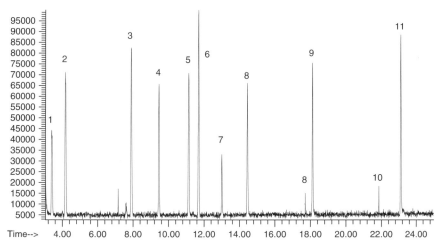

FIGURE 4.10. GC-MS chromatogram of EPA method 524.2 volatiles extracted using HS-SDME. Split (10:1) injection. Extracting solvent: tetradecane (1.3 μL). Conditions and peak identities as in Figure 4.3.

clearly shows that equilibration time is decreased and extraction efficiency improved by efficient sample mixing, and thus mixing or stirring should be maximized to decrease equilibration time. A limitation to about 600 rpm is necessary for SDME, however, since higher stir rates may lead to dislodging of the drop from the needle. An orbital mixer requires a special mixing program that agitates the sample only when the sample is not exposed to the drop, since movement of the needle would again lead to dislodgement of the drop.

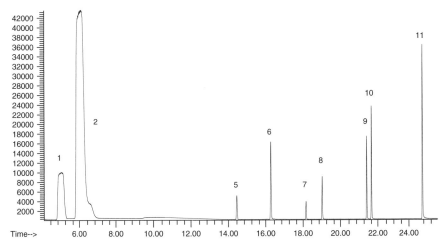

FIGURE 4.11. GC-MS chromatogram of EPA method 524.2 volatiles extracted using HS-SDME. Splitless injection. Extracting solvent: toluene (1.3 μL). Conditions and peak identities as in Figure 4.3.

4.9.2 Ionic Strength

Most analytes with octanol–water partition coefficients (K_{ow} values) less than 1000 (such as gasoline components such as benzene and toluene and the chlorination disinfection by-products) are extracted better with increasing salt concentrations (ionic strength). For these analytes, salt concentration should be maximized (near saturation) if extraction efficiencies need to be maximized. There are few valid literature examples of analytes that have decreased extraction efficiencies with increased salt concentrations.[39,40] Analytes with large (K_{ow} greater than 1000) octanol–water partition coefficients (such as the PAHs and PCBs) are little affected by salt concentration, and adding salt is not necessary unless it is done to decrease the aqueous solubility of the extracting solvent, which may be an important factor for lower–boiling extraction solvents. Headspace extractions can be performed without stirring if the sample is agitated before extraction (e.g., using a vortex device) to ensure sample-headspace equilibrium. These effects are illustrated for the extraction of residual solvents in pharmaceuticals in Figure 4.12 (also see Section 7.7).

4.9.3 Extraction Temperature and Extraction Time

Temperature plays an important role in influencing the diffusion of the analyte across the water–organic, water–air, and air–organic extractant phase boundaries. However, temperature also affects the K_{ow} and K_{aw} (dimensionless Henry's law constant) values. Most experimental results indicate that temperatures up to 40 to 50°C can decrease the equilibration time for SDME and HS-SDME for higher–boiling and polar analytes.[41] Low-boiling analytes, however, may be negatively affected by increases in temperature.[42] This is because extraction time and temperature are interrelated. The longer the extraction time, the more likely the drop

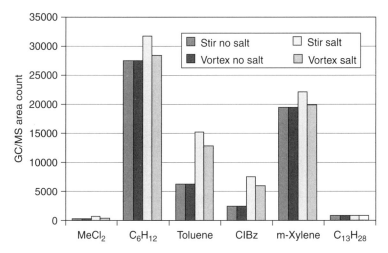

FIGURE 4.12. HS-SDME of USP method 467 residual solvents in pharmaceuticals, followed by GC/MS: comparison of agitation method and ionic strength effect on extraction efficiency. MeCl$_2$: methylene chloride; C$_6$H$_{12}$: cyclohexane; Clbz: chlorobenzene; C$_{13}$H$_{28}$: tridecane.

will also increase in temperature and thus the drop-air or drop-water partition coefficient will decrease, leading to lower extraction efficiencies. Some experiments have been carried out using a cooled needle and extractant drop, to increase

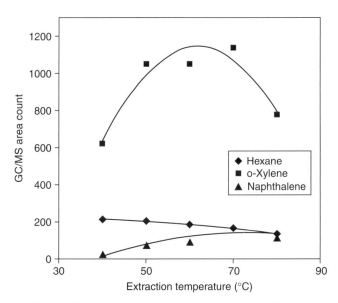

FIGURE 4.13. Effect of extraction temperature on extraction efficiency for HS-SDME extraction of 0.005 µL of unleaded gasoline deposited on wood in a 2-mL vial, followed by GC-MS SIM analysis; inlet split: 10:1; extraction solvent: tetradecane (1.3 µL); extraction time: 7 min; volume injected: 1 µL.

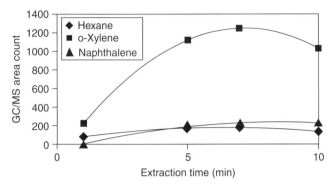

FIGURE 4.14. Effect of extraction time on extraction efficiency for HS-SDME extraction of 0.005 μL of unleaded gasoline deposited on wood in a 2-mL vial. Extraction temperature: 50°C. Other extraction and analytical conditions as in Figure 4.13.

the partition in favor of the drop, thus increasing extraction efficiency.[25,26] The effects of temperature and extraction time are illustrated in Figures 4.13 and 4.14 for three components present in gasoline, extracted from water with tetradecane and analyzed by GC-MS (also see Section 7.8).

4.9.4 Chemical Effects

There have also been successful attempts to favor partitioning into the extraction solvent using chemical means. This includes extraction of acidic or basic chemicals with basic or acid extraction solutions, respectively, when using HF(3)ME.[43] The use of a chemical additive to a HS-SDME extraction solvent to form a derivative of the extracted analyte which is less volatile has also been used for the analysis of amines[44] and aldehydes and ketones.[45] The use of derivatization in solvent microextraction is discussed in detail in Section 5.5 as well as in Sections 6.5.5 and 6.5.6.

4.10 CALCULATION EXAMPLES FOR SDME

It is widely known that solvent volume, solvent lipophilicity, hydrogen bonding, salt concentration, and stirring rate are a few of the important variables that affect traditional macroscale extractions. The same is true for microscale extraction methods such as SME and SPME. However, even a cursory reading of the literature indicates that when designing experiments, many researchers are not careful to take into account the established theory of extractions. We now use the general theory for SME extractions to calculate the expected amount of analyte to be extracted for SDME, headspace SME, dispersive liquid–liquid solvent microextraction, and static headspace extraction, and the effect on each of ionic strength.

It is important to realize that SME is not normally an exhaustive extraction technique; for most analytes it is an equilibrium technique or an intermediate technique. As we shall see, the time required for the equilibrium to be established depends on several factors, including the volumes of the organic extractant and the sample being extracted, the rates of transfer of analyte between phases and the distribution ratio of analyte between phases. The rate constant (k) for the equilibrium is given by the following equations:

$$\frac{1}{\beta_{oo}} = \frac{1}{\beta_o} + \frac{K_{ow}}{\beta_w} \tag{4.1}$$

$$k = \frac{A_i\beta_{oo}[K_{ow}(V_o/V_w) + 1]}{V_o} \tag{4.2}$$

where A_i = interfacial area between the organic and aqueous layers
β_{oo} = overall mass transfer coefficient for the organic phase (cm/s)
β_o = mass transfer coefficient for the organic phase (cm/s)
β_w = mass transfer coefficient for the aqueous phase (cm/s)
V_o = volume of the organic extractant
V_w = volume of the aqueous phase
K_{ow} = distribution ratio between the organic and aqueous phases

Here K_{ow} is the 1-octanol/water distribution ratio, a useful first estimate value, since K_{ow} values are easily obtained from the literature or calculated for values that are not available.[2] In addition, partition coefficients for common extraction solvents usually differ from K_{ow} values by less than an order of magnitude. For example, the K_{ow} value for toluene is 460, the corresponding solvent–water partition coefficient of toluene for hexane is 680, and the coefficient for chloroform is 2700.[46]

Inspection of these equations tells us that the time required for the system to come to equilibrium will be minimized (k maximized) if A_i, β_o, and β_w are maximized and V_w is minimized. *In other words, we want to increase mass transfer of analyte between the water and extracting organic drop and keep the volume of the water to a minimum, determined by the limits of detection for the concentration of the analyte for the method.*

Experimentalists typically try to increase method sensitivity by using larger sample volumes (10 to 40 mL) and increasing temperature. As will be seen in the example calculations below, increasing sample volumes may have little effect on sensitivity and prove deleterious, due to increased equilibration times. Increasing temperature can increase the mass transfer coefficients, but it must be remembered that solubilities of the analyte in the organic and aqueous phases (i.e., K_{ow}) and solubilities and volatilities of the organic phase can also be affected. Appropriate extraction temperatures need to be determined experimentally, although some generalities will be outlined here. In some cases, increasing temperature can have a negative impact on the amount of analyte extracted.[42] One major influence on the mass transfer coefficients is mixing or stirring of the aqueous phase. In general, the stirring rate should be maximized, taking into account the stability of the organic extracting drop. Typical stirring rates range from 400 to 1500 rpm.

At equilibrium or near equilibrium, the following equation derived from the mass balance condition applies:

$$C_o = \frac{K_{ow} C_w^0}{1 + K_{ow}(V_o/V_w)} \qquad (4.3)$$

where C_o is the concentration of the analyte at equilibrium in the organic extractant and C_w^0 is the initial concentration of the analyte in the extracted (aqueous) phase. The equation can be rearranged to represent the amount of analyte (n) extracted into the organic extractant at equilibrium as follows:

$$n = \frac{K_{ow} V_o C_w^0 V_w}{K_{ow} V_o + V_w} \qquad (4.4)$$

As might be expected, these equations show that the larger the K_{ow} value for an analyte, the larger the amount of analyte that will be extracted. But how do the other variables affect n? A few example calculations for three chemicals—benzene, naphthalene, and pyrene—will illustrate this. For all of the following example calculations, for the sake of simplicity and comparison, we assume that the system has reached equilibrium, the use of 1-octanol as the extracting solvent, and identical stirring rates and temperatures. Figure 4.15 shows a spreadsheet in which equation (4.4) is used to calculate the amount of analyte extracted.

Consider the examples below for an aqueous solution with a concentration of 10 ng/mL benzene, naphthalene, or pyrene with no salt addition. The amounts of extracted analytes are calculated using equation (4.4) and listed in Table 4.2.

Interestingly, this illustrates the fact that for analytes with relatively small K_{ow} values (less than 1000), such as benzene, increasing the volume of the water sample

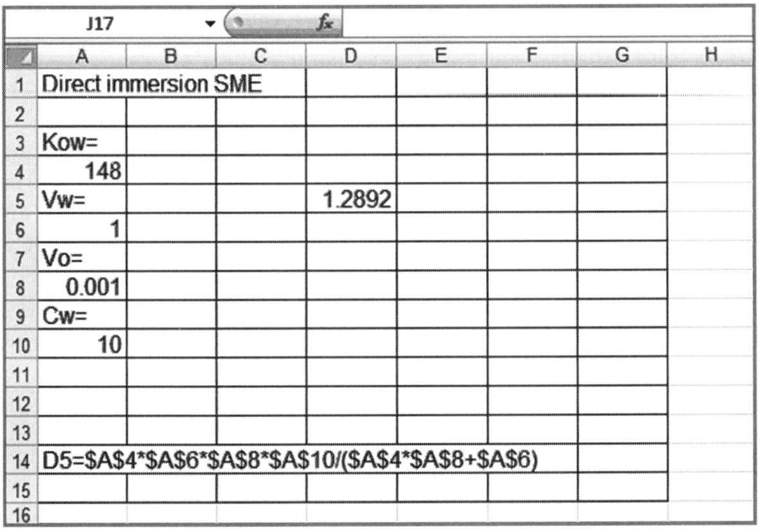

	J17		f_x					
	A	B	C	D	E	F	G	H
1	Direct immersion SME							
2								
3	Kow=							
4	148							
5	Vw=			1.2892				
6	1							
7	Vo=							
8	0.001							
9	Cw=							
10	10							
11								
12								
13								
14	D5=$A$4*$A$6*$A$8*$A$10/($A$4*$A$8+$A$6)							
15								
16								

FIGURE 4.15. Spreadsheet for calculating the amount of analyte extracted in SDME.

TABLE 4.2   Amount of Analyte Extracted by SDME Using 1-Octanol Under Equilibrium Conditions

Analyte	V_w (mL)	V_o (μL)	K_{ow}	n (ng)
Benzene	1	1	148	1.29
	10	1	148	1.45
	20	1	148	1.46
	1	2	148	2.28
	1	3	148	3.07
Naphthalene	1	1	2,140	6.82
	10	1	2,140	17.6
	20	1	2,140	19.3
Pyrene	1	1	135,000	9.93
	10	1	135,000	93.1
	20	1	135,000	174

has very little effect on the amount of analyte extracted while increasing the equilibration time, since this decreases the water mass transfer coefficient, β_w.

On the other hand, increasing the drop size has a nearly directly proportional effect on the amount of analyte extracted. The practical limitation, of course, is the stability of larger drops, especially when stirring the sample. This limitation can be overcome by using hollow fiber–protected microextraction, which is limited in turn by the volume of the extract that can be injected for analysis.

For naphthalene, some 68% of the analyte is extracted using a 1-mL sample. Although more sample is available for analysis using larger sample volumes, only 18% and 10% of the analyte are actually extracted from 10- and 20-mL volumes, respectively, representing only approximately a threefold increase in the amount of analyte injected. Again, however, it will take longer to extract the analyte using larger water samples.

For pyrene, the large K_{ow} value results in a nearly exhaustive extraction when using only 1 mL of water sample. Increasing the sample volume to 10 or 20 mL again increases the amount of analyte available for analysis, this time up to 20-fold.

In summary, the sample volume chosen for extraction will depend on the amount needed for quantification by the method and the K_{ow} values for the extracted analytes. When the K_{ow} values are greater than 1000, a sample size of 1 to 2 mL will yield most of the analyte possible without adding salt. However, larger sample volumes can be used effectively, especially for very dilute solutions less than 100 pg/mL. The equilibration times will be longer, however.

4.11   CALCULATION EXAMPLES FOR DLLME AND HFME

Both DLLME and HFME are variations of SDME and follow the same basic extraction principles, with the exception that a single drop is not used and the volumes of the extracting phase are larger. Thus, the same extraction equation can be used to calculate theoretical amounts of analyte extracted. There are, however, some important "complications" to these two methods that make a simple calculation more complex. A two-phase HF(2)ME method is most similar to SDME, with the exceptions that the volume of the organic extracting phase is 3 to 20 μL, rather than 1 to 2 μL, and there is additional organic phase within the walls of the fiber in addition to the solvent in the lumen of the fiber. Thus, there is also interaction of water and analyte with the polymer. However, as a first approximation we ignore this interaction.

For many three-phase HF(3)ME extractions, the process is nearly, if not completely exhaustive, given enough time for the system to transfer the analytes, which may take 30 to 90 min. This type of extraction is often used to extract acidic or basic chemicals from the sample. Ionization of these chemicals is suppressed by an appropriate pH of the sample solution. Next, they pass through the organic barrier and are reionized in the lumen of the fiber, due to the pH of the acceptor solution, and are not able to pass back out through the organic barrier within the walls of the fiber. In some cases the technique is used to extract nonpolar chemicals, such as PAHs, using a water-soluble binding agent. In these cases, the calculations are essentially the same as for the HF(2)ME examples. The same would also hold for the analogous technique of liquid–liquid–liquid microextraction.

For DLLME there is a more serious complication: the water-soluble dispersing solvent, which may range from 10 to 30% of the volume of the water sample. This co-solvent cannot be ignored in calculating the amount of analyte that will be extracted into the water-insoluble layer, which may also actually contain some unknown amount of dispersing solvent and water. However, we ignore these complications for the following illustrative calculations. We use the K_{ow} for 1-octanol for comparative purposes here. The actual distribution coefficient for a particular solvent, such as CCl_4, would be different. Again, we assume that the analytes are extracted from an aqueous solution with a concentration of 10 ng/mL benzene, naphthalene, and pyrene with no salt added. The amounts of analytes extracted are listed in Table 4.3.

As we can see, while more benzene is available in the extract, only 23% and 60%, respectively, of the total amount of benzene present in the sample is extracted. If a large-volume injection were used, this would provide 10 to 20 times as much extract for analysis as SDME. If we keep things equal, however, and analyze 1 μL of the extract by GC, only 1.14 and 0.596 ng of benzene are actually chromatographed, versus 1.29 ng for SDME. However, if the extracting solvent were

TABLE 4.3 Amounts of Analytes Extracted by DLLME and HF(2)ME Using 1-Octanol Under Equilibrium Conditions

Analyte	V_w (mL)	V_o (μL)	K_{ow}	n (ng)
Benzene	5	10	148	11.4
	5	50	148	29.8
Naphthalene	5	10	2,140	40.5
	5	50	2,140	47.8
Pyrene	5	10	135,000	49.8
	5	50	135,000	50.0

evaporated and replaced by acetonitrile, as much as 29.8 ng would be available for chromatography by reversed-phase HPLC.

For naphthalene, the amounts extracted are now 81% and 96%, respectively, of the total amount of naphthalene present in the sample. If we keep things equal and analyze 1 μL of the extract by GC, only 4.05 and 0.956 ng of naphthalene are actually chromatographed, versus 1.29 ng for SDME. However, if the extracting solvent were evaporated and replaced by acetonitrile, as much as 47.8 ng would be available for chromatography by reversed-phase HPLC.

For pyrene, essentially all of the analyte present in the sample is extracted. If we keep things equal and analyze 1 μL of the extract by GC, 5.0 ng of analyte would be chromatographed versus 1.29 ng for SDME. However, if the extracting solvent were evaporated and replaced by acetonitrile, as much as 50 ng would be available for chromatography by reversed-phase HPLC.

Therefore, these techniques are especially worth considering for GC analysis of nonpolar analytes that have very large K_{ow} values, especially for HPLC analysis, if the analyte would not be lost during evaporation of the extracting solvent. If HF(3) ME were used, however, there is a potential for extracting essentially all of the analyte for analysis, provided that enough time is allowed for the system to equilibrate.

Examples of DLLME and HF-SME with added salt would provide essentially the same results as for SDME with salt, as in the following section.

4.12 CALCULATION EXAMPLES FOR THE EFFECT OF IONIC STRENGTH ON SDME

It has long been known that the addition of salts to an aqueous solution decreases the solubilities of analytes in water and increases the amount of analyte extracted into an organic layer. This salting-out effect varies with the type of salt and the structure of the analyte. The most commonly used salts are sodium chloride and sodium sulfate. Sodium sulfate is preferred for some chemicals which might undergo substitution reactions with the chloride.[47] It turns out that the potential salting-out effect is greater for sodium sulfate than for sodium chloride, but taking the molar masses and solubilities of the two salts into consideration, the practical salting-out effects are nearly the same. Thus, we limit our discussion to sodium chloride. A detailed analysis of the salting-out effect is covered in a highly recommended text by Schwarzenbach et al.[48] and recent publications.[40,49] The effect of increased ionic strength on the organic–water distribution coefficient (we will use K_{ow}) is given by

$$K_{ow(\text{salt})} = K_{ow} \times 10^{+S[\text{salt}]} \qquad (4.5)$$

where $K_{ow(\text{salt})}$ is the K_{ow} value corrected for the addition of salt, S the Setschenow constant for NaCl and a specific organic analyte, and [salt] the molar concentration of the added NaCl.

As we have seen, the K_{ow} value is very important for determining the amount of analyte extracted during SME. Since the $K_{ow(\text{salt})}$ value increases exponentially with salt concentration, this will affect the extraction process significantly. The

Setschenow constant (S), while specific for each analyte and salt, can be estimated for NaCl from the family to which the analyte belongs:

halogenated C1 and C2 compounds	~ 0.2–0.3
alkanes	~ 0.25
benzene derivatives	~ 0.2–0.3
phenols, anilines	~ 0.15
polycyclic aromatic hydrocarbons	~ 0.27–0.35
aldehydes and ketones	~ 0.2–1.0
PCBs	~ 0.3–0.4

A very good estimate for S can be calculated rapidly using the following equation, which requires only knowledge of the K_{ow} value for the chemical.[50]

$$S = 0.04(\log K_{ow}) + 0.114 \qquad (4.6)$$

Using this equation, the resulting Setschenow constants can be calculated: benzene = 0.20, naphthalene = 0.25, and pyrene = 0.32.

The molar concentrations of NaCl at various weight percent concentrations in water are

$$10\% \ NaCl = 1.71 \ M$$
$$20\% \ NaCl = 3.42 \ M$$
$$30\% \ NaCl = 5.13 \ M$$
$$35\% \ NaCl = 5.98 \ M \ (\sim saturated \ solution)$$

As an example, the K_{ow} value for benzene in a 35% (weight/volume) NaCl solution is:

$$K_{ow(35\%NaCl)} = K_{ow} \times 10^{+S[NaCl]} = 148 \times 10^{+0.20 \times 5.98} = 2320$$

Evidently, the K_{ow} for benzene is increased more than 15-fold in a saturated salt solution, resulting in more than a five fold increase in the extraction efficiency, as we shall see.

It should be pointed out that the Setschenow constant value for sodium sulfate (Na_2SO_4), another common salting-out agent, is two to three times that for NaCl, and thus $1/2$ to $1/3$ mols of Na_2SO_4 will have the same effect as 1 mol of NaCl.[48] However, since the formula weight of Na_2SO_4 is 142 g/mol versus 58.5 g/mol for NaCl, the mass of the two salts required would be nearly equivalent.

At NaCl concentrations close to saturation, the total volume of the solution is increased by as much as 15%, and this decrease in analyte concentration should be taken strictly into account. We ignore the volume change for all but one of the following examples.

Consider the direct SME extraction of a 10-ng/mL solution of benzene (S = 0.20), naphthalene (S = 0.25), and pyrene (S = 0.32) in 1 mL of water with 1 μL of organic solvent (1-octanol) at various concentrations of NaCl (Table 4.4).

As we can see, for analytes that have small K_{ow} values (less than 1000), such as benzene, the amount of analyte extracted from a saturated salt solution will be

TABLE 4.4   Amounts of Analyte Extracted by SDME Using 1-Octanol Under Equilibrium Conditions with Various Amounts of Salt Added

Analyte	% NaCl				
	0%	10%	20%	30%	35%
Benzene					
K_{ow}	148	325	715	1570	2320
n (ng)	1.29	2.45	4.17	6.11	6.99
Naphthalene					
K_{ow}	2140	5730	15,300	41,000	66,900
n (ng)	6.82	8.51	9.39	9.76	9.98
Pyrene					
K_{ow}	135,000	476,000			
n (ng)	9.93	10.0			

larger than that from a solution with no added salt (five times more for benzene). Addition of salt can, however, affect the diffusion constants as well, and some low-molecular-weight polar chemicals have been reported to have lowered extraction efficiencies with the addition of salt, so this is one parameter that should be checked during method development.[39,40]

For analytes with K_{ow} greater than 1000, such as naphthalene, nearly exhaustive extraction can be achieved, although in the case of naphthalene, about 68% extraction is possible even without adding salt. For analytes with very large K_{ow} values, such as pyrene, there is nearly exhaustive extraction even without the addition of salt, and there is no real benefit to the addition of salt, even for saturated solutions, unless salt addition is needed to reduce the water solubility of the extracting solvent.

4.13   CALCULATION EXAMPLES FOR HS-SDME

The formula describing the amount of analyte extracted at equilibrium, derived from the mass balance condition for a three-phase system, is

$$n = \frac{K_{ow} V_o C_w^0 V_w}{K_{ow} V_o + K_{aw} V_a + V_w} \tag{4.7}$$

Headspace SME involves the equilibrium between three phases: liquid (aqueous), headspace above the water, and the organic extractant drop. To account for the headspace equilibrium, two new terms are added to the formula: K_{aw}, the air–water distribution coefficient, better known as the dimensionless Henry's law constant (air and water concentrations given in moles/liter) and V_a, the volume of the headspace.

Inspection of the formula shows that in addition to the parameters important for SDME, the amount of analyte extracted will be maximized if K_{aw} and V_a are minimized. The dimensionless Henry's law constant ranges from small to very

	J24			f_x					
	A	B	C	D	E	F	G	H	I
1	Headspace SME								
2									
3	Kow=								
4	148								
5	Vw=			3.5005					
6	10								
7	Vo=								
8	0.003								
9	Cw=								
10	10								
11	Kaw=								
12	0.224								
13	Va=								
14	10								
15									
16	D5=$A$4*$A$6*$A$8*$A$10/($A$4*$A$8+$A$12*$A$14+$A$6)								
17									
18									

FIGURE 4.16. Spreadsheet for calculating the amount of analyte extracted in HS-SDME.

small for most compounds: K_{aw} for benzene is 0.224 and K_{aw} for pyrene is 0.000436. The minimum headspace volume is often determined by the practical limitation of requiring enough headspace for the syringe needle and microdrop while stirring the solution. For a 1.8-mL vial and a 20-mL vial, this amounts to about 0.3 to 0.5 mL and about 5 mL, respectively.

Figure 4.16 shows a spreadsheet in which equation (4.7) is used to calculate the amount of analyte extracted. Consider the examples in Table 4.5 for various head-space volumes and aqueous volumes with an analyte concentration of 10 ng/mL and no salt addition.

Thus, the HS-SDME extraction results parallel those for SDME, yielding slightly less analyte in the extract. Again, for analytes with small K_{ow} values, such as benzene, larger solution volumes do not yield significantly greater amounts of extract while increasing the equilibration time, but increasing the volume of the organic extractant does result in significant improvement in extraction yield.

For analytes with K_{ow} larger than about 1000, such as naphthalene, HS-SDME extracts almost as much analyte as the SDME method.

Theoretically, when the K_{ow} value is very large (e.g., pyrene), HS-SDME can extract the same amount of analyte from water as can SDME. Experimental results are limited for molecules with vapor pressures lower than for pyrene, however, and although theory (see Section 3.3.4.1) predicts that equilibrium between water and headspace should be a fast process, experimental results with PAHs (see Section 7.9) do not bear this out. PAH molecules larger than pyrene are, in fact, extracted poorly using both SDME and HS-SDME compared to lower-molecular-weight molecules. This may be due to slower diffusion of larger molecules through the

TABLE 4.5 Amounts of Analyte Extracted by HS-SDME Using 1-Octanol Under Equilibrium Conditions with No Salt Added

Analyte	V_w (mL)	V_a (mL)	V_o (µL)	K_{ow}	K_{aw}	n (ng)
Benzene	1	0.8	1	148	0.224	1.12
	10	10	1	148	0.224	1.19
	10	5	1	148	0.224	1.31
	20	20	1	148	0.224	1.20
	20	5	1	148	0.224	1.39
	1	0.8	2	148	0.224	2.01
	1	0.8	3	148	0.224	2.74
Naphthalene	1	0.8	1	2,140	0.0182	6.78
	10	10	1	2,140	0.0182	17.4
	20	20	1	2,140	0.0182	19.0
Pyrene	1	0.8	1	135,000	0.000436	9.93
	10	10	1	135,000	0.000436	93.1
	20	20	1	135,000	0.000436	174.2

water sample and/or into the solvent. Therefore, efficient solution stirring and possibly dynamic extraction may be necessary to reach equilibrium within a reasonable time.

4.14 CALCULATION EXAMPLES FOR THE EFFECT OF IONIC STRENGTH ON HS-SDME

As is the case for direct-immersion SME, the addition of salt to a water sample affects the K_{ow} parameters. Thus, K_{ow} will increase exponentially with salt concentration [see equation (4.5)].

For headspace extraction, the Henry's law constant is also influenced by the addition of salt. The equation is analogous to that for K_{ow} using the same Setschenow salting-out constant (S).[3]

$$K_{aw(\text{salt})} = K_{aw} \times 10^{+S[\text{salt}]} \tag{4.8}$$

Since K_{aw} will also increase exponentially with salt concentration, it is even more important when using the salting-out effect to minimize the headspace volume term to maximize the amount of analyte extracted into the microdrop [see equation (4.7)]. The salting-out effect is most important for analytes with small K_{ow} values ($K_{ow} < 1000$) and relatively large dimensionless Henry's law constants ($K_{aw} > 0.05$), as illustrated by the sample calculations below.

Consider examples with 1 mL of water, and 0.8 mL of headspace, and a 1-µL 1-octanol microdrop (Table 4.6).

Now let us see what happens to the amount of benzene extracted by HS-SDME if we take into account the change in molar concentration of salt, sample volume, and headspace volume when high concentrations of salt are used. Conditions: 35% added NaCl, sample volume increased to 1.15 mL due to added salt and headspace

TABLE 4.6 Amounts of Analyte Extracted by HS-SDME Using 1-Octanol Under Equilibrium Conditions with Various Amounts of Salt Added

Analyte	% NaCl				
	0%	10%	20%	30%	35%
Benzene					
K_{ow}	148	325	715	1570	2320
K_{aw}	0.224	0.493	1.08	2.37	3.52
n (ng)	1.12	1.89	2.77	3.52	3.78
Naphthalene					
K_{ow}	2140	5730	15,300	41,000	66,900
K_{aw}	0.0182	0.0487	0.0974	0.349	0.569
n (ng)	6.78	8.46	9.34	9.69	9.80
Pyrene					
K_{ow}	135,000	476,000			
K_{aw}	0.000436	0.00154			
n (ng)	9.93	9.98			

volume decreased to 0.65 mL. Since the water volume is increased, the NaCl concentration is decreased, resulting in $K_{ow(NaCl)} = 1620$ and $K_{aw(NaCl)} = 2.46$:

$$n = \frac{1620 \times 0.001\,\text{mL} \times 10\,\text{ng/mL} \times 1.15\,\text{mL}}{1620 \times 0.001\,\text{mL} + 2.46 \times 0.65\,\text{mL} + 1.15\,\text{mL}} = 4.26\,\text{ng}$$

Thus, although it is valid to estimate the amount of analyte extracted when not taking into account the increase in aqueous volume (by some 15%) by the addition of 35% NaCl, it can be seen that correcting for the volume changes results in a larger amount of analyte extracted, due to the resulting decreased headspace volume and smaller K_{aw}.

Consider the examples below for extraction of benzene with various sample and headspace volumes and a 1-μL 1-octanol microdrop. The changes in volumes of the sample and headspace are ignored.

a. *Conditions:* 35% NaCl added, 10-mL sample, and 10-mL headspace, $K_{ow(NaCl)} = 2320$ and $K_{aw(NaCl)} = 3.52$:

$$n = \frac{2320 \times 0.001\,\text{mL} \times 10\,\text{ng/mL} \times 10\,\text{mL}}{2320 \times 0.001\,\text{mL} + 3.52 \times 10\,\text{mL} + 10\,\text{mL}} = 4.88\,\text{ng}$$

This can be compared to 3.78 ng extracted for a 1-mL sample with 0.8 mL headspace and 35% NaCl added and to 1.19 ng extracted for a 10-mL sample and 10 mL of headspace with no salt added.

b. *Conditions:* 35% NaCl added, 10-mL sample, and 5-mL headspace, $K_{ow(NaCl)} = 2320$ and $K_{aw(NaCl)} = 3.52$:

$$n = \frac{2320 \times 0.001\,\text{mL} \times 10\,\text{ng/mL} \times 10\,\text{mL}}{2320 \times 0.001\,\text{mL} + 3.52 \times 5\,\text{mL} + 10\,\text{mL}} = 7.75\,\text{ng}$$

Thus, there is about 60% more analyte extracted by decreasing the headspace by half and a sixfold increase compared to not adding salt.

An inspection of Table 4.6 reveals that for analytes with large K_{ow} values ($K_{ow} > 1000$), such as naphthalene, although adding salt does increase the amount extracted, the increase is negligible. It is also evident that salt addition is not needed for essentially exhaustive extraction of pyrene. For analytes with very large K_{ow} values, the addition of salt increases the extracted amount only by an insignificant amount and is not necessary unless increased ionic strength is used to decrease the extracting solvent solubility in water.

4.15 CALCULATION EXAMPLES FOR STATIC HEADSPACE EXTRACTION

We provide here calculations for standard static headspace analysis for comparison to SME. An equation for the amount of analyte extracted at equilibrium into the headspace above a water solution can be derived analogously as for SME, from the mass balance condition in a two-phase system:

$$n = \frac{K_{aw} V_a C_w^0 V_w}{K_{aw} V_a + V_w} \tag{4.9}$$

4.15.1 Benzene: Static Headspace at Equilibrium

a. *Conditions:* 1 mL of water, 0.8 mL of headspace, $K_{aw(benzene)} = 0.224$, no salt added

$$n = \frac{0.224 \times 0.8\,\text{mL} \times 10\,\text{ng/mL} \times 1\,\text{mL}}{0.224 \times 0.8\,\text{mL} + 1\,\text{mL}} = 1.52\ \text{ng}$$

b. *Conditions:* 10 mL of water, 10 mL of headspace, $K_{aw(benzene)} = 0.224$, no salt added

$$n = \frac{0.224 \times 10\,\text{mL} \times 10\,\text{ng/mL} \times 10\,\text{mL}}{0.224 \times 10\,\text{mL} + 10\,\text{mL}} = 18.3\ \text{ng}$$

If we assume that a maximum of 10% of the headspace can be withdrawn and analyzed, for a 1-mL sample 0.152 ng would be injected, and for the 10-mL sample (more typical static conditions) 1.83 ng would be analyzed, compared to 1.29 ng for SDME and 1.12 ng for HS-SDME for 1-mL samples.

4.15.2 Naphthalene: Static Headspace at Equilibrium

a. *Conditions:* 1 mL of water, 0.8 mL of headspace, $K_{aw(naphthalene)} = 0.0182$, no salt added

$$n = \frac{0.0182 \times 0.8\,\text{mL} \times 10\,\text{ng/mL} \times 1\,\text{mL}}{0.0182 \times 0.8\,\text{mL} + 1\,\text{mL}} = 0.14\ \text{ng}$$

b. *Conditions:* 10 mL of water, 10 mL of headspace, $K_{aw(\text{naphthalene})} = 0.0182$, no salt added

$$n = \frac{0.0182 \times 10\,\text{mL} \times 10\,\text{ng/mL} \times 10\,\text{mL}}{0.0182 \times 10\,\text{mL} + 10\,\text{mL}} = 1.79\,\text{ng}$$

If we assume that a maximum of 10% of the headspace can be withdrawn and analyzed, for a 1-mL sample 0.014 ng would be injected, and for the 10-mL sample (more typical static conditions) 0.179 ng would be analyzed, compared to 6.82 ng for SDME and 6.78 ng for HS-SDME for 1-mL samples.

4.15.3 Pyrene: Static Headspace at Equilibrium

a. *Conditions:* 1 mL of water, 0.8 mL of headspace, $K_{aw(\text{pyrene})} = 0.000436$, no salt added

$$n = \frac{0.000436 \times 0.8\,\text{mL} \times 10\,\text{ng/mL} \times 1\,\text{mL}}{0.000436 \times 0.8\,\text{mL} + 1\,\text{mL}} = 0.0035\,\text{ng}$$

b. *Conditions:* 10 mL of water, 10 mL of headspace, $K_{aw(\text{pyrene})} = 0.000436$, no salt added

$$n = \frac{0.000436 \times 10\,\text{mL} \times 10\,\text{ng/mL} \times 10\,\text{mL}}{0.000436 \times 10\,\text{mL} + 10\,\text{mL}} = 0.044\,\text{ng}$$

If we assume that a maximum of 10% of the headspace can be withdrawn and analyzed, for a 1-mL sample 0.00035 ng would be injected, and for the 10-mL sample (more typical static conditions) 0.0044 ng would be analyzed, compared to 9.93 ng for SDME and HS-SDME with 1-mL of water.

Thus, static headspace extraction yields favorable results for low-boiling analytes with a limited solubility in water and relatively large K_{aw} values, but SME yields better results for analytes with a limited solubility in water having higher boiling points, larger K_{ow} values, and small K_{aw} values.

Static headspace is often performed at elevated temperatures, since this increases the K_{aw} value significantly for many analytes, as illustrated for pyrene.

c. *Conditions:* 10 mL of water, 10 mL of headspace, 50°C, $K_{aw(\text{pyrene, 50°C})} = 0.00263$, no salt added

$$n = \frac{0.00263 \times 10\,\text{mL} \times 10\,\text{ng/mL} \times 10\,\text{mL}}{0.00263 \times 10\,\text{mL} + 10\,\text{mL}} = 0.26\,\text{ng}$$

Although still much less is available for analysis than for HS-SDME, increasing the temperature from 25°C to 50°C has increased the amount available for headspace analysis almost sixfold. Increasing the temperature may or may not have a positive effect on HS-SDME, since the temperature increase would also increase the air/1-octanol partition coefficient (K_{ao}), decreasing the amount dissolved in the drop. This would be counterbalanced if the syringe needle and thus drop were cooled, which could be accomplished by using a thermoelectric needle cooling attachment and/or using dynamic extraction to shorten the extraction time.

4.16 CALCULATION EXAMPLES FOR SOLVENT SOLUBILITY

We end this discussion with examples for solvent solubility in water. It is often forgotten that solvents, such as ether and ethyl acetate, commonly used for macroextractions, are actually quite soluble in water. Even solvents such as chloroform and toluene, which are considered to be insoluble in water, have significant aqueous solubilities when used for SME. Water solubility, as well as K_{ow} and K_{aw}, is affected by ionic strength, and salt addition can be used to decrease the extracting solvent solubility while enhancing K_{ow} and K_{aw}.

An equation describing the effect of added salt on solubility[48] is given by

$$C_{w(\text{salt})} = C_w \times 10^{-S[\text{salt}]} \tag{4.10}$$

Notice the negative sign before the Setschenow constant value. Thus, solubility decreases exponentially with salt concentration. The results for five possible solvents are listed in Table 4.7, which lists water solubilities for five extracting solvents in mg/mL and μL/mL.

Thus, if 1 μL of chloroform were used for SDME extraction of 1 mL of water, the chloroform would dissolve completely. Even at a 35% salt concentration, chloroform would not be an appropriate solvent. Most of the carbon tetrachloride, toluene, and 1-octanol would dissolve also. The dissolution does not happen instantaneously, but if an extraction is carried out over a period of 30 min or more (static SDME), there is a great risk of losing most of the drop and mistakenly withdrawing water into the syringe. This would, in turn, risk injecting water and any dissolved salts into the GC at the end of the extraction. This problem is even worse if extraction is performed at elevated temperatures, since not only do the solubilities increase, but the evaporation rate does as well. Even adding 10% salt significantly reduces the solubilities of these solvents, however. Using a saturated salt solution reduces the solubilities to acceptable levels, and it is thus recommended that nearly saturated salt conditions be employed for SME, if possible, when using an extracting solvent with moderate solubility such as 1-octanol. This may be necessary,

TABLE 4.7 Extracting Solvent Solubilities in Water and Salt Solutions

Solvent	Solubility (mg/mL water) 0% NaCl	Solubility (μL/mL water) 0% NaCl	Solubility (μL/mL water) 10% NaCl	Solubility (μL/mL water) 35% NaCl
CHCl$_3$	8.40	5.60	2.70	1.20
CCl$_4$	0.83	0.52	0.22	0.09
Toluene	0.56	0.65	0.27	0.12
1-Octanol	0.54	0.65	0.26	0.10
Tetradecane	0.00000090	0.0000012	0.0	0.0

even though it is not required when extracting analytes with large K_{ow} values, such as the PAHs. Conversely, when using a solvent with a very low solubility in water, such as tetradecane, salt addition is not necessary unless analytes with small K_{ow} values are to be extracted.

REFERENCES

1. Schwarzenbach, R. P.; Gschwend, P. K.; Imboden, D. M., *Environmental Organic Chemistry*, 2nd ed., Wiley-Interscience, Hoboken, NJ, 2002, pp. 213–244.
2. Estimation program interface (EPI) suite, http://www.epa.gov/opptintr/exposure/pubs/episuite.htm.
3. Schwarzenbach, R. P.; Gschwend, P. K.; Imboden, D. M., *Environmental Organic Chemistry*, 2nd ed., Wiley-Interscience, Hoboken, NJ, 2002, pp. 181–212.
4. Psillakis, E.; Kalogerakis, N., Developments in single-drop microextraction. *Trends Anal. Chem.* 2002, *21* (1), 53–63.
5. Xu, L.; Basheer, C.; Lee, H. K., Developments in single-drop microextraction. *J. Chromatogr. A* 2007, *1152* (1–2), 184–192.
6. Ouyang, G.; Zhao, W.; Pawliszyn, J., Kinetic calibration for automated headspace liquid-phase microextraction. *Anal. Chem.* 2005, *77* (24), 8122–8128.
7. Lambropoulou, D. A.; Konstantinou, I. K.; Albanis, T. A., Recent developments in headspace microextraction techniques, for the analysis of environmental contaminants in different matrices. *J. Chromatogr. A* 2007, *1152* (1–2), 70–96.
8. He, Y.; Lee, H. K., Liquid-phase microextraction in a single drop of organic solvent by using a conventional microsyringe. *Anal. Chem.* 1997, *69* (22), 4634–4640.
9. Wang, Y.; Kwok, Y. C.; He, Y.; Lee, H. K., Application of dynamic liquid-phase microextraction to the analysis of chlorobenzenes in water by using a conventional microsyringe. *Anal. Chem.* 1998, *70* (21), 4610–4614.
10. Saraji, M., Dynamic headspace liquid-phase microextraction of alcohols. *J. Chromatogr. A* 2005, *1062* (1), 15–21.
11. Hou, L.; Shen, G.; Lee, H. K., Automated hollow fiber–protected dynamic liquid- phase microextraction of pesticides for gas chromatography-mass spectrometric analysis. *J. Chromatogr. A* 2003, *985* (1–2), 107–116.
12. Huang, S.; Huang, S., Dynamic hollow fiber protected liquid phase microextraction and quantification using gas chromatography combined with electron capture detection of organochlorine pesticides in green tea leaves and ready-to-drink tea. *J. Chromatogr. A* 2006, *1135* (1), 6–11.
13. Kokosa, J. M., Automation of liquid phase microextraction, U.S. patent 7,178,414 B1, Feb. 20, 2007.
14. Chromsys, LLC, Alexandria, VA.
15. Pawliszyn, J.; Pedersen-Bjergaard, S., Analytical microextraction: current status and future trends. *J. Chromatogr. Sci.* 2006, *44* (6), 291–307.
16. Ouyang, G.; Zhao, W.; Pawliszyn, J., Automation and optimization of liquid-phase microextraction by gas chromatography. *J. Chromatogr. A* 2007, *1138* (1–2), 47–54.
17. Rezaee, M.; Assadi, Y.; Milani Hosseini, M. R.; Aghaee, E.; Ahmadi, F.; Berijani, S., Determination of organic compounds in water using dispersive liquid–liquid micro-extraction. *J. Chromatogr. A* 2006, *1116* (1–2), 1–9.

18. Berijani, S.; Assadi, Y.; Anbia, M.; Milani Hosseini, M. R.; Aghaee, E., Dispersive liquid–liquid microextraction combined with gas chromatography–flame photometric detection: very simple, rapid and sensitive method for the determination of organophosphorus pesticides in water. *J. Chromatogr. A* 2006, *1123* (1), 1–9.

19. Kozani, R. R.; Assadi, Y.; Shemirani, F.; Milani Hosseini, M. R.; Jamali, M. R., Part-per-trillion determination of chlorobenzenes in water using dispersive liquid–liquid microextraction combined with gas chromatography–electron capture detection. *Talanta* 2007, *27* (2), 387–393.

20. Farajzadeh, M. A.; Bahram, M.; Jönsson, J. Å., Dispersive liquid–liquid microextraction followed by high performance liquid chromatography–diode array detection as an efficient and sensitive technique for determination of antioxidants. *Anal. Chim. Acta* 2007, *591* (1), 69–79.

21. Battle, R.; Nerin, C., Application of single-drop microextraction to the determination of dialkyl phthalate esters in food simulants. *J. Chromatogr. A* 2004, *1045* (1–2), 29–35.

22. Liu, J. F.; Jiang, G. B.; Chi, Y. G.; Cai, Y. Q.; Zhou, Q. X.; Hu, J. T., Use of ionic liquids for liquid-phase microextraction of polycyclic aromatic hydrocarbons. *Anal. Chem.* 2003, *75* (21), 5870–5876.

23. Peng, J. F.; Liu, J. F.; Jiang, G. B.; Tai, C.; Huang, M. J., Ionic liquid for high temperature headspace liquid-phase microextraction of chlorinated anilines in environmental water samples. *J. Chromatogr. A* 2005, *1072* (1), 3–6.

24. Nazarenko, A. Y., Liquid-phase headspace micro-extraction into a single drop. *Am. Lab.*, 2004, *36* (16), 30–33.

25. Yamini, Y.; Hosseini, M. H.; Hojaty, M.; Arab, J., Headspace solvent microextraction of trihalomethane compounds into a single drop. *J. Chromatogr. Sci.* 2004, *42* (1), 32–36.

26. Yamini, Y.; Hojaty, M.; Hosseini, M. H.; Shamsipur, M., Headspace solvent microextraction: a new method applied to the preconcentration of 2-butoxyethanol from aqueous solutions into a single microdrop. *Talanta* 2004, *62* (2), 265–270.

27. Sigma-Aldrich Catalog, 2007, Tetradecane, 172456.

28. Zhu, L.; Zhu, L.; Lee, H. K., Liquid–liquid–liquid microextraction of nitrophenols with a hollow fiber membrane prior to capillary liquid chromatography. *J. Chromatogr. A* 2001, *924* (1–2), 407–414.

29. Zhao, L.; Zhu, L.; Lee, H. K., Analysis of aromatic amines in water samples by liquid–liquid–liquid microextraction with hollow fibers and high-performance liquid chromatography. *J. Chromatogr. A* 2002, *963* (1–2), 239–248.

30. Zhu, L.; Ee, K. H.; Zhao, L.; Lee, H. K., Analysis of phenoxy herbicides in bovine milk by means of liquid–liquid–liquid microextraction with a hollow-fiber membrane. *J. Chromatogr. A* 2002, *963* (1–2), 335–343.

31. Kuuranne, T.; Kotiaho, T.; Pedersen-Bjergaard, S.; Rasmussen, K. E.; Leinonen, A.; Westwood, S.; Kostiainen, R., Feasibility of a liquid-phase microextraction sample cleanup and liquid chromatographic/mass spectrometric screening method for selected anabolic steroid glucuronides in biological samples. *J. Mass Spectrom.* 2003, *38* (1), 16–26.

32. Sarafraz-Yazdi, A.; Es'haghi, Z., Comparison of hollow fiber and single-drop liquid-phase microextraction techniques for HPLC determination of aniline derivatives in water. *Chromatographia* 2006, *63* (11–12), 563–569.

33. Pedersen-Bjergaard, S.; Rasmussen, K. E., Liquid–liquid–liquid microextraction for sample preparation of biological fluids prior to capillary electrophoresis. *Anal. Chem.* 1999, *71* (14), 2650–2656.

34. Hou, L.; Lee, H. K., Dynamic three-phase microextraction as a sample preparation technique prior to capillary electrophoresis. *Anal. Chem.* 2003, *75* (11), 2784–2789.

35. Choi, K.; Kim, Y.; Chung, D. S., Liquid-phase microextraction as an on-line preconcentration method in capillary electrophoresis. *Anal. Chem.* 2004, *76* (3), 855–858.

36. Gil, S.; Fragueiro, S.; Lavilla, I.; Bendicho, C., Determination of methylmercury by electrothermal atomic absorption spectrometry using headspace single-drop micro-extraction with in situ hydride generation. *Spectrochim. Acta, B* 2005, *60* (1), 145–150.

37. Xia, L.; Hu, B.; Jiang, Z.; Wu, Y.; Chen, R.; Li, L., Hollow fiber liquid phase micro-extraction combined with electrothermal vaporization ICP-MS for the speciation of inorganic selenium in natural waters. *J. Anal. At. Spectrom.* 2006, *21* (3), 362–365.

38. Zeini Jahromi, E.; Bidari, A.; Assadi, Y.; Milani Hosseini, M. R.; Jamali, M. R., Dispersive liquid–liquid microextraction combined with graphite furnace atomic absorption spectrometry: ultra trace determination of cadmium in water samples. *Anal. Chim. Acta* 2007, *585* (2), 305–311.

39. Almeida, M. B.; Alvarez, A. M.; De Miguel, E. M.; Del Hoyo, E. S., Setchenow coef-ficients for naphthols by distribution method. *Can. J. Chem./Rev. Can. Chim.* 1983, *61* (2), 244–248.

40. Ni, N.; El-Sayed, M. M.; Sanghvi, T.; Yalkowsky, S. H., Estimation of the effect of NaCl on the solubility of organic compounds in aqueous solutions. *J. Pharm. Sci.* 2000, *89* (12), 1620–1625.

41. Kaykhaii, M.; Nazari, S.; Chamsaz, M., Determination of aliphatic amines in water by gas chromatography using headspace solvent microextraction. *Talanta* 2005, *65* (1), 223–228.

42. Li, X.; Xu, X.; Wang, X.; Ma, L., Headspace single-drop microextraction with gas chromatography for determination of volatile halocarbons in water samples. *Intern. J. Environ. Anal. Chem.* 2004, *84* (9), 633–645.

43. Ho, T. S.; Pedersen-Bjergaard, S.; Rasmussen, K. E., Recovery, enrichment and se-lectivity in liquid-phase microextraction. Comparison with conventional liquid–liquid extraction. *J. Chromatogr. A* 2002, *963* (1–2), 3–17.

44. Deng, C.; Li, N.; Wang, L.; Zhang, X., Development of gas chromatography-mass spectrometry following headspace single-drop microextraction and simultaneous deri-vatization for fast determination of short-chain aliphatic amines in water samples. *J. Chromatogr. A* 2006, *1131* (1–2), 45–50.

45. Deng, C.; Yao, N.; Li, N.; Zhang, X., Headspace single-drop microextraction with in-drop derivatization for aldehyde analysis. *J. Sep. Sci.* 2005, *28* (17), 2301–2305.

46. Schwarzenbach, R. P.; Gschwend, P. K.; Imboden, D. M., *Environmental Organic Chemistry*, 2nd ed., Wiley-Interscience, Hoboken, NJ, 2002, pp. 1197–1208.

47. Code of Federal Regulations (CFR) 40, Part 136, revised as of July 1, 1995, Appendix A to Part 136: Methods for Organic Chemical Analysis of Municipal and Industrial Was-tewater, method 551.1: determination of chlorination disinfection byproducts, chlori-nated solvents and halogenated pesticides/herbicides in drinking water by liquid–liquid extraction and gas chromatography with electron-capture detection, revision 1.0.

48. Schwarzenbach, R. P.; Gschwend, P. K.; Imboden, D. M., *Environmental Organic Chemistry*, 2nd ed., Wiley-Interscience, Hoboken, NJ, 2002, pp. 133–180.

49. Johnson, S. R.; Zheng, W., Recent progress in the computational predictions of aqueous solubility and absorption. *AAPS J.* 2006, *8* (1), E27–E40.

50. Ni, N.; Yalkowsky, S. H., Prediction of Setschenow constants. *Int. J. Pharm.* 2003, *254* (2), 167–172.

CHAPTER 5

METHOD DEVELOPMENT IN SOLVENT MICROEXTRACTION

5.1 INTRODUCTION

In this chapter we discuss the practical approach to SME method development. It should be remembered that if we are faced with a specific analytical task that may call for the use of SME, a good first step is to check the Excel worksheets on the CD-ROM included with this book to see if the needed SME procedure has already been developed. Each file is a compilation of all the SME procedures available for a particular SME mode: headspace, direct immersion, dispersive, continuous flow, and hollow fiber–protected two- and three-phase, and includes the information on sample matrix, analyte(s) determined, extraction conditions, and analytical characteristics of the procedure. If the procedure is not available, we have to follow the stages in SME method development outlined in Figure 5.1. We should remember that in some cases not all the steps have to be completed, especially the first three, because we may have previous experience with particular analytes and/or matrices, and the analytical literature may provide some information as well. The discussion that follows focuses on solvent microextraction from aqueous matrices, since at present they are the most common applications of SME.

Solvent Microextraction: Theory and Practice, By John M. Kokosa, Andrzej Przyjazny, and Michael A. Jeannot
Copyright © 2009 John Wiley & Sons, Inc.

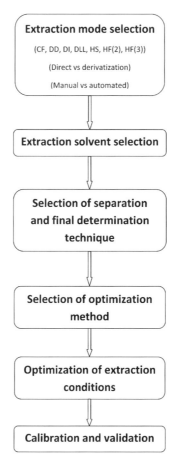

FIGURE 5.1. Steps in SME method development.

5.2 EXTRACTION MODE SELECTION

Three factors must be considered when selecting a solvent microextraction mode: (1) sample matrix, (2) analyte volatility, and (3) analyte polarity (i.e., affinity to the matrix). More detailed guidelines on selection of the optimal SME mode for the task at hand are provided in sections 4.2 and 6.3. A summary of the selection criteria is shown in Figure 5.2, and a flowchart facilitating the selection process is provided in Figure 5.3. Briefly, for volatile analytes, headspace SME should be tried first. This mode is the most versatile of all variants of SME, since it can be used for gaseous, liquid, and solid samples. Headspace SME can be implemented in two different modes: single-drop or hollow fiber–protected. The single-drop approach is simpler, suitable for occasional use, and easily automated. The hollow fiber approach has a number of advantages, including elimination of solvent drop instability, larger drop volumes, and increased interfacial area between the solvent

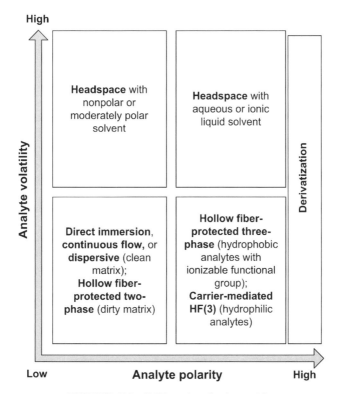

FIGURE 5.2. SME mode selection guide.

and the sample. However, this approach is more difficult to automate, and fibers must be individually prepared.

Nonpolar and moderately polar semivolatiles can be extracted from clean matrices by direct immersion (DI) (single drop, drop-to-drop, or floating organic drop), continuous flow (CF), or dispersive microextraction (DLLME). Single-drop SME is the simplest of the three modes and requires only a microsyringe, pure organic solvent, and a magnetic stirrer, so this mode is appropriate for occasional use, and the single-drop mode can be automated. Continuous-flow SME requires more equipment, such as a pump, but provides the highest sensitivity, as it always operates in the equilibrium extraction mode. The two most important features of dispersive SME are short extraction time and high sensitivity resulting from a very high interfacial area between the solvent and the sample. On the other hand, this mode cannot be automated. Therefore, dispersive liquid–liquid microextraction is best suited for rapid occasional analyses. For solvent microextraction from complex and dirty matrices, three modes are available: drop-to-drop (DD), hollow fiber–protected two-phase [HF(2)], and membrane-assisted solvent extraction (MASE). Drop-to-drop SME is the preferred mode when sample volume is limited (e.g., blood plasma, as it requires only about 10 to 25 μL of sample) (see Figure 2.1). The other advantages of drop-to-drop SME are short extraction times, typically 5 to 10 min,

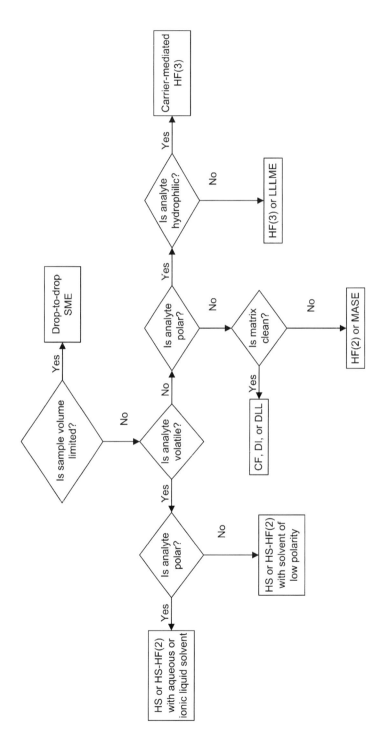

FIGURE 5.3. Flowchart for SME mode selection. CF, continuous flow; DI, direct immersion; DLL, dispersive liquid–liquid; HF(2), hollow fiber–protected two-phase; HF(3), hollow fiber–protected three-phase; HS, headspace; MASE, membrane-assisted solvent extraction; LLLME, liquid–liquid–liquid microextraction.

and no need for sample pretreatment, even for complex samples such as body fluids. Hollow fiber–protected two-phase SME uses small volumes of solvent and provides excellent sample cleanup and high sensitivity. Since the hollow fibers used in this mode are disposable, there is no problem with fiber fouling and analyte carryover effects. The shortcomings of HF(2)ME include relatively long extraction times, typically 15 to 60 min, although in some cases extending to several hours, the lack of commercially available automated equipment, and a lack of commercially prepared fibers. Membrane-assisted solvent extraction provides very high sensitivity and low detection limits, often in the range 1 to 10 ng/L by being ordinarily coupled to large-volume injection (LVI) gas chromatography. The fully automated membrane-assisted solvent extraction system is available commercially from Gerstel GmbH (Mühlheim, Germany). Two drawbacks of MASE technique are that it generally uses more organic solvent than other SME modes (800 μL vs. several microliters), and that a new membrane bag has to be preconditioned prior to the first extraction.

Polar semivolatiles fall into two categories: they are either hydrophobic compounds with an ionizable functional group, such as acid –COOH or base –NH$_2$, or are hydrophilic compounds highly soluble in water. Examples of the former category are phenols and aromatic amines, while the latter category includes, for example, drugs such as practolol, ranitidine, or codeine, which have high solubility in water at low pH values. Hydrophobic compounds with ionizable functional groups can be extracted by one of the three-phase SME modes, either hollow fiber–protected three-phase [HF(3)] or liquid–liquid–liquid microextraction (LLLME). Both techniques offer comparable enrichment factors, detection limits, and extraction times, and both are relatively simple and inexpensive. HF(3)ME is better suited for complex and dirty matrices and limited sample volumes (e.g., blood plasma samples), but it requires preliminary cleaning of the fiber by sonication in a solvent, such as acetone. As was the case with HF(2)ME, one drawback of three-phase microextraction techniques is a relatively long extraction time. This is understandable in view of the fact that the entire mass transfer of the analyte takes place in the liquid phase, where diffusion coefficients are low, and involves two processes: the transfer from the sample to the solvent and back-extraction from the solvent to the acceptor phase. If short extraction times are of primary importance (high sample throughput), electromembrane extraction (EME) can be used. In EME, typical extraction times are about 5 min, but the technique requires additional equipment: two electrodes and a dc power supply. Hydrophilic semivolatiles call for carrier-mediated HF(3)ME. In this technique, a relatively hydrophobic ion-pairing reagent (called a carrier) with acceptable water solubility forms ion pairs with the analytes. The ion-pair complexes are then extracted into the organic phase, followed by back-extraction into an aqueous acceptor phase. In terms of analytical performance, carrier-mediated HF(3)ME is similar to HF(3)ME except for the need to use an ion-pairing agent.

For some classes of analytes, such as basic drugs, more specific guidelines for selection of membrane-based SME mode have been developed.[1] The mode selected depends on computed solubility and log D for the analytes, where D is the distribution ratio defined by equation (3.13). To this end, drug solubilities at pH 2 and log D values at pH 13 are computed using computer programs from Advanced Chemistry Development, Inc. (Toronto, ON, Canada): ACD/Solubility DB and ACD/Log D DB, respectively. The optimum SME mode: HF(2), HF(3), or carrier-mediated HF(3) is then selected depending on the solubility and log D, as shown in Figure 5.4.[1] Based on this strategy, the extractability of new compounds may be readily predicted to speed up method development.

After selecting the extraction mode, the next decision we have to make is whether to choose exhaustive or equilibrium extraction. Some sample preparation techniques, such as solid-phase extraction or purge and trap, operate exclusively as exhaustive extraction methods. Others, including solid-phase microextraction and solvent microextraction, can be used in either an exhaustive or an equilibrium mode. A detailed discussion of this topic can be found in Chapters 3 and 4. Here we provide only a summary and conclusions. The equilibrium concentration of analyte

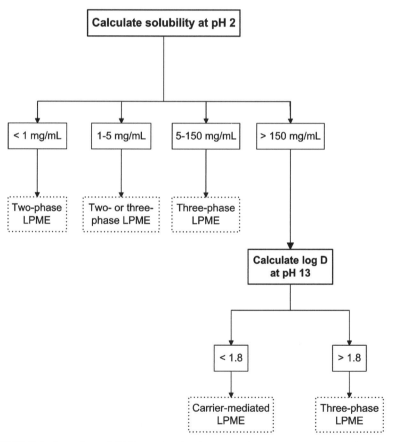

FIGURE 5.4. Overview of SME mode selection. (Reprinted with permission from Ref. 1; copyright © Wiley-VCH Verlag GmbH & Co.)

in the organic phase as a function of the initial analyte concentration in the aqueous phase is given by the following expression [equation (3.3)]:

$$C_o = \frac{KC_w^0}{1 + K(V_o/V_w)} \tag{5.1}$$

where C_o and C_w^0 are the equilibrium and initial analyte concentrations in the organic and aqueous phases, respectively, K is the equilibrium distribution constant of analyte between the organic and water phases, and V_o and V_w are the volumes of the organic and aqueous phase (sample), respectively. An inspection of equation (5.1) reveals that the extraction mode depends on the values of K and the phase ratio, V_o/V_w. When the analyte has a relatively low affinity for the extraction solvent (relatively small K value) and the sample volume is large (the phase ratio is very small) as is the case in continuous-flow SME, the second term in the denominator of equation (5.1) is very small and can be neglected, and the equation is reduced to $C_o \approx KC_w^0$. Under these conditions, a negligible amount of analyte is removed from the sample and solvent microextraction operates in equilibrium mode. The other limiting case occurs when K is so large and/or the phase ratio is relatively large (the sample volume is limited) that the second term in the denominator of equation (5.1) is much greater than unity, and C_o becomes $C_w^0(V_w/V_o)$. This is the case of exhaustive extraction, which can take place for analytes with a very high affinity for the extraction solvent when the sample volume is limited. Because the organic solvent volume is usually very small, exhaustive extraction is often not observed. In many cases, an intermediate extraction mode between the two extremes is encountered, where the amount of analyte extracted is significant (not negligible) but far less than 100%. In solvent microextraction, only one mode, continuous-flow SME, ensures implementation of equilibrium extraction automatically. In all the other modes, equilibrium extraction can be accomplished by proper selection of the organic solvent and the phase ratio. Equilibrium extraction is usually simpler, more selective, and applicable to field sampling. On the other hand, exhaustive extraction does not require surrogates, since all the analytes are transferred to the extracting solvent and there is no need to compensate for variations in matrix composition and their effect on distribution coefficients.

5.3 STATIC VS. DYNAMIC EXTRACTION

In static solvent microextraction, the solvent is not deliberately agitated (although stirring the sample solution still causes convection in both phases, especially in direct-immersion mode), while in dynamic SME both a small volume of sample and the acceptor phase are in motion, the latter forming a thin film of organic solvent (see Section 3.3.3.4). Static SME is simpler in implementation, because it does not require additional equipment, such as a syringe pump, a variable-speed stirrer motor, or an autosampler. It is therefore recommended for SME method development and for occasional use. If it turns out that the extraction time optimized in the static mode is exceedingly long, (e.g., over 60 min), dynamic SME may be a better option. The dynamic mode speeds up mass transfer, reducing the extraction time, but it does not

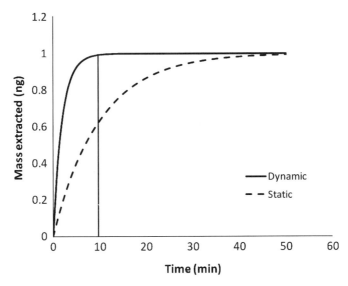

FIGURE 5.5. Extraction time profile of a hypothetical analyte extracted under identical conditions using static and dynamic mode.

improve the extraction efficiency. In both extraction modes, the amount of extracted analytes remains the same provided that the equilibrium between the two phases (donor and acceptor) has been reached. This is illustrated in Figure 5.5, which shows a hypothetical analyte extracted by static and dynamic SME under identical conditions (sample volume, solvent type and volume, temperature, ionic strength, pH, etc.). It is evident that in the dynamic mode, equilibrium is reached much faster (after 10 min), whereas the static mode requires at least 50 min to reach equilibrium. Notice, however, that in both cases the amount extracted at equilibrium is the same (1 ng). As we shall see later in the chapter, however, in practical SME procedures it is not necessary to reach equilibrium, and a shorter extraction time than equilibration time can often be used, as long as two conditions are met: (1) the extraction time is carefully controlled for solutions, both standards and samples, and (2) the slope of the extraction time profile has reached a small value. For example, in Figure 5.5, static extraction could be terminated after 30 min rather than 50 min.

It should be evident from the discussion above that the best uses of dynamic SME are in automated procedures for handling a large number of samples, where high sample throughput is essential. The reduction in extraction time due to dynamic extraction is achieved in most SME modes, including single-drop headspace, hollow fiber–protected headspace, direct immersion, and membrane-based techniques: HF (2) and HF(3), but this advantage should be weighed against increased complexity and the cost of equipment required for dynamic solvent microextraction.

5.4 SELECTION OF MANUAL VS. AUTOMATED EXTRACTION

Every manual mode of solvent microextraction is characterized by extreme simplicity and a very low cost of operation, which in our view makes this sample preparation technique so attractive in the first place. Therefore, a manual SME setup is a good starting point for any optimization of microextraction parameters, since it requires only small volumes of very pure organic solvents and common laboratory equipment, such as a microsyringe, sample vials, a magnetic stirrer, a pH meter, and a stopwatch or timer, and the instrumentation needed for the final determination (e.g., a chromatograph). For membrane-based techniques, inexpensive and disposable hollow fibers are also necessary. Manual mode can be sufficient if SME procedures are used only occasionally and the cost of dedicated, automated equipment is unjustified. On the other hand, laboratories handling routine analyses of a large number of samples should be interested in purchasing automated systems taking advantage of SME. At present, two fully automated solvent microextraction systems are available commercially. One of them is the dedicated membrane-assisted solvent extraction system that was mentioned in Section 5.2. The other system is a CTC CombiPal autosampler (Zwingen, Switzerland) which, used in combination with a gas chromatograph, is more versatile and can be employed in several SME modes. Using the associated Cycle Composer software, the auto-sampler can be programmed to implement static and dynamic modes of headspace and direct immersion as well as static hollow fiber–protected two-phase SME.[2] In addition to the advantage of unattended operation, automated SME provides better repeatability of the results, although an experienced operator working with solutions having analyte concentrations on the order of µg/L can achieve relative standard deviations less than about 7% in manual mode. The scope of SME applications would be greatly expanded, especially in the area of clinical and forensic analysis, if the automated systems for HF(3)ME and electromembrane extraction (EME) techniques became available.

5.5 SELECTION OF DIRECT VS. DERIVATIZATION SME

Most analytes can be extracted directly from the sample matrix by SME, and this should be the preferred way of using solvent microextraction. There are several cases, however, when derivatization can either make possible or improve an analytical procedure. Derivatization in SME has two major objectives:

1. To convert nonextractable or poorly extractable analytes into derivatives that have much larger partition coefficients between the sample and the acceptor phase (solvent) and therefore improved extractability. Examples of analytes nonextractable by direct SME include ionic species, such as metal ions,[3–29] metalloid ions,[30–35] and inorganic anions,[36–38] as well as other inorganic[39,40] and organometallic compounds.[41–44] Polar analytes are usually poorly extracted by typical organic solvents used in SME unless aqueous solutions or ionic liquids are employed as extractants (see Chapter 6 for more details). As indicated in Figure 5.2, all polar analytes, both volatile and semivolatile, are amenable to derivatization. These include amines,[45–51] carbonyl compounds: aldehydes and ketones,[52–61] phenols and chlorophenols,[62–72] organic acids,[73–80]

amino acids[81,82] and peptides,[83] diisocyanates,[84] and anabolic steroids.[85] To this end, derivatization is carried out either prior to extraction (preextraction derivatization) or concurrently with extraction (extraction–derivatization).

2. To make analytes compatible with the detection method used and to improve the selectivity and sensitivity of final determination. In this case, derivatization follows solvent microextraction (postextraction). For example, gas chromatographic determination of some analytes, especially those polar and thermally unstable, can result in irreversible adsorption, thermal decomposition in the injection port, peak tailing, and poor sensitivity. Derivatization of extracted analytes can improve their volatility and thermal stability, bringing about better GC separation. Furthermore, properly selected derivatizing agents can significantly improve sensitivity of the analytical procedure by providing increased response of the GC detector. It is well known that the electron capture detector is highly sensitive (and selective to some extent) to halogenated compounds. This explains the fact that carbonyl compounds are often derivatized with *O*-2,3,4,5,6-(pentafluorobenzyl)hydroxylamine hydrochloride or 2,4,6-trichlorophenylhydrazine, and amines are frequently derivatized with pentafluorobenzaldehyde.

A substantial number of published SME procedures involved derivatization, and the combination of microextraction and derivatization was recently reviewed.[86,87] We should remember, however, that there are problems that can limit the extent of use of derivatization reactions in SME. They include, for example, lack of reagents suitable for functional groups of interest, additional time required to complete chemical reaction, variability in reaction yield resulting in inferior repeatability of results, high cost of some derivatizing reagents, and possible interferences. In general, derivatization complicates SME procedures and should be regarded as a last resort and not the first choice of analytical chemists. Classification of derivatization approaches utilized in SME is depicted in Figure 5.6.

5.5.1 Preextraction Derivatization

In the preextraction technique, the derivatizing agent is first added to the vial containing the sample. The derivatives are subsequently extracted by SME and introduced

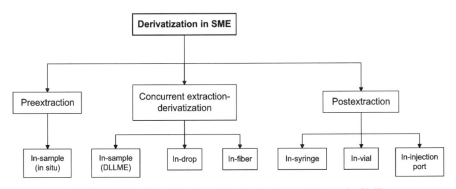

FIGURE 5.6. Classification of derivatization techniques in SME.

into the analytical instrument. This approach, called *in-sample* or *in situ derivatization*, has been used most commonly in SME,[3,4,10–15,17,18,21,23–25,28,31,33–35,38–41,44,45, 50–54,59–62,68–70,72–75,78,81–83,88] although it is not as rapid as concurrent extraction–derivatization. Almost all SME procedures for the determination of metal ions took advantage of in situ derivatization. This approach has also been applied to extract phenols, amines, carbonyl, and organometallic compounds.

5.5.2 Concurrent Extraction–Derivatization

The most efficient and effective derivatization strategy is concurrent extraction and derivatization. This approach reduces derivatization time and simultaneously improves analyte extractability. It can also be used in field sampling. Concurrent extraction–derivatization is implemented in three ways. The first one involves *in-sample extraction combined with simultaneous derivatization*. This strategy is utilized almost exclusively with dispersive liquid–liquid microextraction[19,20,29,30,46,65,89] and homogeneous liquid–liquid extraction.[7–9] In this procedure, a solution containing both the extracting solvent and derivatizing agent in a disperser solvent is rapidly injected into the sample solution (see Figure 2.2). As a result of high interface area between the sample, derivatizing agent, and extracting solvent, derivatization and extraction are rapid. Typically, extraction times are in the range 2 to 20 min.

The second and third ways of implementing concurrent extraction–derivatization are used in single-drop (both direct immersion[6,16,26,27,32,67] and headspace[10,31,33,34,36,37,43,47,48,52,55–58,77,84] modes) and hollow fiber–based SME and are called *in-drop*[6,10,16,26,27,31–34,36,37,43,47,48,52,55–58,67,77,84] and *in-fiber*[22,73,85] *derivatization*, respectively. In this case, the derivatizing agent is dissolved directly in the extracting solvent, which is exposed to the sample. When the analytes are extracted, they are converted into derivatives with high solubility in the organic solvent. It should be noted that the process is no longer equilibrium but exhaustive extraction, since the derivatives will be collected in the solvent as long as extraction continues and the derivatizing agent is present in excess.

In some cases of simultaneous extraction–derivatization, one reactant (usually derivatizing agent) is hydrophobic, while the other (typically analyte) is hydrophilic. To promote the transport of analyte from the aqueous sample into the extracting organic solvent containing the derivatizing reagent, a phase-transfer catalyst is added. The catalysts used in SME have always been ion-pairing agents. This approach is called *phase-transfer catalysis* (PTC).[87] Phase-transfer catalysis has been used to derivatize 11-nor-Δ^9-tetrahydrocannabinol-9-carboxylic acid,[73] phenolic endocrine disruptors,[67] and phenols.[66]

5.5.3 Postextraction Derivatization

Postextraction derivatization is carried out after the extraction process has been completed. In this approach, the derivatizing agent is added to the organic extract containing the analyte(s) either directly in a microsyringe[71,90] (*in-syringe derivatization*) or transferred to a vial[49,76] (*in-vial derivatization*). Derivatized analytes are then introduced into the analytical instrument. Alternatively, post-extraction derivatization can be performed in the injection port of a gas chromatograph[63,64,79,80,91] (in-injection port derivatization or simply, in-port). Several different strategies are available in this case. In the first one,[79,80] analytes with acid or base properties are extracted as ion pairs, followed by decomposition at the high

TABLE 5.1   Summary of Derivatization Reactions Used in SME for Selected Classes of Analytes

Analytes	Derivatization Mode	Reaction Type	SME Mode[a]	Detection Technique[a]
Phenols,	In-sample	Acetylation	SDME, HF(2)	GC-MS
chlorophenols	Concurrent (in-sample)	Acetylation	DLLME	GC-ECD
	In-drop	Tosylation	SDME	GC-MS
	In-drop	Esterification	SDME	GC-FID
	In-syringe	Silylation	SDME	GC-MS
	In-injection port	Silylation	HF(2)	GC-MS
Amines	In-sample	Schiff base reaction	HF(2)	GC-MS
	In-sample	Iodination	SDME	GC-FID
	Concurrent (in-sample)	Schiff base reaction	DLLME	GC-MS
	In-drop	Schiff base reaction	HS-HF(2)	GC-MS
	In-drop	Schiff base reaction	HS	GC-MS
	In-vial	Tosylation	HF(3)	HPLC-UV
Carbonyl	In-sample	Hydrazone formation	SDME	HPLC-UV
compounds	In-sample	Oxime formation	HS	GC-MS
(aldehydes,	In-drop	Oxime formation	HS	GC-MS
ketones)	In-drop	Hydrazone formation	HS	GC-MS
Acids: organic,	In sample	Alkylation	HF(2)	GC-MS
haloacetic,	In-sample	Esterification	HS-HF(2)	GC-ECD
alkylphosphonic,	In-drop	Silylation	HS	GC-MS
acidic herbicides	In-syringe	Silylation	SDME	GC-MS
	In-vial	Pentafluorobenzylation	SDME	GC-MS
	In-injection port	Ion pair decomposes to butyl ester	HF(2)	GC-MS
Amino acids	In-sample	Conversion to N(O,S)-ethoxycarbonyl amino acid ethyl esters	SDME	GC-MS

[a] SDME, single-drop microextraction; HS, headspace; HS-HF(2), headspace hollow fiber–protected two-phase; DLLME, dispersive liquid–liquid microextraction; HF(2), hollow fiber–protected two-phase; HF (3), hollow fiber–protected three-phase; GC, gas chromatography; MS, mass spectrometry; FID, flame ionization detector, ECD, electron capture detector; HPLC-UV, high-performance liquid chromatography with ultraviolet detection.

temperature of the GC injection port to form volatile by-products and the alkyl derivatives of the target compounds. No other reagent is added after the extraction of analyte ion pairs. In the second approach,[63,64] the analytes extracted are introduced into the GC injection port, followed by injection of the derivatizing agent, and held there for 2 min in order to complete the reaction. In the third approach,[91] after analytes are extracted into the organic solvent, the derivatizing agent is drawn into the same microsyringe that contains the extracted analytes, and the mixture is injected into the GC, where the derivatization reaction takes place in the hot injection port. The objective of postextraction technique is mainly to improve selectivity and sensitivity of the detection method and to reduce interferences. It is also used when the derivatizing agent decomposes in the presence of sample matrix (i.e., water).

Some analytical procedures involving SME require not one but two derivatization reactions.[10,31,33,34,52,73] The first reaction is carried out in the sample, and the second takes place in the extracting solvent (in-drop or in-fiber). The most common application of this approach includes the determination of certain metals or metalloids by first converting them into volatile hydrides by reaction with sodium borohydride, followed by sequestration of the hydride in an aqueous solution of Pd(II) using headspace single-drop SME.[10,31,33]

A summary of complexation reactions used for the extraction of metal ions is provided in Chapter 6 (Table 6.3). A compilation of derivatization reactions used for other important classes of analytes is summarized in Table 5.1. A brief inspection of Table 5.1 reveals that the most common final determination technique for derivatized analytes is GC-MS, which indicates that one of the reasons for derivatization (at least for the postextraction approach) is to take advantage of the high resolution afforded by capillary gas chromatography as well as the superior sensitivity and analyte identification ability of mass spectrometry. More details on derivatization of various classes of analytes are provided in Chapter 6.

5.6 EXTRACTION SOLVENT SELECTION

Selection of the optimum extraction solvent is one of the most important tasks in method development in SME. There are some general requirements that the solvent must meet irrespective of the solvent microextraction mode employed. Solvent properties both required and desired include:

1. Good selectivity for target analytes.
2. High extraction efficiency of analytes (i.e., large partition coefficients between the extraction solvent and sample). In other words, polarity of the solvent should match that of analytes ("like dissolves like").
3. Immiscibility with water. The solvent should be insoluble in water or at least have very low solubility. This requirement does not always apply in headspace SME if aqueous solutions are used for extraction of analytes.

4. Compatibility with the final determination technique: for example, with the mobile phase in HPLC and capillary electrophoresis, as well as selective detectors, such as the electron capture detector (ECD), which is incompatible with oxygenated and halogenated solvents.

5. Good chromatographic behavior: the solvent peak must be separated from the analyte peaks.

6. Low volatility to minimize solvent losses during extraction.

7. Low cost and ready availability.

8. High purity: no contaminants interfering with the determination must be present.

9. Low toxicity.

In addition, various SME modes have their own specific requirements for the solvent. In drop-based techniques, direct immersion and continuous flow, a low rate of drop dissolution is desirable. Furthermore, good drop stability in sample solution is a must. In single-drop direct-immersion SME, the stability of the organic drop depends on three factors:[92,93] (1) upward floating force ($F_f = V_o \rho_w g$), (2) downward gravity ($F_g = V_o \rho_o g$), and (3) adhesion forces (F_a) resulting from surface tension (σ), where V_o, ρ_o, ρ_w, and g are the volume of the organic solvent drop, the density of the organic solvent, the density of the sample solution, and the acceleration due to gravity, respectively. Therefore, the stability of the organic solvent drop suspended at the tip of the microsyringe is affected by the surface tension and the densities of both the organic solvent and the sample solution. Under experimental conditions, σ is constant. Stable drops are formed when the organic solvent used has a high surface tension and the difference $F_f - F_g$ is less than F_a (i.e., when the densities of water and organic solvent are similar). When $F_f - F_g$ is greater than F_a as a result of increased drop volume V_o (since $\rho_w > \rho_o$), the organic drop begins to move slowly upward away from the tip outlet.[92]

In continuous-flow SME, in addition to the three forces described above, there exist three additional forces:[94] the collision force (F_c), coming from the molecular momentum of the moving sample solution, the intermolecular force (F_v) (van der Waals force) between the molecules of the solvent drop and of the poly(ether ether ketone) (PEEK) tubing or stainless steel syringe needle, and drag (F_d). F_f, F_d, and F_c are directed upward and try to push the solvent drop away from the tip of the tubing, whereas F_g, F_a, and F_v are downward forces which ensure that the solvent drop is retained at the tip of the tubing or syringe needle. Depending on the force equilibrium, the solvent drop is retained at either the tubing outlet or the needle tip (when $F_g + F_a + F_v \geq F_f + F_c + F_d$) or it is detached from the tubing or needle tip when $F_g + F_a + F_v < F_f + F_c + F_d$.

In headspace SME, the primary requirements for the solvent are high boiling point and low vapor pressure, to avoid drop evaporation during extraction. This eliminates certain solvents from consideration even though they may meet other criteria, such as high extraction efficiency or good selectivity. For example, although toluene is one of the preferred extraction solvents in direct-immersion and hollow fiber–protected two-phase SME, it should be avoided in headspace SME, since toluene drops were found to evaporate completely when exposed to air within 4 min[95] or 2 min.[96] Other than that, good drop stability is also required, which implies a high surface tension of the solvent.

In dispersive liquid–liquid microextraction, the extracting solvent has to have a density higher than water in order to be separated from the sample by centrifugation, while the disperser solvent has to be miscible with the extraction solvent and water. Typical disperser solvents used in DLLME include acetone, methanol, acetonitrile, and ethanol. In liquid–liquid–liquid and directly suspended droplet microextraction, the solvent has to have a density lower than water, so that it forms a separate layer on top of the sample, into which the microdrop of the acceptor solution can be immersed. A recent paper[97] discusses a combination of dispersive and directly suspended droplet modes, which makes use of 2-dodecanol as an extracting solvent, which has a density lower than that of water. The analytes are extracted by means of dispersive liquid–liquid microextraction, but the extract is separated from the sample by solidification of a floating organic droplet in an ice bath. This technique avoids the use of chlorinated solvents, most of which are considered to be hazardous.

In hollow fiber–based SME (both two- and three-phase), the solvent has to be compatible with the fiber, so that it is easily immobilized in the fiber pores and does not leak. Since almost all fiber-based SME applications make use of Accurel porous polypropylene hollow fiber, which is hydrophobic, the solvent has to have a low polarity matched to that of the fiber to effectively wet the walls of the pores of the hollow fiber. In addition, because mass transfer through the solvent layer in the fiber is the diffusion-controlled rate-determining step, it is desirable that the solvent have low viscosity to improve diffusion coefficients of the analytes in the solvent and the kinetics of mass transfer. In addition, in hollow fiber–protected three-phase SME, the solubility of analytes in the solvent should be higher than that in the donor phase but lower than that in the acceptor phase. It should be pointed out that in HF(3)ME variations in volume of extracting solvent do not affect analyte enrichment. Electromembrane extraction calls for a solvent that has a certain polarity or water content to provide sufficient electrical conductance.

Most SME procedures described in the analytical literature use a purely empirical approach to solvent selection. Several different organic solvents immiscible with water are selected and used to extract target analytes. The one with the largest analyte enrichment factors and the chromatographic peak not overlapping with the analyte peaks is then used in the next steps of method development. A compilation of all solvents used in different modes of SME along with the classes of analytes they were used to extract is provided in Table 5.2. As we can see, the following solvents are used most often:

continuous flow	carbon tetrachloride
direct immersion	hexane, toluene
headspace	1-octanol, hexadecane, dodecane, decane

TABLE 5.2 SME Solvent Selection Guide

Solvent[a]	Mode Used[a,b]	Analytes Extracted
Aliphatic hydrocarbons		
Hexane	**SD**, CF, HF(2), HF(3)	Trihalomethanes (THMs), carbonyl compounds, organochlorine and organophosphorus pesticides (OCPs, OPPs), organohalogen compounds, volatile organosulfur compounds, phthalate esters, polychlorinated biphenyls (PCBs), polybrominated diphenyl ethers (PBDEs), aromatic hydrocarbons
Heptane	HF(3)	Monosubstituted phenols
Octane	SD, HF(2), HF(3)	Polycyclic aromatic hydrocarbons (PAHs), analgesic drug, organometallic compounds
Isooctane	SD, HF(2)	Iodine (derivatized), carbamates, OPPs, chlorobenzenes, PCBs
Nonane	HF(2)	Halogenated anisoles
Decane	**HS**, HF(2)	Organotin compounds, aliphatic amines, aldehydes, ketones, PBDEs, derivatized volatile organic acids
Undecane	HF(2), HF(3)	PBDEs, dinitrophenols, antiparasitic drug
Dodecane	HS, HF(2), HF(3)	Volatile plant components, chlorobenzenes, antiepileptic and antidepressive drugs, *n*-alkanes, insecticide, antimicrobial agent
Tridecane	HS	Volatile organochlorine compounds
Tetradecane	HF(2)	OCPs
Hexadecane	**HS**	Essential oil components, gasoline components, aromatic hydrocarbons, volatile organic compounds
Heptadecane	HS	Volatile organic compounds (VOCs), essential oil components
Cyclopentane	HF(2)	Volatile organochlorine compounds
Cyclohexane	HS, **HF(2)**	Essential oil components, OCPs, OPPs, triazine herbicides, PAHs, PCBs

Chlorinated hydrocarbons

Dichloromethane	SD	Volatile flavor components, chemical warfare agents (CWAs), herbicide
Chloroform	**SD**, CF, HF(2)	Drugs of abuse, derivatized inorganic species, derivatized amino acids, phenolic endocrine disruptors, hydrophobic biomolecules, surfactant, antibiotic, alkaloid, volatile flavor components, CWAs
Carbon tetrachloride	SD, **DLL, CF,** HF(2)	OPPs, triazine herbicides, CWAs, aromatic amines, derivatized inorganic species, phenols, antioxidants, antidepressant drugs, phthalate esters, organotin compounds, organosulfur pesticides
1,2-Dichloroethane	SD	Analgesic, herbicide, CWAs
1,1,2,2-Tetrachloroethane	SD, DLL	Analgesic, PBDEs, carbamate pesticides
Tetrachloroethene	DLL	PAHs, drugs
1,1,1-Trichloroethane	DLL	Organophosphorus esters
Chlorobenzene	**DLL**	OPPs, triazine herbicides, chlorophenols, chlorobenzenes, anilines, phthalate esters
1,2-Dichlorobenzene	DLL	Metal ions (derivatized)
1,2,4-Trichlorobenzene	HF(3)	Chlorophenols

Aromatic hydrocarbons

Benzene	CF, HF(3)	PAHs, local anaesthetics
Toluene	**SD**, HS, CF, **HF(2)**, HF(3)	Chlorobenzenes, phenols, PAHs, OCPs, OPPs, triazine herbicides, antifouling agents, aldehydes, ketones, derivatized inorganic species, nitroaromatic explosives, surfactant, aromatic amines, PCBs, PBDEs, chloroacetanilide herbicides, nitrobenzenes, antidepressant drugs, phthalate esters, PAHs, endocrine disruptors, primary amines, haloethers, essential oil components, fungicides, illicit drugs, organomercury compounds, peptides, and proteins

(Continued)

117

TABLE 5.2 *Continued*

Solvent[a]	Mode Used[a,b]	Analytes Extracted
o-Xylene	SD, HS, HF(2)	Fungicides, alcohols, low-molecular-weight drugs, trihalomethanes (THMs), hydroxycarbonyls, organosulfur pesticides
m-Xylene	SD, HF(2)	Alkaloid
p-Xylene	SD, HS, HF(2)	Antimicrobial agents, OCPs, VOCs, essential oil components
p-Cymene	HS	Esters
Phenylhexane	HF(3)	Metal ions (following derivatization)
Alcohols		
Benzyl alcohol	SD, **HS**, HF(3)	Residual solvents, aniline derivatives, aromatic compounds (BTX), ether, VOCs, aliphatic and aromatic amines, aldehyde, chlorocarbons, alcohol
1-Butanol	HS	PAHs, carbonyl compounds (following in-drop derivatization)
1-Pentanol	HS	Mononitrotoluenes
1-Hexanol	HS, HF(3)	Odorants, organic acid
1-Heptanol	HF(3)	Acidic drugs
1-Octanol	SD, **HS**, **HF(2)**, **HF(3)**	PAHs, quinone, triazine herbicides, hydrophilic drugs, bioactive plant components, aldehydes, volatile flavor components, volatile halocarbons, haloanisoles, aromatic hydrocarbons (BTEX), alcohols, haloacetic acids, volatile residual solvents, THMs, mycotoxin, benzodiazepines, OCPs, carbamate pesticides, chlorobenzenes, chlorophenols, fatty acids, acidic herbicides, diuretics, food dyes, antimalarial drug, acidic and basic drugs, aromatic amines, amino alcohols, phenols, anabolic steroid glucuronides, cough suppressant, peptides, *Strychnos* alkaloids, nitrophenols, nonsteroidal antiinflammatory drugs (NSAIDs)
1-Nonanol	HS, HF(2), HF(3)	OCPs, weakly basic drugs, drugs of abuse, hydroxyaromatic compounds
1-Undecanol	SD	PAHs, OPPs, fat-soluble vitamins, metal ions (following derivatization)
Ethylene glycol	HS	Alcohols
Glycerol	HS	Alkaloid

Esters

Ethyl acetate	SD, HF(2), HF(3)	Drugs of abuse, OCPs, pyrethroids, aniline derivatives, phenols
Butyl acetate	SD, HS, HF(2)	Triazine herbicides, phenol and chlorophenols, aromatic compound, benzodiazepines
Amyl acetate	CF, SD	Phenols
Hexyl acetate	SD	Phenols (silylated), organic acids, acidic herbicides
Dodecyl acetate	HF(3)	Antidepressant drugs
o-Dibutyl phthalate	SD, HS	Local anaesthetics, short-chain fatty acids
Tributyl phosphate	SD, HF(3)	Metal ions (following derivatization), alkaloid
Triolein	HS	Aromatic hydrocarbons (BTEX)

Ethers

Anisole	SD, HS	Derivatized inorganic species
Dihexyl ether	HF(2), **HF(3)**	Anabolic steroid metabolites, chlorophenols, nitrophenols, antidepressant drugs, drugs of abuse, haloacetic acids, phenoxyacetic acids, basic drugs, NSAIDs, hydroxyaromatic compounds, metalions (after derivatization), pesticide residues, short-chain fatty acids, aniline derivatives, antimalarial drug, polyamines
2-Nitrophenyl octyl ether	HF(3)	Basic drugs

Ketones

Acetone	HS (with water and/or glycerol)	Alkaloids
2-Heptanone	HF(2)	Degradation product of NSAID
6-Undecanone	HF(3)	Phosphate ester
Benzoylacetone	CF	Metal ions
Diisobutyl ketone	HF(3)	Sulfonamides

(Continued)

119

TABLE 5.2 *Continued*

Solvent[a]	Mode Used[a,b]	Analytes Extracted
Ionic liquids		
1-Butyl-3-methylimidazolium hexafluorophosphate	SD, HS	Hydrocarbons, phenols, chlorophenols, chloroanilines, OCPs, chlorobenzenes, NSAIDs
1-Hexyl-3-methylimidazolium hexafluorophosphate	SD, HS, DLL	Alkylphenols, ketone, chlorobenzenes, OPPs, pyrethroids, metal ions (after derivatization)
1-Octyl-3-methylimidazolium hexafluorophosphate	SD, HS, HF(3)	Derivatized aldehyde, chlorophenols, PAHs, aromatic compounds, phthalates, phenols, aromatic amines, herbicides, organotin and organomercury compounds
Miscellaneous		
Water	HS	Inorganic and organomercury species (solution of derivatizing agents), amphetamines (solution of phosphoric acid), alcohol, PAHs (water saturated with cyclodextrin), chlorophenols (water mixed with acetonitrile), phenols (NaOH solution), diisocyanates (solution of dibutylamine)
Pyridine	HS	Derivatized inorganic species (in mixture with benzyl alcohol)
N-Methylpyrrolidone	HS	Volatile residual solvents
N,N-Dimethylformamide	HS	Volatile sulfur compounds
Acetonitrile	HS	Chlorophenols (in mixture with water)
Carbon disulfide	DLL	THMs
4-Nitro-m-xylene	HF(3)	Sulfonamides
1-Isopropyl-4-nitrobenzene	HF(3)	Basic drugs
Silicone oil AR 20	HF(3)	Antidepressant drugs

[a] The most common solvents and the modes in which they are used are shown in boldface type.

[b] CF, continuous flow; SD, single drop; DLL, dispersive liquid–liquid; HS, headspace; HF(2), hollow fiber–protected two-phase; HF(3), hollow fiber–protected three-phase.

dispersive	carbon tetrachloride, chlorobenzene
hollow fiber–protected two-phase	1-octanol, toluene, cyclohexane
hollow fiber–protected three-phase	1-octanol, dihexyl ether

These solvents can be tried first. If they are unsuitable for the task at hand, a more systematic approach is needed. The procedure recommended for solvent selection in SME is to choose several (five or six) common organic solvents meeting the required criteria listed at the beginning of this section but providing a broad range of polarities as described by Rohrschneider polarity scale,[94,98,99] log K_{ow},[100] or dielectric constant[101] and to perform extraction of the analytes. For example, one might choose isooctane (polarity index 0.1), cyclohexane (0.2), toluene (2.4), xylene (2.5), butyl acetate (4.0), and chloroform (4.4).[94] The solvent providing the best enrichment factor and lack of interference in the final determination is then selected for the SME procedure. To facilitate the initial choice of solvents for method development, Table 5.3 summarizes the physicochemical properties of common solvents relevant in SME. Most of the data were taken from a paper by Barwick,[102] but numerous other online sources were also used. For practical reasons, water solubility of solvents, S, is expressed in microliters of solvent per milliliter of sample at 20°C. In this way it is easy to estimate the maximum solvent loss due to dissolution during microextraction. For example, the solubility of toluene is 0.55 μL/mL, which means that if we use a 1-mL aqueous sample and a 1-μL toluene drop in direct-immersion SME, then at equilibrium (i.e., if extraction is long enough) 0.55 μL of the toluene will dissolve and only 0.45 μL will be left. In practice, this is unlikely to happen because the dissolution process is very slow, but we should be aware of the fact that after most DI extractions, at least some of the solvent will be gone. It is therefore recommended that solvent dissolution be taken into consideration during SME method development, as well as the fact that the solvent volume drawn back into the microsyringe following extraction is less than the volume extruded from the microsyringe in the beginning of SME.

5.7 SELECTION OF FINAL DETERMINATION METHOD

Solvent microextraction is compatible with a large number of chromatographic and spectroscopic methods of analysis. The extraction process results in a substantial sample cleanup but does not guarantee that only the analytes and none of the interferences are extracted. Therefore, in most SME applications, a powerful separation method has to be coupled with the detection technique. As a result of its superior selectivity and sensitivity, capillary gas chromatography is the preferred method of separation and final determination of volatile and semivolatile organic analytes. All the SME modes using organic solvents as acceptor solutions are compatible with GC, which means that the extracts can be introduced directly into the GC injection port. Although an important group of extracting solvents, ionic liquids, often used with HPLC, is incompatible with GC, a simple interface for

TABLE 5.3 Summary of Physical Properties of Solvents Used in SME

Solvent[a]	Density (g/cm^3)	B.p. (m.p.)[b] (°C)	Dielectric Constant (ϵ)	Viscosity (cP at 20°C)	Surface Tension (g/s^2)	Vapor Pressure (kPa at 25°C)	Dipole Moment (D)	S[c]	log K_{ow} (log D)[d]	P[e]	δ[f]
Aliphatic hydrocarbons											
Hexane	0.65	69	1.8	0.372	18.4	17	0.08	0.02	3.94	0.1	~8
Heptane	0.68	98.4	1.92	0.408	20.21	6.1	~0	0.0044	4.66	0.2	~8
Octane	0.70	126	1.95	0.706	21.8	1.87	0	0.0009	5.15	0.1	7.16
Isooctane	0.69	99.2	1.94	0.48	16.2	5.2	0	Insoluble	5.83	0.1	7
Decane	0.73	174.1	2.0	1.08	22.43	0.17	0	1.2×10^{-5}	5.98	0.3	7.19
Undecane	0.74	196	2.0	1.08	19.23	0.075	0	5.4×10^{-6}	6.6	—	7.8
Dodecane	0.75	216	2.0	1.82	25	0.028	0	4.0×10^{-6}	7.13	—	7.8
Tridecane	0.76	235.4	2.0	2.948	26.1	0.011	0	6.2×10^{-6}	7.66	—	8.0
Tetradecane	0.76	253.7	2.0	2.29	26.5	0.009 (35°C)	0	2.9×10^{-6}	7.20	—	7.9
Hexadecane	0.77	287 (18)	2	3.34	27.6	0.0004	0	Insoluble	8.25	—	8.01
Heptadecane	0.78	302.1 (22)	—	4.21	22.77	0.00025	0	Insoluble	9.79	—	8.41
Cyclopentane	0.74	49	1.97	0.44	23	42.4	0	0.22 (25°C)	3.0	0.1	8.10
Cyclohexane	0.78	81	2.02	0.98	24.99	13.1	0.3	0.070	4.15	0.2	8.2
Chlorinated hydrocarbons											
Dichloromethane	1.33	40	8.9	0.41 (25°C)	27.36	46.5	1.14	12	1.25	3.4	9.6
Chloroform	1.48	61	4.8	0.57	27.2	21.1	1.15	5.54	1.97	4.4	9.3
Carbon tetrachloride	1.59	77	2.24	0.965	26.9	11.94	0	0.50	2.64	1.7	8.6
1,2-Dichloroethane	1.25	83.5	10.6	0.84	35.14	11.6	1.75	6.50	1.48	—	9.78

Aromatic hydrocarbons and *Alcohols* (continued table)

1,1,2,2-Tetrachloroethane	1.54	146.5	7	1.501	32.9	1.07	1.5	1.9	2.39	—	10.2
Tetrachloroethene	1.63	121.1	2.4	0.89 (25°C)	31.74	2.46	0	0.092	3.40	3.40	9.29
1,1,1-Trichloroethane	1.34	74	7.2	4	25.4	13.3	1.78	0.37	2.48	—	8.57
Chlorobenzene	1.1	132	5.6	7.68	33	1.47	1.55	0.18	2.89	2.7	9.7
1,2-Dichlorobenzene	1.309	180.5	9.9	1.324 (25°C)	36.61	0.4 (30°C)	2.27	0.076	3.43	—	10.05
1,2,4-Trichlorobenzene	1.454	214.4 (16.9)	3.98	32.9 (0°C)	39.1	0.067	0.67	0.020	4.02		
Aromatic hydrocarbons											
Benzene	0.88	80	2.28	0.604 (25°C)	28.22	12.7	2.27	2.0	2.13	2.7	9.19
Toluene	0.86	110	2.38	0.590	29.71	3.79	0.36	0.55	2.73	2.40	8.91
o-Xylene	0.88	144	2.545	0.812	30.10	0.7	1.3	0.19 (25°C)	3.12	—	8.85
m-Xylene	0.86	139	2.35	0.62	28.90	0.8	1.3	0.19 (25°C)	3.20	—	8.85
p-Xylene	0.86	138	2.27	0.34 (30°C)	29.0	0.9	1.3	0.21 (25°C)	3.15	2.4	8.85
p-Cymene	0.857	177	2.25	1.023	28.10	0.19	0.19	0.027 (25°C)	4.1		
Phenylhexane	0.867	226	—	1.655	29.47	0.017	0.37	Insoluble	5.34		
Alcohols											
Benzyl alcohol	1.05	205	13.1	4.43 (30°C)	39.0	0.02	1.7	33.3	1.10	5.7	11.5
1-Butanol	0.81	118	17.5	2.55 (25°C)	26.3 (10°C)	0.96	1.66	77.8 (25°C)	0.88	3.9	11.4
1-Pentanol	0.829	137.9	13.9	3.42 (25°C)	24.97	0.27	1.65	32.6	1.41	—	10.9
1-Hexanol	0.832	156	13.33	4.53 (25°C)	25.73	0.067 (21°C)	1.417	7.1	2.03	—	10.7
1-Heptanol	0.837	176	12.1	5.86 (25°C)	26.47	0.016	1.41	1.2	2.62		

(Continued)

123

TABLE 5.3 *Continued*

Solvent[a]	Density (g/cm³)	B.p. (m.p.)[b] (°C)	Dielectric Constant (ϵ)	Viscosity (cP at 20°C)	Surface Tension (g/s²)	Vapor Pressure (kPa at 25°C)	Dipole Moment (D)	S[c]	log K_{ow} (log D)[d]	P[e]	δ[f]
1-Octanol	0.83	195	10.3	6.49 (30°C)	26.35 (35°C)	0.0106	1.40	0.65	3.0	3.4	10.3
1-Nonanol	0.8419	215	8.6	9.72 (25°C)	28	0.0032	1.41	0.15	3.77		
1-Undecanol	0.8298	245 (19)	—	—	30.1	0.00024 (20°C)	1.67[g]	0.023	4.72		
Ethylene glycol	1.11	197	37.7	26 (15°C)	48.4	0.0080 (20°C)	2.28	Miscible	−1.36	6.9	17.0
Glycerol	1.261	290 (18)	41.01	954 (25°C)	64.8	0.000027	2.617	Miscible	−1.76	—	21.1
Esters											
Ethyl acetate	0.90	77	6.08	0.428 (25°C)	23.6	12.1	1.78	87.8	0.73	4.3	8.9
Butyl acetate	0.875	126	5	0.63 (30°C)	24.8 (0°C)	2.0	1.84	7.8	1.78	4.0	8.6
Amyl acetate	0.876	149	4.81	0.87 (25°C)	12	0.53	1.9	1.9	2.30	—	8.5
Hexyl acetate	0.8902	169.2	4.42	1.08	27	0.133	1.86	0.57	2.96	—	8.6
Dodecyl acetate	0.865	265	—	2.81 (35°C)	29.6	0.00013	—	0.00046	6.1	—	8.2
o-Dibutyl phthalate	1.05	340	6.58	16.63	34	0.000016	2.82	0.012	4.9	—	8.3
Tributyl phosphate	0.982	289	6.45	3.88	27.79	0.0008	2.36	6.1	4.27	—	8.8
Triolein	0.9146	385.2	3.109	39.88 (37°C)	35.8	5.16×10^{-7}	—	Insoluble	−0.59		
Ethers											
Anisole	0.9961	154	4.3	1.05	36.18 (15°C)	0.47	1.2	1.04	2.11	3.8	9.7
Dihexyl ether	0.8035	223	—	1.706	25.61	0.0197	0	Insoluble	5.23		

2-Nitrophenyl octyl ether	1.041	351.2	24	12.8	37.9	0.00001125	6.33	Insoluble	5.45	5.4	9.6
Ketones											
Acetone	0.79	56	21.1	0.32	22.8	30.8	2.7	Miscible	−0.24		
2-Heptanone	0.8111	151	11.658	0.714	26.17	0.49	2.6	5.3	1.98		
6-Undecanone	0.831	228	2.005	1.95 (15°C)	27.4	0.0107	—	0.060	4.09		
Benzoylacetone	1.073	262	3.8	—	38.7	0.00147	2.40[g]	0.36	2.52		
Diisobutyl ketone	0.806	165	9.9	1.03	23.92	0.16 (20°C)	1.88	0.62	2.66		
Ionic liquids											
1-Butyl-3-methyl-imidazolium hexafluorophosphate	1.3673	349[h]	11.4	393	48.8	Very low	Ions	13.7	−0.69	—	14.56
1-Hexyl-3-methyl-imidazolium hexafluorophosphate	1.29366	417[h]	8.9	586	43.4	Very low	Ions	5.8	—	—	13.98
1-Octyl-3-methyl-imidazolium hexafluorophosphate	1.22	416[h]	—	682	36.5	Very low	Ions	1.6	—	—	13.59
Miscellaneous											
Water	1	100	78	0.01	72.8	3.168	1.85	—	—	10.2	23.4
Pyridine	0.983	115	12.4	0.97	38	2.77	2.2	Miscible	0.65	5.3	10.6

(Continued)

TABLE 5.3 *Continued*

Solvent[a]	Density (g/cm³)	B.p. (m.p.)[b] (°C)	Dielectric Constant (ϵ)	Viscosity (cP at 20°C)	Surface Tension (g/s²)	Vapor Pressure (kPa at 25°C)	Dipole Moment (D)	S[c]	log K_{ow} (log D)[d]	P[x,e]	δ[f]
N-Methyl-pyrrolidone	1.033	202	33	1.65 (25°C)	40.7	0.867	4.1	Soluble	-0.54	6.7	11.24
N,N-Dimethyl-formamide	0.94	153	37	0.82	35.2	0.493	3.8	Soluble	-1.01	6.4	11.8
Acetonitrile	0.78	82	37.5	0.35	29.04	9.73	3.44	Miscible	-0.34	6.2	12.1
Carbon disulfide	1.26	46.2	2.64	0.363	32.25	48	0.08	1.7	1.94	1.0	9.9
4-Nitro-m-xylene	1.135	244	—	—	40.7	0.00644	—	0.12	2.87		
1-Isopropyl-4-nitrobenzene	1.096	236.4	—	—	38.8	0.0097	—	Insoluble	3.29		
Silicone oil AR 20	1.01	—	2.7	20	—	—	—	Insoluble			

Source: Ref. 102.

[a]Most commonly used solvents are shown in boldface type.

[b]Melting points are listed only for those solvents that can freeze at ambient temperatures, thereby causing problems with a syringe.

[c]S, water solubility in μL solvent per mL sample at 20°C.

[d]K_{ow}, octanol–water partition coefficient.

[e]P' Rohrschneider polarity index.

[f]δ Hildebrand solubility parameter.

[g]In benzene.

[h]Decomposes.

the direct introduction of ionic liquids into the gas chromatograph has been developed recently[103,104] (see Figure 6.11). Mass spectrometric (MS) detection coupled to GC offers very high sensitivity and analyte identification and discrimination, which makes it the method of choice for the analysis of complex environmental and biological samples. Alternatively, flame ionization detection is a highly sensitive and universal technique suitable for virtually all organic analytes. For very complex samples containing specific classes of analytes, such as organochlorine pesticides, selective detectors are widely used. These include electron capture, nitrogen–phosphorus, and flame photometric detectors. It should be pointed out, however, that for samples containing large amounts of chlorinated compounds, GC-MS is preferred to ECD, because halogen-selective detectors such as ECD cannot discriminate between target analytes and interferences.

High-performance liquid chromatography (HPLC) and capillary electrophoresis (CE) are effective in the separation and determination of nonvolatile and thermally labile analytes. Typically, the two techniques use ultraviolet/visible light (UV/VIS) detection, which is inferior in terms of sensitivity compared to the detection methods used in GC. An exception is the LC-MS or LC-MS-MS combination, which has detection limits comparable to GC-MS.[85,105–113] It should be noted, however, that HPLC in its most common reverse-phase mode and CE are directly compatible with the acceptor solutions that are at least partially aqueous: that is, those utilized in hollow fiber–protected three-phase, liquid–liquid–liquid microextraction, and headspace SME with an aqueous solvent. If the SME procedure calls for an organic solvent immiscible with water, but HPLC or CE are preferred as the detection methods, an extra step of solvent exchange and extract reconstitution is required.[98,114–116]

Although mass spectrometry is a common detection system used with GC and HPLC for solvent microextraction applications, a direct combination of SME and MS has recently gained attention as well.[117–125] As a rule, SME is combined with matrix-assisted laser desorption/ionization (MALDI) MS. The small extracting solvent volume is compatible with MALDI-MS sampling requirements.

Following solvent microextraction, inorganic analytes, such as metal and metalloid ions, are commonly detected by one of the spectroscopic techniques: UV/VIS spectrophotometry,[3,7,9,22,23,25] spectrofluorimetry,[82,126] electrothermal atomic absorption spectrometry (ET AAS),[4–6,8,10,18–20,26,27,29–34,43,89,127–130] and inductively coupled plasma (ICP)–optical emission spectroscopy,[15] or ICP–mass spectrometry.[28,35,129] UV/VIS detection offers inexpensive instrumentation, spectrofluorimetry has very low detection limits, while ET AAS and ICP techniques both provide very high selectivity and sensitivity.

5.8 SELECTION OF EXTRACTION OPTIMIZATION METHOD

In analytical chemistry, the term *optimization* means discovering conditions that provide the best possible response, such as enrichment factor, detector signal, and chromatographic resolution. Traditionally, in the SME method, development optimization has been carried out by varying one parameter while keeping all the other parameters at a constant level and monitoring the response (analyte enrichment factor or chromatographic peak area). This optimization technique, called *one-variable-at-a-time* (OVAT), has two major disadvantages: It is time consuming, and it does not account for the presence of interactions among various parameters being optimized. For example, if we optimize an SME procedure at ambient temperature by OVAT, set all other extraction parameters constant and vary the extraction time, we will arrive at a value when the mass extracted–time profile levels off, which will provide us with the optimum extraction time. However, when the extraction time is then fixed at the optimum value and we optimize the extraction temperature next, if the optimum temperature happens to be elevated, the extraction time should be reoptimized, because extraction time and temperature are not independent variables. Extraction kinetics depends strongly on the temperature: the rate of extraction will be improved at higher temperatures and thus the optimum extraction time will be reduced, although the partition coefficient will become less favorable and the amount of analyte extracted will decrease.

Experimental design involves identification of the parameters (factors) that may affect the response, planning the experiments and using statistical analysis for the evaluation of the effects of the factors.[131–135] There are two basic approaches to experimental design: simultaneous and sequential. In *simultaneous design*, one or more experimental responses are recorded for a set of experiments performed in a systematic way in order to predict the optimum and the interaction effects, generally using regression analysis.[132] The data obtained can be modeled; therefore, a description (equation) is obtained relating the variation of the response with factors that can be used for subsequent optimization. For example, the equation can relate analyte enrichment factor to SME variables: solvent volume, temperature, extraction time, ionic strength and pH of the sample, stirring rate, and so on. The equation can then be used to find the maximum enrichment factor and the corresponding optimum values of experimental variables. In contrast, sequential design does not allow obtaining a model of the response. *Sequential designs* are useful when the experimenter has only limited knowledge about how far the optimum region is from the starting experimental point to obtain guidance and direction for the next optimization experiments.

The most common simultaneous design used in analytical chemistry is *response surface methodology* (RSM), a collection of mathematical and statistical techniques based on the fit of a polynomial equation to the experimental data, which must describe the behavior of a data set with the objective of making statistical predictions. It is used when a response or responses of interest are influenced by several variables. The objective of RSM is to optimize the levels of these variables simultaneously to achieve the best system performance and to investigate the interactions between the variables.[134] The most commonly used RSM techniques include central composite designs (CCD), star design, Doehlert design, Box–Behnken design, D-optimal design, three-level factorial design, and mixture design.

Prior to applying the response surface methodology, screening experiments are performed to find which experimental variables influence the response significantly. *Factorial designs* are among the most common screening experiments used, and two-level factorial designs are particularly useful plans for evaluation of the effects of variables on response and the interactions between them. Full factorial designs suffer from one disadvantage: If a number of factors (variables) is large, the number of experiments required can become excessively large, because the number of experiments in two-level factorial designs is related to the number of factors, k, through 2^k. Accordingly, in practice, only a fraction of full factorial design is often performed, generating *fractional factorial designs*.[135] A particular type of fractional factorial design is the Plackett–Burman design, which assumes that the interactions between variables can be ignored completely. It is typically used for screening experiments in which only the main effects are evaluated by performing a reduced number of experiments.

Among sequential designs, *simplex design* is the approach used most often. In this case, a stepwise strategy is followed and the experiments are performed one by one. Disadvantages are related to a large number of experiments that can be required to find the optimum, the lack of information about the interactions among factors, and the impossibility of finding an exact optimum value.

Close to 20 solvent microextraction method development procedures have taken advantage of experimental design, all but one using simultaneous design strategies.[34,84,100,136–149] A summary of those procedures is compiled in Table 5.4, which provides information on target analytes, sample matrix type, SME mode, and the final determination technique, the variables optimized, and the type of experimental design methodology used. It should be noted that in some cases the optimization procedure included solvent selection, which is typically performed at an earlier stage in SME method development.

5.9 OPTIMIZATION OF EXTRACTION CONDITIONS

The next step in SME method development involves optimization of extraction conditions, such as sample, headspace, and solvent volume, acceptor solution volume in three-phase modes, time, temperature, pH, ionic strength, agitation method and rate, fiber type and length, as well as plunger motion rate, dwelling time, and number of cycles in dynamic SME. A summary of typical ranges of values of extraction parameters that have been used in various SME modes is compiled in Table 5.5. The table provides a good starting point when developing a new microextraction procedure. The selection of some extraction parameters was discussed in Chapter 4 and is covered in Chapter 6. Brief guidelines for optimization of extraction parameters are provided below.

5.9.1 Optimization of Sample Volume

Detailed calculations involving the amount of analyte extracted for various sample volumes, organic solvent volumes, analyte distribution coefficients, and ionic

TABLE 5.4 Applications of Experimental Design in SME

Analytes	Matrix	SME Mode/ Detection Method[a]	SME Variables Optimized[b]	Optimization Technique
Diisocyanates	Air	HS/UPLC-MS-MS	**Drop volume, temperature, time,** relative position, vial size, mixing effect, stirring rate, concentration of derivatizing agent	Type III screening, Box–Behnken modeling, modified simplex optimization
Dialkyl phthalate esters	Food simulants	SD/GC-FID	**Time,** temperature, ionic strength, **stirring rate,** drop volume, sample mass, sampling depth	Plackett–Burman screening, circumscribed central composite optimization
Mononitrotoluenes	Wastewater	HLLE/GC-FID	Sample volume, phase separator reagent concentration, **extracting solvent volume,** consolute solvent volume	4-Factor, 3-level Box–Behnken design
Copper(II) ion	Water	DLLME/AAS	**Dispersion solvent volume, extracting solvent volume, sample volume,** ionic strength, pH	Circumscribed central composite design
Selenium	Water	HS/ET AAS	**Pd(II) concentration in drop,** NaBH$_4$ concentration, volume of NABH$_4$, **time,** sample volume, HCl concentration	Fractional factorial design screening
Dinitrophenols	Water	HF$_{(3)}$/HPLC-UV	**Sample volume, fiber length,** stirring rate	Three-variable Doehlert matrix design
2,4,6-Trichloroanisole, 2,4,6-tribromoanisole	Wine	HS/GC-ECD	**Temperature, time,** pH, **ionic strength,** matrix type	Fractional factorial design, Box–Behnken design
Pesticides	Water	USAEME/GC-MS	**Time, type of solvent, ionic strength,** phase volume ratio	Multifactorial screening
Essential oil components	Vapor, liquid food simulants	HS and SD/GC-MS	Vial volume, spiking volume, **preheating time,** drop volume, **extraction time, temperature, drop position, filling (%), stirring rate, ionic strength**	Plackett–Burman screening, response surface modelling

Analyte	Matrix	Method	Relevant factors[b]	Design
Fat-soluble vitamins	Water, urine, fruit juice	DSD/HPLC-UV	Temperature, **stirring rate,** ionic strength, **time,** drop volume	Orthogonal array design (fractional factorial design)
Short-chain fatty acids	Blood plasma	HS/GC-FID	**Temperature, time, ionic strength, sample volume**	Orthogonal array design (fractional factorial design)
Trihalomethanes	Water	SD/GC-ECD	**Drop volume, time, ionic strength**	Factorial design
Benzophenone-3	Urine	SD/HPLC-UV	Drop volume, sample volume, pH, **ionic strength,** solvent type, **time, stirring rate**	Plackett–Burman screening design, circumscribed central composite design optimization
Chlorobenzenes	Water	MW-HS/HPLC-UV	Drop volume, **sample volume,** stirring rate, ionic strength, **time, solvent type,** microwave power, length of Y-shaped glass tube	Plackett–Burman screening design, mixed-level factorial design
Chlorobenzenes	Water	HS/HPLC-UV	Drop volume, sample volume, **temperature, time, ionic strength,** stirring rate	Plackett–Burman screening design, central composite design optimization
NSAIDs	Wastewater	HF(3)/HPLC-UV	**Solvent type, HCl concentration in donor phase, NaOH concentration in acceptor phase, stirring rate, time, sample ionic strength**	Mixed-level orthogonal array design
Metacrate	Water	DLLME/HPLC-UV	Sample volume, extracting solvent volume, disperser solvent volume, time, ionic strength	Orthogonal array design, circumscribed central composite design

Source: Data from Refs. 34, 84, 100, and 136–149.

[a] SD, single drop; DLLME, dispersive liquid–liquid microextraction; DSD, directly suspended droplet; HLLE, homogeneous liquid–liquid extraction; HS, headspace; HF(3), hollow fiber–protected three-phase; MW, microwave-assisted; USAEME, ultrasound-assisted emulsification-microextraction; NSAIDs, nonsteroidal antiinflammatory drugs; HPLC, high-performance liquid chromatography; UPLC, ultraperformance liquid chromatography; MS, mass spectrometry; UV, ultraviolet detection; AAS, atomic absorption spectrometry; ET, electrothermal; FID, flame ionization detector; ECD, electron capture detector.

[b] Relevant factors are shown in boldface type.

TABLE 5.5 Summary of Typical Ranges of Values of Extraction Parameters in Various SME Modes[a]

Extraction Parameter	Drop-Based SME[b]					Membrane-Based SME[c]					Dispersive SME[d]	
	CF	SD	HS	DD	DSD	LLLME	HF(2)	HF(3)	MASE	MMLLE	DLL	HLLE
Solvent volume (µL)	1.5–4	1–4	1–4	0.5–2	7–20	100–400	1–20	1–21	800–1000	11–125	8–52	20–200
Sample volume (mL)	n/a	2–40	1–40	7–30 µL	1–20	0.5–6	1–15	1–200	15	1–40	5–10	2.5–40
Sample flow rate (mL/min)	0.1–1.5	n/a	n/a	n/a	n/a	n/a	n/a	n/a	n/a	0.1–1.0	n/a	n/a
Headspace volume (mL)	n/a	n/a	1–50	n/a	n/a	n/a	n/a	n/a	n/a	n/a	n/a	n/a
Disperser solvent volume (mL)	n/a	n/a	n/a	n/a	n/a	n/a	n/a	n/a	n/a	n/a	0.5–2.5	1–6
Acceptor solution volume (µL)	n/a	n/a	n/a	n/a	n/a	1–5	n/a	5–100	n/a	n/a	n/a	n/a
Extraction time (min)	5–15	5–60	4–60	5–10	5–60	6–45	3–60	5–60	30–60	30–50	1–30	3–30
Sample temperature (°C)	amb.	25–70	20–90	amb.	25–65	amb.–50	amb.–100	amb.–30	35–55	50–80	amb.–80	amb.

Solvent temperature (°C)	amb.	amb.	−6–amb.	amb.	ice bath–amb.	amb.	−1–amb.	amb.	35–55	50–80	n/a	amb.
Sample pH	1.5–6	2–14	1–12	8–10	2–12	1–14	1–13	1–14	2–6	n.r.	0.7–14	1–9
Ionic strength (%NaCl)	none–15	none–30	none–37.5	none	none–17.5	none–20	none–33	none–20	33	none	none–30	none–16
Agitation method and rate (rpm)	n/a	200–1300 (m)	300–1500 (m)	none	500–1250 (m)	250–1250 (m)	240–2000 (m,v)	100–1500 (m,v)	750 (v)	n/a	1000–6000 (c)	1800–3500 (c)
Fiber type and length (cm)	n/a	n/a	Q 3/2 Accurel, 1.3–6.5	n/a	n/a	n/a	Q 3/2 Accurel, 1–11	Q 3/2 Accurel, 1–25	PP bag, 2–4	PP membr. 6.1–22	n/a	n/a
Dynamic Mode												
Plunger motion rate (μL/min)	n/a	18–240	1–84	160	n/a	n/a	18–90	18–40	n/a	n/a	200	n/a
Dwelling time (s)	n/a	1–5	5	6	n/a	n/a	4–8	3–10	n/a	n/a	n.r.	n/a
Number of cycles	n/a	15–100	25–80	20	n/a	n/a	10–90	20–240	n/a	n/a	8	n/a

^{a} amb., ambient; PP, polypropylene; membr, membrane; n/a, not applicable; n.r., not reported; (c), centrifuged; (m), magnetically stirred; (v), vibrated.
^{b} CF, continuous flow; SD, single drop; HS, headspace; DD, drop-to-drop; DSD, directly suspended droplet.
^{c} LLLME, liquid–liquid–liquid microextraction; HF(2), hollow fiber-protected two-phase; HF(3), hollow fiber-protected three-phase; MASE, membrane-assisted solvent extraction; MMLLE, microporous membrane liquid–liquid extraction.
^{d} DLL, dispersive liquid–liquid microextraction; HLLE, homogeneous liquid–liquid extraction.

strength were performed in Sections 4.10 to 4.14 for direct immersion, headspace, dispersive, and hollow fiber–protected SME. This section therefore provides a summary of these considerations. An inspection of Table 5.5 reveals that typical sample volumes used in SME vary from 1 to 40 mL (except for drop-to-drop and hollow fiber–protected three-phase modes). The volume of the sample depends primarily on the distribution coefficient of the analyte(s). This is illustrated in Figure 5.7, depicting the dependence of the amount of analyte extracted on sample volume for various organic solvent volumes in SDME for three analytes with widely different octanol–water distribution coefficients: benzene ($K_{ow} = 148$), naphthalene ($K_{ow} = 2140$), and pyrene ($K_{ow} = 135,000$). It is evident that for analytes with K_{ow} values up to about 200, such as benzene, maximum sensitivity (which is related to the amount of analyte extracted) is reached for small sample volumes, on the order of several milliliters, and a further increase does not improve sensitivity. It should also be noted that for both benzene and naphthalene, the amount of analyte extracted is directly proportional to organic solvent volume. For analytes with a moderate distribution coefficient of about 2000, such as naphthalene, an increase in sample volume from 1 mL to 20 mL results in about a threefold sensitivity improvement. Further increases in sample volume do not significantly affect the amount of analyte extracted. In addition, an increase in sample volume has a disadvantage: It results in an increased equilibration time, since the rate constant for the extraction process, k, is inversely proportional to the sample volume [see equation (3.30) and (4.2)]. Therefore, the sample volume chosen must be a compromise between the sensitivity and analysis time requirements. For semi-volatile analytes with very large distribution coefficient values, such as pyrene, the amount of analyte extracted, and therefore sensitivity, are almost directly proportional to sample volume over a wide range of volumes (Figure 5.7C). Notice also that for pyrene the amount of analyte extracted is much less dependent on organic solvent volume for V_o values larger than 1 μL. Again, in this case the choice of sample volume has to be a compromise between sensitivity and extraction time. In practice, we should consider the analyte levels in the sample and detection limits of the final determination method. The sample volume selected must allow us to extract the amount of analyte that can be quantitated by the instrumental method of choice.

5.9.2 Optimization of Headspace Volume

Headspace SME involves a three-phase system: sample–headspace–solvent. Consequently, at equilibrium, analytes will be present in all three phases instead of being partitioned between the sample and the solvent only, as is the case in direct immersion SME. Therefore, headspace SME has inferior sensitivity, expressed as the amount of analytes extracted, compared to SDME, particularly for very volatile analytes, which will tend to accumulate in the headspace (see Section 4.15 for detailed calculations). This loss of sensitivity depends strongly on the headspace volume and distribution constant values, as shown in Figure 5.8, which illustrates the dependence of the amount of analyte extracted on sample volume for various headspace volumes in headspace SME for the same three analytes that were used in

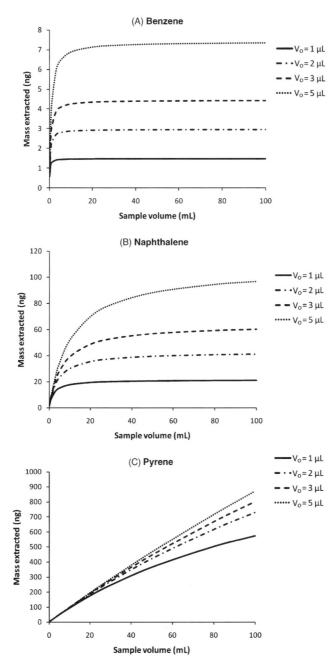

FIGURE 5.7. Dependence of mass of analyte extracted on sample volume for various organic drop volumes in direct-immersion SDME for (A) benzene, octanol–water distribution coefficient $K_{ow} = 148$; (B) naphthalene, $K_{ow} = 2140$; and (C) pyrene, $K_{ow} = 135,000$. Initial analyte concentration in water is assumed to be $C_w^0 = 10$ ng/mL.

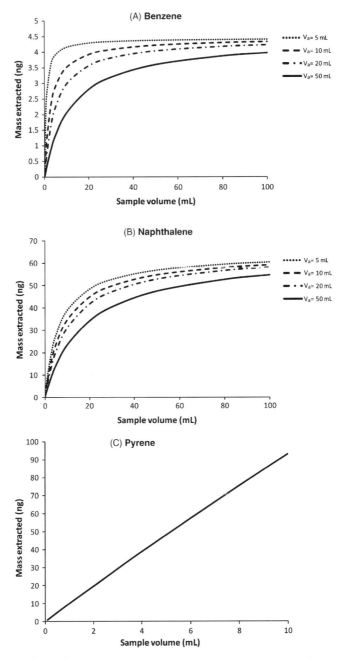

FIGURE 5.8. Dependence of mass of analyte extracted on sample volume for various headspace volumes in HS-SDME for (A) benzene, octanol–water distribution coefficient $K_{ow} = 148$, air–water partition coefficient (Henry's law constant) $K_{aw} = 0.224$; (B) naphthalene, $K_{ow} = 2140$, $K_{aw} = 0.0182$; and (C) pyrene, $K_{ow} = 135{,}000$, $K_{aw} = 0.000436$. Initial analyte concentration in water is assumed to be $C_w^0 = 10$ ng/mL and organic drop volume $V_o = 3$ μL.

Figure 5.7. In this case, the 1-octanol volume was fixed at $V_o = 3$ µL, and the following values of air–water distribution constant, K_{aw}, were used: 0.224 (benzene), 0.0182 (naphthalene), and 0.000436 (pyrene). It is clearly evident that for relatively volatile analytes such as benzene and, to a smaller extent, naphthalene, which have water–organic solvent partition coefficients between 100 and about 2000, and relatively large air–water distribution constants, the headspace volume determines the amount of analyte extracted. To obtain high-sensitivity headspace extraction, the headspace volume should be kept low (see Figure 5.8A and B). For example, for benzene, the amounts of analyte extracted at equilibrium from a 20-mL aqueous sample using 3 µL of 1-octanol for 5-, 10-, 20-, and 50-mL headspace volumes are equal to 4.30, 3.91, 3.56, and 2.80 ng, respectively. For comparison, the amount of benzene extracted by SDME under these conditions ($V_w = 20$ mL, $V_o = 3$ µL) is 4.34 ng. It also follows from Figure 5.8A and B that for small headspace volumes (say, 5 mL), maximum sensitivity is already reached for sample volumes of about 20 mL for benzene and about 50 mL for naphthalene, and a further increase in V_w does not bring about significant improvements in the amount of analytes extracted. This explains why, in practice, the headspace volumes vary from 1 to 50 mL (Table 5.5).

Semivolatile analytes such as pyrene represent quite different behavior (Figure 5.8C). In this case, the amount of analyte extracted does not depend on the headspace volume (the curves for V_a equal to 5-, 10-, 20-, and 50-mL overlap), and there is a direct proportionality between the amount of analyte extracted and the sample volume. The amount of analyte extracted in headspace SME is very close to that extracted by SDME. It should be kept in mind, however, that semivolatile analytes with very small K_{aw} coefficients also have very low diffusion coefficients, particularly in the liquid phase. As a result, equilibration times for semivolatiles can be long. To reduce extraction times, we can optimize the headspace capacity: that is, the amount of analyte contained in the headspace, which is equal to $K_{aw}V_a$. If the amount of analyte extracted is small compared to headspace capacity (say, less than 5%), analyte extraction will take place almost exclusively from the headspace, the amount of analyte that will have to be transferred from the sample to the headspace is small, and the equilibrium between the sample and the headspace will, for all practical purposes, not be disturbed. This results in rapid extraction (taking only several minutes), since the diffusion coefficients of analytes in the gaseous phase are much larger than those in the liquid phase. The headspace capacity can be maximized by increasing either headspace volume or K_{aw} values, or both. The air–water distribution constants can conveniently be increased by raising the temperature or adjusting sample ionic strength and/or pH. Since increasing headspace volume reduces the amount of analyte extracted, the choice of V_a will be a compromise between the sensitivity of the procedure and extraction time.

5.9.3 Optimization of Solvent Volume

The selection of solvent volume is determined by three factors: the amount of analyte that has to be extracted to allow quantitative analysis (sensitivity), the requirements of SME mode employed, and the volume of the extract that can be introduced into the instrument used for final determination. The amount of analyte extracted at equilibrium depends on the solvent volume in every SME mode. This relationship is represented by equations (4.4) and (4.7) for direct immersion and headspace SME, respectively, and illustrated in Figures 5.9 and 5.10 for three analytes varying in volatility: benzene, naphthalene, and pyrene, assuming that $V_w = 10$ mL, $C_w^0 = 10$ ng/mL, and $V_a = 5$ mL. In SDME, for volatile analytes that have relatively small K_{ow} values, such as benzene, the first term in the denominator of equation (4.4) ($K_{ow}V_o$) is small in comparison with V_w and can be neglected. Equation (4.4) then becomes $n = K_{ow}V_oC_w^0$; that is, the amount of analyte extracted is directly proportional to the octanol–water distribution constant, the solvent volume, and the analyte concentration in the sample, and extraction takes place in the equilibrium mode. This case is illustrated in Figure 5.9A, where the relationship between the amount of analyte extracted and solvent volume for benzene ($K_{ow} = 148$) is almost linear over solvent volumes ranging from 0.1 to 20 μL. The other limiting case occurs for semivolatile analytes, such as pyrene, with very large K_{ow} values. In this case, the first term in the denominator of equation (4.4) is large in comparison with V_w, which can then be neglected, so that the equation becomes $n = C_w^0V_w$. This is the case for exhaustive extraction, which can be observed for pyrene for solvent volumes greater than 1 μL (Figure 5.9C), when the entire amount of analyte present in the sample is extracted and a further increase in solvent volume does not result in any improvements in sensitivity. Naphthalene has the K_{ow} value intermediate between benzene and pyrene. For this analyte, an intermediate extraction mode should be expected (Figure 5.9B). For small solvent volumes (up to 2 μL), the equilibrium mode is observed, with the amount of analyte extracted being almost directly proportional to the solvent volume. For large solvent volumes (greater than 20 μL), the amount of analyte extracted becomes almost independent of the solvent volume (i.e., exhaustive extraction takes place).

The dependence of the amount of analyte extracted on solvent volume for the same three analytes in headspace SME is shown in Figure 5.10. For all three analytes, the dependence observed is virtually identical to that for SDME. This illustrates an important point: Under equilibrium conditions, the amount of analyte extracted is independent of the location of solvent drop in the system (in solution or in headspace). The headspace role is to serve as a barrier between the sample and the extracting solvent.

In conclusion, from the point of view of SME sensitivity, the largest possible solvent volumes should be used for extraction of analytes with relatively small distribution constant values (such as benzene); semivolatile analytes (pyrene) do not benefit from solvent volumes larger than a few microliters, and for analytes with intermediate distribution constant values, sensitivity can be improved by increasing the solvent volume up to a certain point.

The second factor to consider when deciding on the solvent volume is SME mode. Single-drop techniques (direct immersion, headspace, continuous flow) are limited by drop stability, and typically, solvent volumes used vary from 1 to 4 μL (Table 5.5). Drop-to-drop microextraction uses smaller volumes, ranging from 0.5 to 2 μL, whereas in directly suspended droplet SME, solvent volumes range from

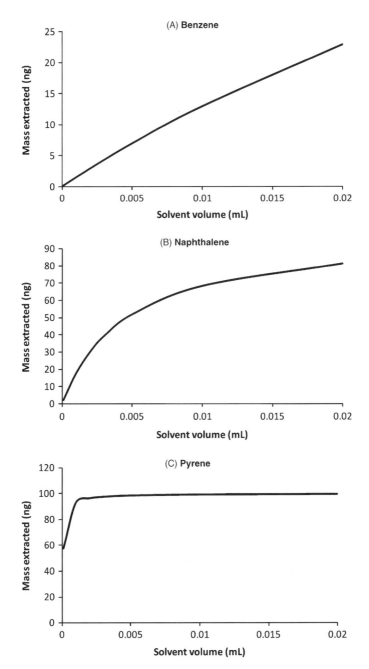

FIGURE 5.9. Dependence of mass of analyte extracted on solvent volume in direct immersion SDME for (A) benzene, octanol–water distribution coefficient $K_{ow} = 148$; (B) naphthalene, $K_{ow} = 2140$, and (C) pyrene, $K_{ow} = 135,000$. Initial analyte concentration in water is assumed to be $C_w^0 = 10$ ng/mL, and sample volume $V_w = 10$ mL.

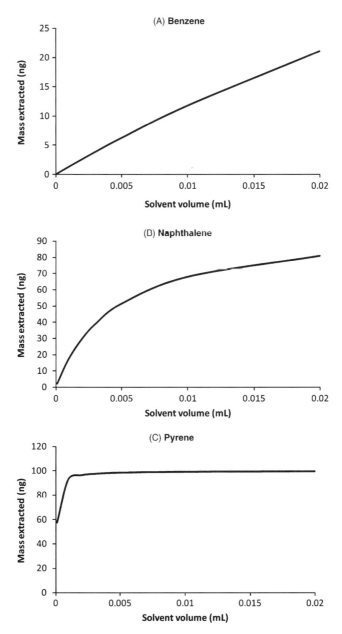

FIGURE 5.10. Dependence of mass of analyte extracted on solvent volume in headspace SDME for (A) benzene, octanol–water distribution coefficient $K_{ow} = 148$, air–water partition coefficient (Henry's law constant) $K_{aw} = 0.224$; (B) naphthalene, $K_{ow} = 2140$, $K_{aw} = 0.0182$; and (C) pyrene, $K_{ow} = 135{,}000$, $K_{aw} = 0.000436$. Initial analyte concentration in water is assumed to be $C_w^0 = 10$ ng/mL, and sample and headspace volumes are $V_w = 10$ mL and $V_a = 5$ mL, respectively.

7 to 20 μL. Dispersive SME modes employ larger volumes: from 8 to 52 μL in dispersive liquid–liquid microextraction to 20 to 200 μL in homogeneous liquid–liquid extraction. In hollow fiber–protected two-phase SME, typical solvent volumes range from 1 to 20 μL, while in membrane-assisted solvent extraction and microporous membrane liquid–liquid extraction these values are 800 to 1000 μL and 11 to 125 μL, respectively. Two membrane-based modes—hollow fiber–protected three-phase and liquid–liquid–liquid ME—stand out among solvent microextraction techniques, because the organic solvent used in them acts not as an extracting medium but as a barrier to separate the donor solution (sample) from the aqueous acceptor solution. Accordingly, the organic solvent volume is larger in liquid–liquid–liquid microextraction than in any other SME mode (except for membrane-assisted solvent extraction), ranging from 100 to 400 μL, while in HF(3) ME the organic solvent volume is usually unspecified (with a few exceptions, when it varies from 1 to 21 μL). Instead, to saturate the fiber pores, the fiber is typically immersed in an organic solvent for a prescribed time, ranging from 2 s to 15 min. The extracting solution in liquid–liquid–liquid microextraction and HF(3) is the aqueous acceptor solution, whose volume ranges from 1 to 5 μL and from 5 to 100 μL, respectively. In dispersive liquid–liquid microextraction techniques, a disperser solvent is used to disperse the extracting solvent in the sample solution. The disperser solvent must be capable of dissolving the extracting solvent and be miscible with water. The disperser solvent volumes in dispersive liquid–liquid microextraction and homogeneous liquid–liquid extraction vary from 0.5 to 2.5 mL and from 1 to 6 mL, respectively.

The last factor to consider when selecting the extracting solvent volume is the instrumental method used for the final determination of analytes. Gas chromatography, which is compatible with all organic solvents, should be the method of choice for volatile and semivolatile analytes. When the extracting solvent is lower boiling than analytes, as is often the case in direct immersion, continuous flow, and HF(2) modes, splitless injection can be used. Higher-boiling solvents, commonly employed in headspace SME, call for split injection (see Section 4.8 for examples). The injection volumes in GC rarely exceed 3 μL, typically ranging from 0.5 to 3 μL, except for membrane-assisted solvent extraction, where large-volume injection (LVI) is generally used. Therefore, unless the SME mode requires larger solvent volumes (e.g., dispersive mode), it is pointless to work with solvent volumes larger than 3 μL. Therefore, to increase the amount of analyte extracted, we should try not to increase solvent volume above 3 μL but to look for ways of increasing the concentration of analyte in the drop, C_o. This can be achieved by increasing the distribution coefficients between the sample and the extractant through temperature adjustment and/or matrix modification.

In the large-volume injection used in membrane-assisted solvent extraction, 100 to 400 μL of the organic extract are introduced through a multipurpose sampler at a slow rate into an injection port cooled to 20 to 30°C. The column head pressure is reduced and the flow rate through the split vent is adjusted to 100 mL/min to purge most of the solvent.[150,151] Next, the vent is closed and the injection port and column temperatures are programmed. This approach results in much improved sensitivities for analytes with boiling points much higher than the extracting solvent used.

High-performance liquid chromatography can handle larger injection volumes, and volumes ranging from 1 to 100 μL of the extract are injected into HPLC instruments. Most of these extracts are aqueous solutions obtained by HF(3)ME, although extracts in some organic solvents, such as 1-octanol or ionic liquids, are also introduced directly through the sampling valve. When using other organic solvents, incompatible with the mobile phase required for the separation of analytes, the extracts are usually evaporated to dryness and redissolved in the mobile phase. Capillary electrophoresis typically calls for aqueous solutions with extractant volumes ranging from 12 nL to 100 μL. The extract volume used in SME combined with visible spectrophotometry is 350 μL. Electrothermal atomic absorption spectroscopy requires from 3 to 20 μL of the extract, while in flame AAS the extract volume is much larger (500 μL).

5.9.4 Optimization of Sample Flow Rate

Only two SME modes: continuous-flow microextraction and microporous membrane liquid–liquid microextraction are based on flowing sample being in contact with the extractant either directly (continuous-flow microextraction) or through a membrane (microporous membrane liquid–liquid microextraction). Two factors have to be considered when selecting the optimum flow rate. The first one is stability of the organic microdrop in continuous-flow microextraction. At high sample flow rates, the microdrop can be dislodged from the tip of PEEK tubing or microsyringe needle, which imposes an upper limit on practical sample flow rates. The second factor is extraction dynamics. An increase in sample flow rate decreases the thickness of the interfacial layer surrounding the solvent droplet, improving mass transfer of analytes and analyte peak areas, and speeding up extraction. However, reduction in peak areas is observed after exceeding a certain flow rate, which can be attributed to too high a linear velocity of the sample solution to allow establishment of extraction equilibrium in the interfacial layer of the two phases: sample and organic.[94] Consequently, sample flow rates used in practice in continuous-flow microextraction and microporous membrane liquid–liquid microextraction range from 0.1 to 1.5 mL/min and from 0.1 to 1.0 mL/min, respectively.

5.9.5 Optimization of Extraction Time

Equations used in Chapter 4 and 5 to calculate the amounts of analytes extracted from aqueous solutions with 1-octanol were based on the assumption that the system (sample-extracting solvent) achieves equilibrium. In practice, however, equilibration times can be prohibitively long, especially for analytes with high molar masses, for which diffusion coefficients in liquids are very small and distribution constants K very large. In those SME techniques where the organic solvent is in direct contact with the sample, including continuous-flow microextraction, SDME, or directly suspended droplet (DSD), long extraction times result in higher solvent loss, due to dissolution. Equilibration time can be reduced by employing dynamic

SME, as discussed in Sections 3.3.3.4 and 5.3. Kinetics of mass transfer in solvent microextraction is described in detail in Section 3.3. Therefore, only a summary is provided here as it applies to extraction time in various SME modes. According to equation (3.30), the rate constant for the extraction process in two-phase systems is directly proportional to A_i, the water–organic interfacial area. Consequently, the shortest equilibration times should be observed for the techniques that offer the highest interfacial contact areas: dispersive and homogeneous microextraction. High extraction rates, and therefore short equilibration times, will also occur when both V_o and V_w are kept small [again, see equation (3.30)], as in drop-to-drop SME.

Increasing the overall mass transfer coefficient $\overline{\beta}$ will increase the rate of extraction proportionally. This can be achieved when the sample is flowing (continuous-flow SME) or by increasing the rate of convection by mechanical agitation (stirring or vibration). Increasing temperature will also improve mass transfer by enhancing the diffusion coefficients of analytes. Hollow fiber–protected two-phase SME should offer shorter equilibration times compared to direct immersion and directly suspended droplet SME by providing both increased interfacial area A_i and the possibility of more vigorous agitation as a result of protection afforded by the hollow fiber.

The longest equilibration times occur in three-phase solvent microextraction techniques: liquid–liquid–liquid, hollow fiber–protected three-phase, and membrane-assisted solvent extraction. In this case, analytes have to be transferred from the sample through a barrier (liquid organic solvent membrane) into the acceptor solution. The overall rate of extraction will then be limited by mass transfer in both the water and organic phases.

When the equilibration time is prohibitively long, a shorter extraction time can be used for quantitative analysis provided that the equilibration time profile (mass of analyte extracted vs. extraction time) levels off. For example, in Figure 5.11, which depicts extraction-time profiles for selected aromatic hydrocarbons extracted from a 1.7% gasoline solution in motor oil by headspace SME following a 15-min equilibration between the sample and the headspace, an extraction time of 2 min could be used, even though the equilibrium between the sample and the extracting solvent clearly has not yet been reached for ethylbenzene and xylenes. If the extraction time is carefully controlled for each standard and sample, the relative error in the mass of analyte extracted will be relatively small. According to equation (3.31), the concentration of analyte in the extracting solvent at any time is directly proportional to the equilibrium concentration provided that the rate constant k and the extraction time are kept constant. The presence of complex sample matrix can affect the equilibration time. For example, if an aqueous sample contains suspended solids that adsorb analytes, the release of analytes from the solid phase could be the rate-determining step in extraction kinetics. Consequently, the optimum extraction time should be redetermined for such complex samples if it was originally optimized for standard solutions in pure water.

Typical extraction times depend on the molar mass of analytes and their volatility and the SME mode selected. In accordance with the discussion above, they are shortest for drop-to-drop (5 to 10 min), continuous flow (5 to 15 min), dispersive

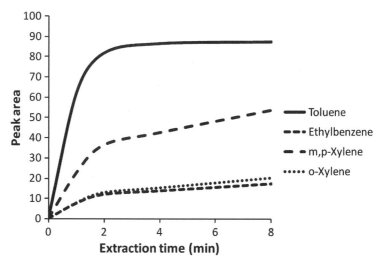

FIGURE 5.11. Extraction time profiles of selected aromatic hydrocarbons extracted from 1.7% gasoline solution in motor oil stirred at 1200 rpm by HS-SDME. Extraction temperature: 23°C.

(1 to 30 min), and homogeneous (3 to 30 min) modes. For other SME modes, they range from 3 to 60 min, depending mostly on the type of analyte(s).

5.9.6 Optimization of Sample and Solvent Temperature

Temperature has a manifold effect on solvent microextraction. It affects both the amount of analyte extracted (sensitivity, through thermodynamics of partition) and extraction kinetics (rate of extraction). In addition, temperature influences the rate of dissolution of organic solvents in direct-immersion methods and the rate of evaporation of extracting solvents in headspace SME. Therefore, unless the SME system is thermostated, temperature fluctuations will deteriorate reproducibility of results. The effect of temperature on distribution constants was discussed in Section 3.2.4. Briefly, the water–organic solvent distribution constant, K_{ow}, depends only weakly on temperature, because the enthalpy of transfer of solute from the aqueous phase to the organic phase, ΔH, is relatively small [see equation (3.11)]. Consequently, in those SME modes where two or three liquids are in contact and no headspace is involved in the partition equilibrium, sample temperature can be increased to improve the rate of extraction without a substantial penalty in the amount of analyte extracted. In practice, extraction temperatures for SDME, directly suspended droplet, liquid–liquid–liquid microextraction, hollow fiber–protected three-phase, membrane-assisted solvent extraction, microporous membrane liquid–liquid extraction, and dispersive modes vary from ambient to 80°C, and in the case of hollow fiber–protected two-phase SME go as high as 100°C (Table 5.5). Elevated sample temperatures are used both in direct-immersion and headspace mode, primarily for semivolatile analytes, for which the mass transfer process between the phases is slow, owing to small diffusion coefficients. For practical reasons, extraction temperatures used in continuous-flow and drop-to-drop SME are ambient. In the former case, an elevated sample temperature in a flowing system would result in rapid dissolution of the organic microdrop; in the latter case the rate of extraction is sufficiently high to make temperature control unnecessary.

In contrast, the effect of temperature on water–air (K_{aw}) and air–solvent (K_{oa}) distribution constants is significant, because ΔH in equation (3.11) is in this case equal to the large heat of vaporization (or condensation) of the solute. Consequently, in headspace SME the temperature has the opposite effect on partition coefficients K_{aw} and K_{oa} as well as on extraction rate. To improve the rate of transfer of analyte into the headspace and to transfer more analyte(s) into the headspace, sample temperature should be relatively high (but less than the boiling point of water). At the same time, to reduce solvent evaporation from the organic solvent in a microdrop or in a fiber and to increase the amount of analyte(s) transferred from headspace to extracting solvent, the solvent temperature should be kept low (the solution process is exothermic: ΔH is negative). This problem has been resolved by using two independently operated temperature-controlling baths: one for heating the sample and the other for cooling the syringe with the solvent. In practice, solvents have been cooled to temperatures ranging from -6 to $4°C$ (see Section 6.4 for more detail). Few other SME modes make use of solvent temperature control; in most cases the solvent has the same temperature as the sample. Solvent and sample temperatures are raised in some membrane-based techniques (membrane-assisted solvent extraction: 35 to 55°C; microporous membrane liquid–liquid microextraction: 50 to 80°C) to ensure rapid mass transfer of analytes through the membrane. In the directly suspended droplet technique, the sample and extract temperatures are lowered following the extraction process by putting the system in an ice bath to facilitate separation of the extract from the sample.

5.9.7 Optimization of pH of Sample and Acceptor Solution

The adjustment of pH of sample and/or acceptor solution can improve the amount of analytes extracted which are acidic or basic. A detailed discussion of this topic is provided in Sections 3.2.5 and 6.5.5. Briefly, in any two-phase SME mode, the sample pH should be modified to suppress ionization of any acidic or basic analyte. To this end, the pH should be lowered so that it is at least 1.5 pH units below the pK for any acidic species. Conversely, the pH should be raised to the value that is at least 1.5 pH units larger than the pK of any basic solute. To assure good reproducibility, best results are obtained when samples containing acidic or basic analytes are buffered. In practice, all common SME modes, including continuous-flow microextraction, direct immersion, dispersive liquid–liquid microextraction, HF(2), HF(3), and HS, make use of sample pH adjustment for acidic and basic analytes, primarily for carboxylic acids, phenols, and organic bases, including drugs of abuse, pharmaceuticals, and alkaloids, but also in the complexation of some metal ions. Typically, for acidic analytes, the sample pH ranges from 0.1 to 3.5, whereas for basic analytes it varies between 10 and 14.

In two three-phase techniques, liquid–liquid–liquid microextraction and HF(3), a third solution (acceptor) is used in addition to sample (donor solution) and organic solvent. Analytes in their nonionized form are extracted from an aqueous solution into an organic solvent and at the same time back-extracted into the aqueous acceptor solution. To ensure accumulation of the analytes in the acceptor solution and to prevent their back-extraction into the organic solvent, the pH of the acceptor solution has to be adjusted to the value that guarantees ionization of the analytes. Therefore, a basic acceptor solution should be used for acidic analytes and an acidic acceptor solution for basic analytes.

5.9.8 Optimization of Ionic Strength

According to Debye theory, the water solubility of charged organic solutes such as simple organic acids and bases will generally increase with increasing ionic strength, since the activity coefficients of ionized species decrease with increasing ionic strength of solution. Consequently, it is essential to convert these analytes into their neutral forms by pH modification, as discussed in Section 5.9.7. The solubility of hydrophobic organic solutes in water usually decreases in the presence of dissolved salts due to the salting-out effect. According to hydration theories,[152] as the concentration of ions in water increases, more and more water molecules are used to hydrate them, and fever water molecules are available to hydrate the organic solute. This leads to an increase in distribution constant K in two-phase SME, and an increase in K_{aw} and K_{ow} in headspace SME, thus improving the amount of analyte extracted and sensitivity of an SME procedure. The effect of ionic strength on the distribution constants is described quantitatively by the empirical Setschenow equation [see equations (3.12) and (4.5) as well as the description of the ionic strength effect on distribution constants in Sections 3.2.4 and 4.12 and examples of calculations in Section 4.12]. Sodium chloride and sodium sulfate are the two salts most commonly used to increase ionic strength of aqueous samples. However, it should be kept in mind that some sample matrices, such as seawater, already contain dissolved salts (3% in seawater). In addition, if a sample pH is adjusted with a buffer solution, the ionic strength also increases. In practice, the amounts of salt (usually NaCl) added vary from none to a maximum of 37% by weight (see Table 5.5). When developing a solvent microextraction procedure, we should consider the nature of analytes and the value of octanol–water partition coefficient in particular, before deciding whether ionic strength of samples has to be increased. Commonly, if the K_{ow} values of analytes are large (semivolatile analytes), the amounts of analytes extracted are in most cases sufficient for quantitative analysis without salt addition. For other analytes, the effect of salt addition on distribution constants has to be determined empirically. For some analytes, an increase in sample ionic strength has been found to result in a decrease in the extraction efficiency.[153–156] This phenomenon was explained tentatively by assuming that in addition to the salting-out effect, the salt dissolved in the aqueous solution may change the physical properties of the Nernst diffusion film and reduce the rate of diffusion of analytes into the organic solvent. In such a case, analytes from samples are extracted without salt addition.

5.9.9 Optimization of Agitation Method and Rate

Only three solvent microextraction techniques do not use sample agitation. In drop-to-drop SME, the volumes of both sample and organic solvent are very small. This ensures rapid mass transfer from the sample to the extractant even in the absence of stirring. Both continuous-flow and microporous membrane liquid–liquid extraction make use of flowing sample, which provides convection in addition to diffusion, thus increasing the mass transfer rate. In all other SME modes, sample agitation improves extraction rate, thus reducing extraction time, especially for higher-molar-mass analytes, which have small values of diffusion coefficients. Agitation of samples increases mass transfer of analytes, particularly in two-phase systems, where the resistance to mass transfer in the aqueous phase for analytes with relatively large distribution constant values is more important as a rate-determining step than the mass transfer resistance in the organic phase, as discussed in Section 3.3.3.1. For drop-based techniques, film theory (see Section 3.3.3.3) predicts that more vigorous stirring results in smaller film thicknesses, causing an increase in the mass transfer coefficient and hence an increased extraction rate for analytes. The two most common agitation techniques employed in SME are magnetic stirring and mechanical vibration. Typically, magnetic stirring is used in drop-based SME modes, where vibration would result in dislodgment of the organic drop from the needle tip of a microsyringe. To prevent drop dislodgment, the stirring rate is limited to no more than 1250 to 1300 rpm. To provide reproducible agitation conditions, the magnetic stirrers used in solvent microextraction should have digital control of the stirring rate. In membrane-based SME modes, organic solvent is protected by a fiber, which allows the use of vibrations in addition to magnetic stirring, as well as more vigorous agitation with rates up to 2000 rpm [HF(2)] or 1500 rpm [HF(3)]. In dispersive techniques (dispersive liquid–liquid microextraction and homogeneous liquid–liquid extraction), characterized by very fast mass transfer owing to a very high interfacial area between the sample and the solvent, extraction times are generally short. Consequently, manual shaking is often sufficient for proper agitation of samples. However, both these techniques use centrifugation for the separation of extract from sample. Centrifugation speeds range from 1000 to 6000 rpm for dispersive liquid–liquid microextraction to 1800 to 3500 rpm for homogeneous liquid–liquid extraction.

Headspace SME techniques (single drop and hollow fiber-protected) for semivolatile analytes require vigorous sample agitation. However, equilibration times between the gaseous phase and the extracting solvent are short and often limited by the diffusion rate of analytes in the acceptor phase. To improve mass transfer in the acceptor phase, dynamic techniques can be applied. However, for volatile analytes such as benzene and toluene, rate of extraction is limited by mass transfer in both the aqueous and organic phases.[157] It is therefore important to enhance mass transfer processes in headspace SME of volatile analytes in both the water and organic phases by magnetic stirring (aqueous sample) and using the dynamic mode (acceptor solution).

The most vigorous agitation technique that has been used in SME is sonication. However, since it generates large quantities of heat that affect sample temperature in an uncontrolled way and may lead to decomposition of analytes, sonication has been used exclusively during preliminary treatment steps for solid samples aimed at release of analytes from solid matrices prior to their isolation and enrichment during microextraction proper.

5.9.10   Selection of Fiber Type and Length

Membrane-based liquid microextraction can be performed in a hollow fiber, a flat membrane module, or a membrane bag.[158] Its advantages include high analyte enrichment and efficient sample cleanup. Membranes used in membrane-based SME have to be hydrophobic and compatible with the organic solvent used. The membranes used are either microporous [hollow fiber–protected headspace, HF(2), HF(3), and microporous membrane liquid–liquid microextraction] or nonporous (membrane-assisted solvent extraction). In microporous membranes the pores are filled with an organic solvent. In nonporous membrane-based SME, the extraction mechanism is different; analytes must diffuse through the membrane to the acceptor phase (see Section 6.5.2 for more details). Most procedures based on porous membranes have used microporous polypropylene fibers, usually Q 3/2 Accurel hollow fibers manufactured by Membrana (Wuppertal, Germany), although other fiber materials, such as poly(vinylidene difluoride), have been tried as well.[159] Polypropylene is highly compatible with a wide range of organic solvents and strongly immobilizes the organic solvent in the pores, which is essential for ensuring that the organic phase does not leak during extraction. The most common type of polypropylene hollow fibers has an inner diameter on the order of 600 to 1200 μm, appropriate for the microliter volumes of the acceptor solution employed in microextraction. Wall thickness is usually 200 μm, since thinner walls do not provide sufficient mechanical stability, while thicker walls result in extended extraction times due to longer diffusion paths.[160] Typically, the pore size is 0.2 μm with a porosity of 70%, which ensures efficient microfiltration, allowing penetration of only small molecules of target analytes through the pores of the hollow fiber and exclusion of high-molar-mass interfering compounds.[161]

Generally, a 1-cm length of the hollow fiber with an inside diameter of 600 μm is capable of immobilizing about 8 μL of solvent.[162] Of that volume, the amount retained in the lumen is about 2.8 μL. Fiber lengths used in membrane-based SME depend primarily on the extract volume needed for quantitative analysis. Two-phase extraction systems utilize an organic solvent as the extractant, which is directly compatible with GC. Since GC analysis requires no more than 1 to 3 μL of the extract, fibers used in hollow fiber–protected headspace and hollow fiber–protected two-phase SME are relatively short, ranging from 1.3 to 6.5 cm in HF-HS-SDME to 1 to 11 cm in HF(2)ME. Three-phase extraction systems make use of aqueous acceptor solutions, which are compatible with HPLC or CE. In this case, larger volumes of the extract (5 to 100 μL) can be used for quantitation. Consequently, longer fibers (1 to 25 cm) are employed in hollow fiber–protected three-phase SME.

Modules with flat-sheet membranes can be utilized in three-phase SME (supported liquid membrane systems) or in two-phase SME (microporous membrane liquid–liquid extraction). The selection of membrane material depends on the solvent microextraction mode used. In three-phase SME, porous PTFE membranes with a thickness of about 60 μm and polyethylene backing have been used to support the extracting organic solvent. In two-phase systems such as microporous membrane liquid–liquid microextraction, thinner porous polyethylene or polypropylene membranes with a thickness of 25.4 μm have been utilized.[163]

Membrane-assisted solvent extraction uses nonporous dense polypropylene membrane bags filled with an organic solvent. Dense polypropylene is resistant to most organic solvents and has good stiffness, so it can withstand even vigorous stirring. To improve the mass transfer kinetics, the membrane thickness is reduced to 30 μm and the extraction temperature is increased to about 40°C. The bag is agitated to facilitate extraction of the analytes into the bag. Typical membrane bags are 4 cm in length and 6 mm in internal diameter.[150,151]

5.9.11 Optimization of Dynamic Mode Parameters

The choice of static versus dynamic solvent microextraction is considered in Section 5.3, and the model for dynamic techniques is discussed in Section 3.3.3.4. Only a summary is provided here. In dynamic solvent microextraction, the organic solvent is contained in the microsyringe, and microliter volumes of the aqueous sample or headspace are repeatedly drawn into and expelled out of the syringe.[92] The formation of a thin film of organic solvent on the inside walls of the syringe in dynamic SME provides a large interfacial area of contact with the aqueous or headspace phase, which improves the rate of mass transfer of analytes into the solvent. Dynamic SME also ensures continuous renewal of the aqueous plug or headspace and the organic film with each cycle, thus inducing convection not only in the sample phase, which could also be achieved by other means (magnetic stirring, vibration), but also in the organic solvent, which would be difficult to accomplish otherwise. Dynamic solvent microextraction can be automated readily.

Dynamic SME has been implemented in both two-phase (direct immersion,[2,92,119,164–168] drop-to-drop,[119] hollow fiber–protected two-phase,[2,47,52,79,169–177] dispersive[178]) and three-phase (headspace,[179–181] hollow fiber–protected headspace,[2,47,174,182] and hollow fiber–protected three-phase[101,183–186]) systems. In general, parameters to be optimized include volume of sample (or headspace over sample) drawn into the syringe and volume of organic solvent (or aqueous acceptor solution), plunger motion rate, dwelling time, and a number of cycles. Typical volumes of sample (donor solution) or headspace drawn into the microsyringe range from 2 to 8 μL, while usual volumes of organic solvent or aqueous acceptor solution vary from 1 to 5 μL. These volumes are limited by the capacity of microsyringes utilized in dynamic solvent microextraction (10 μL). If we assume that instantaneous equilibrium is established between the aqueous sample plug and the organic film, complete mixing of the phases takes place after each cycle, and the overall composition of the phases does not change substantially after each cycle, the amount of analyte extracted should be directly proportional both to the volume of aqueous sample drawn into the syringe and to the number of cycles *n*, when *n* is small.[92] In practice, the volume of sample drawn into the syringe is limited by the syringe capacity, while the number of cycles is limited by the solubility of organic solvent in water, by the required sample throughput, and by the fact that the amount of analyte extracted during each cycle actually decreases with the number of cycles because the organic film is not renewed completely each time. In practice, the number of cycles ranges from 8 to 240 (see Table 5.5). The rate of plunger motion is generally high for all but one dynamic SME mode, to reduce extraction time and ranges from 18 to 240 μL/min (0.3 to 4 μL/s). The only exception is headspace SME, where the plunger motion rate is 1 to 84 μL/min. A dwell time of several seconds is used in the beginning and at the end of each cycle to achieve equilibrium. Since the analyte is transferred from the aqueous (or gaseous) phase into a very thin film of organic solvent (or aqueous acceptor solution), the process is very fast. Consequently, dwelling times are typically short, ranging from 1 to 10 s (Table 5.5).

5.9.12 Analytical Characteristics of SME Procedures and Quantitative Analysis

Optimization of extraction parameters concludes SME method development. The next step involves selection of the quantitative analysis method, as described in Section 6.3. Briefly, for relatively clean sample matrices such as tap water, a calibration curve (external calibration) method is suitable. More complex matrices, such as lake water or physiological fluids, call for the standard addition method, either single-point or (preferably) multipoint, which compensates for the matrix effect. To eliminate the effect of variability in extraction and/or instrumental conditions on analytical results, most SME procedures make use of the internal standard method or its variant, isotopic dilution (if mass spectrometry is employed as the detection technique). Once the quantitation method has been selected, we have to characterize the analytical procedure developed. Analytical characteristics typically include limit of detection and/or limit of quantitation, sensitivity, linear dynamic range, accuracy, precision, selectivity, sample throughput, and robustness. The last step usually involves validation of the procedure.

5.9.12.1 Limit of Detection   The *limit of detection* (LOD) is defined as the lowest concentration level that yields the response statistically different from a blank. The *method detection limit* (MDL) is the lowest concentration of analyte that a method can detect reliably in a sample or a blank, and the *instrument detection limit* (IDL) is the smallest signal above background noise that an instrument can detect reliably.[187] In most cases, method detection limits are the same as instrument detection limits only if there is no transfer of analyte from the sample to a second phase (such as organic solvent).[188] In solvent microextraction, the MDL is usually much lower than the IDL because SME by design isolates and enriches analytes from the sample matrix. The most common technique of estimating the method detection limits is based on determining repeatedly the response to solutions with a known low concentration of analyte (preferably lower than 10 times the MDL) and finding the concentration yielding the response equal to three times the standard deviation of the determinations. In practice, the detection limit can be calculated from the following formula: $LOD = 3s/m$, where s is the standard deviation of a low-concentration sample and m is the slope of the calibration curve. The method detection limit can also be calculated from the instrument detection limit if the phase ratio, V_o/V_w, is known and the percent recovery R (expressed as a decimal fraction) has been determined experimentally:[188]

$$MDL = \frac{IDL}{R} \frac{V_o}{V_w} \qquad (5.2)$$

Limits of detection of SME procedures for some commonly determined analytes are provided in Tables 6.2 through 6.4. The LOD values for other analytes can be found in the files provided on the CD-ROM included with this book.

5.9.12.2 Limit of Quantitation

The *limit of quantitation* (LOQ) is the level of analyte above which quantitative results may be obtained with a specified degree of confidence.[187] Usually, the concentration corresponding to a response being 10 times the standard deviation of a low-concentration sample is taken as the limit of quantitation: $LOQ = 10s/m$, where s and m have the same meaning as above.

5.9.12.3 Linear Dynamic Range

The LOQ is often used to define the lower limit of the *linear dynamic range* (LDR), a range that extends from the limit of quantitation to the upper limit of linearity where the response deviates from a straight line. A common measure of linearity is the square of the correlation coefficient, r^2, which is sometimes called the *coefficient of determination*. r^2 must be very close to 1 to represent a linear fit. Commonly, a value of 0.995 or better is considered acceptable. A useful analytical procedure should have an LDR of at least two orders of magnitude. An inspection of the summaries of SME modes on the CD-ROM accompanying this book reveals that most solvent microextraction procedures developed thus far meet this criterion. In a few cases, the procedures have been found to have a linear range of just one order of magnitude, but it should be remembered that even in those cases, correct quantitative analysis can be carried out with nonlinear calibration curves as long as a sufficiently large number of standard solutions were used during calibration. Typically, however, the LDR extends for two to four orders of magnitude (two to five for direct immersion) for the SME modes other than those membrane-based. For HF(2)ME and HF(3)ME, the linear range is usually limited to one to three orders of magnitude. After determining the linear dynamic range for standard solutions, which are typically prepared in deionized or distilled water, the LDR should be confirmed with real sample matrices.

5.9.12.4 Sensitivity and Enrichment Factor

Sensitivity is usually defined as the ability of a method to discriminate between small differences in analyte concentration and is determined as the slope m of the calibration curve (calibration sensitivity). A convenient measure of sensitivity of an analyte isolation and enrichment procedure, such as solvent microextraction, is the enrichment factor, E_f, which is defined as the ratio of analyte concentration in the final extract to its initial concentration in the original sample: $E_f = C_o/C_w$. For systems at equilibrium, the enrichment factor can be calculated by rearranging equation (3.3) or (3.8) for two- and three-phase systems, respectively, to arrive at equations (5.3) (two-phase system) and (5.4) (three-phase system):

$$E_f = \frac{C_o}{C_w^0} = \frac{1}{1/K + V_o/V_w} \tag{5.3}$$

$$E_f = \frac{1}{1/K_{ow} + K_{aw}V_a/K_{ow}V_w + V_o/V_w} \tag{5.4}$$

where the symbols have the same meaning as in Section 3.2.1. It is clear that the enrichment factor improves with an increase in the organic–water distribution constant of an analyte, K_{ow}, and the phase ratio V_o/V_w. It also increases with a decrease in the air–water distribution constant, K_{aw}, and in the phase ratio V_a/V_w. This confirms conclusions reached in Sections 5.9.2 and 5.9.3 that for maximum sensitivity we should use small headspace volumes and a small ratio of solvent to sample volume. In practice, the enrichment factors for solvent microextraction procedures range from 2 to 27,000, the lowest corresponding to direct immersion (7 to 875) and the highest to hollow fiber–protected three-phase SME (2 to 27,000) (see the summary files on the CD-ROM). It should be kept in mind that analyte enrichment is not the only objective of solvent microextraction. SME is also used to isolate analyte(s) from a matrix and to replace an original matrix with one compatible with the detection method. Therefore, the enrichment factor should not be the only criterion considered when developing an analytical procedure involving SME.

5.9.12.5 Precision *Precision* describes the degree to which data generated from replicate or repetitive measurements differ from one another. It is expressed as the variance, standard deviation, or relative standard deviation (coefficient of variation). The precision is considered at three levels with different repetition conditions: repeatability, intermediate precision, and reproducibility.

- *Repeatability*, also called *intra-assay precision* or *intra-day precision*, is evaluated by analyzing a homogeneous sample several times over a short period of time within a laboratory by the same analyst using the same instrument and equipment. The repeatability indicates how reproducible the analytical procedure can be.
- *Intermediate precision*, also called *ruggedness* or *inter-day precision*, expresses the precision of observed values obtained from multiple samplings of a homogeneous sample by different analysts on different instruments on different days in the same laboratory.

- *Reproducibility*, or *interlaboratory precision*, is the most general measure of precision observed when aliquots of the same sample are analyzed in different laboratories.

The precision of an analytical procedure involving solvent microextraction is affected by variations in all the extraction parameters that have been discussed so far: sample, solvent, and headspace volumes, temperature, time (if equilibrium has not been reached), agitation rate, sample matrix, pH, ionic strength, fiber length, diameter, wall thickness, and pore size, as well as plunger motion rate and dwelling time in dynamic mode. The precision of an analytical procedure can also deteriorate as a result of analyte losses through adsorption on the walls of containers or PTFE-coated stirring bars. This problem is especially serious for nonpolar semivolatiles (see Section 6.5.4). The capacity of an analytical procedure to remain unaffected by small but deliberate variations in method parameters is called the *robustness*. Robustness provides an indication of a method's reliability during normal use in a variety of laboratories. The robustness test can be a part of method validation.

So far, only intra-assay and intermediate precision, expressed as relative standard deviations (RSDs), have been reported in the analytical procedures involving SME. Relative standard deviations for SME procedures typically vary from 1 to 30%, although for the majority of samples and analytes they are less than 10%. For example, in procedures making use of headspace solvent microextraction, out of 97 RSD values reported, 57 were less than 10% and only 40 exceeded 10%, but only for some analytes.

5.9.12.6 *Accuracy*

Accuracy, or *trueness*, is a measure of the bias of observed values obtained by an analytical procedure. Accuracy is expressed as the difference between the average value obtained from a large number of determinations and the true value or accepted reference value. Accuracy cannot be evaluated unless the true value or reference value is known. If standard reference materials are available, bias can be estimated from the results of their analysis. When reference materials are not available, accuracy is often estimated from the recovery of spiked samples. To be valid, recoveries of spiked standards must be determined in the same matrix as the sample, since matrix effects can cause wide variability in recoveries, especially for organic compounds. Because recovery often varies with concentration, the spike and analyte concentrations should be as close as practical.[187] Since very few SME procedures employ the exhaustive mode, absolute recovery, defined as the ratio of the amount of analyte extracted to the amount of analyte added to the sample, is of little use. *Relative recovery*, defined as the ratio of the concentrations found in the investigated matrix to those in distilled or deionized water spiked with the same amounts of analytes, is generally used instead of absolute recovery. The recovery values close to 100% indicate the lack of matrix effect and good accuracy of the procedure. A solvent microextraction procedure is regarded as accurate if the recoveries of analytes present in an original sample match those introduced with a spike. This condition is verified using certified reference materials or by comparing with standard methods known to be accurate. Typical relative recovery values in various SME modes, determined by spiking samples with known amounts of analytes, range from 70 to 120%, indicating reasonably good accuracy in most cases.

5.9.12.7 Selectivity The *selectivity* of an analytical method is its ability to measure accurately the concentration of analyte free from interference due to other matrix components, impurities, or degradation products. Interferences arise from two sources: components that are inherent in the sample (matrix) and contaminants (artifacts) that have been introduced during the analytical process.[187] To separate target analyte(s) from interferences and artifacts, SME procedures are coupled with powerful separation and determination techniques (see Section 5.7 for more details). In addition, the microextraction procedures themselves can provide substantial sample cleanup, especially headspace and fiber-based SME techniques, which can prevent high-molar-mass interferences from being extracted. Despite this, some interferences may still affect extraction yields. For example, dissolved organic matter, humic acids in particular, often lower analyte recoveries from natural water samples, and proteins in physiological fluids bind some drugs, which have to be released prior to their extraction. Consequently, some samples require preliminary preparation, such as filtration, dilution, protein precipitation, or pH adjustment, which precedes solvent microextraction proper. This is especially true for solid samples (see Section 6.4).

5.9.12.8 Sample Throughput *Sample throughput* is defined as the number of samples that can be analyzed per unit of time, typically per hour. Sample throughput is determined by the time required to complete all steps of an analytical procedure after samples have been collected and brought to the laboratory, including preliminary sample preparation, extraction, and final determination. Optimization of extraction time in solvent microextraction is discussed in Section 5.9.5. Times of final determination of extracted analytes are very diverse. Spectroscopic techniques are rapid and usually require no more than a few minutes. Chromatographic and capillary electrophoretic methods are more time consuming, because they combine separation of analytes with their quantitative determination and range from several minutes to more than an hour. Sample throughput can be improved by using high-speed gas chromatography or ultra performance liquid chromatography. Another consideration is the time between preparation of standards or sample collection and extraction, which should be minimized to reduce changes in the matrix and/or the analytes, as a result of, for example, adsorption of hydrophobic analytes on the surface of containers. In addition, time between microextraction and final determination should be minimized to avoid analyte losses through their evaporation to the ambient air.

5.9.12.9 Verification and Validation *Verification* is the general process used to decide whether a method under consideration is capable of producing accurate and reliable data. *Validation* is an experimental process involving interlaboratory studies or using other methods or the use of reference materials to evaluate the suitability of methodology for its intended purpose.[187] So far, no interlaboratory

TABLE 5.6   Tap Water Concentrations of Trihalomethanes (in µg/L) Determined by EPA Method 551.1 and a Headspace Solvent Microextraction – GC-ECD Procedure in the Manual and Automated Mode (Average of Three Determinations)

Method	$CHCl_3$	$CHBrCl_2$	$CHBr_2Cl$	$CHBr_3$
EPA 551.1	6.8	4.7	2.0	0.11
HS-SDME				
Manual	6.7	5.1	2.1	0.20
Automated	6.7	4.9	1.9	0.16

studies involving solvent microextraction have been reported. However, validation by comparing the results of quantitative analysis with certified values obtained for standard reference materials having similar matrices and target analytes as the samples of interest has been performed for a number of SME procedures.[15,16,18,28,43,89,95,189] SME procedures have also been validated by comparing their results with those obtained by other extraction techniques. In the majority of cases, solvent microextraction has been compared with SPME,[39,51,63,64,76,80,95,153,169,171,190–200] although other extraction techniques, including liquid–liquid extraction,[48, 63, 76, 95, 109, 196, 201] solid-phase extraction,[76,109,153,202] and stir bar sorptive extraction,[203] have also been used as a reference. An example of validation of a headspace solvent microextraction—a GC-ECD procedure for the determination of trihalomethanes in tap water samples developed by the authors of this book (see Section 7.5) by comparison with the standard EPA Method 551.1, based on liquid–liquid extraction using methyl *tert*-butyl ether or pentane and GC-ECD determination—is provided in Table 5.6. It is evident that the HS-SDME procedure yields results that are in good agreement with the standard EPA method.

REFERENCES

1. Pedersen-Bjergaard, S.; Rasmussen, K. E.; Brekke, A.; Ho, T. S.; Halvorsen, T. G., Liquid-phase microextraction of basic drugs: selection of extraction mode based on computer calculated solubility data. *J. Sep. Sci.* 2005, *28* (11), 1195–1203.
2. Ouyang, G.; Zhao, W.; Pawliszyn, J., Automation and optimization of liquid-phase microextraction by gas chromatography. *J. Chromatogr. A* 2007, *1138* (1–2), 47–54.
3. Baghdadi, M.; Shemirani, F., Cold-induced aggregation microextraction: a novel sample preparation technique based on ionic liquids. *Anal. Chim. Acta* 2008, *613* (1), 56–63.
4. Carasek, E.; Tonjes, J. W.; Scharf, M., A liquid–liquid microextraction system for Pb and Cd enrichment and determination by flame atomic absorption spectrometry. *Quim. Nova* 2002, *25* (5), 748–752.
5. Dadfarnia, S.; Salmanzadeh, A. M.; Shabani, A. M. H., A novel separation/preconcentration system based on solidification of floating organic drop microextraction for determination of lead by graphite furnace atomic absorption spectrometry. *Anal. Chim. Acta* 2008, *623* (2), 163–167.

6. Fan, Z.; Zhou, W., Dithizone–chloroform single drop microextraction system combined with electrothermal atomic absorption spectrometry using Ir as permanent modifier for the determination of Cd in water and biological samples. *Spectrochim. Acta B* 2006, *61* (7), 870–874.

7. Ghiasvand, A. R.; Mohagheghzadeh, E., Homogeneous liquid–liquid extraction of uranium(VI) using tri-*n*-octylphosphine oxide. *Anal. Sci.* 2004, *20* (6), 917–919.

8. Ghiasvand, A. R.; Moradi, F.; Sharghi, H.; Hasaninejad, A. R., Determination of silver (I) by electrothermal-AAS in a microdroplet formed from a homogeneous liquid–liquid extraction system using tetraspirocyclohexylcalix[4]pyrroles. *Anal. Sci.* 2005, *21* (4), 387–390.

9. Ghiasvand, A. R.; Shadabi, S.; Mohagheghzadeh, E.; Hashemi, P., Homogeneous liquid–liquid extraction method for the selective separation and preconcentration of ultra trace molybdenum. *Talanta* 2005, *66* (4), 912–916.

10. Hashemi, P.; Rahimi, A.; Ghiasvand, A. R.; Abolghasemi, M. M., Headspace microextraction of tin into an aqueous microdrop containing Pd(II) and tributyl phosphate for its determination by ETAAS *J. Braz. Chem. Soc.* 2007, *18* (6), 1145–1149.

11. Igarashi, S.; Ide, N.; Takagai, Y., High-performance liquid chromatographic–spectrophotometric determination of copper(II) and palladium(II) with 5,10,15,20-tetrakis(4*N*-pyridyl)porphine following homogeneous liquid–liquid extraction in the water–acetic acid–chloroform ternary solvent system. *Anal. Chim. Acta* 2000, *424* (2), 263–269.

12. Igarashi, S.; Ide, N.; Takahata, K.; Takagai, Y., HPLC spectrophotometric determination of metal–porphyrin complexes following a preconcentration method by homogeneous liquid–liquid extraction in a water/pyridine/ethyl chloroacetate ternary component system. *Bunseki Kagaku.* 1999, *48* (12), 1115–1122.

13. Igarashi, S.; Takahashi, A.; Ueki, Y.; Yamaguchi, H., Homogeneous liquid–liquid extraction followed by X-ray fluorescence spectrometry of a microdroplet on filter-paper for the simultaneous determination of small amounts of metals. *Analyst* 2000, *125* (5), 797–798.

14. Kumemura, M.; Korenaga, T., Quantitative extraction using flowing nano-liter droplet in microfluidic system. *Anal. Chim. Acta* 2006, *558* (1–2), 75–79.

15. Li, L.; Hu, B., Hollow-fibre liquid phase microextraction for separation and preconcentration of vanadium species in natural waters and their determination by electrothermal vaporization-ICP-OES. *Talanta* 2007, *72* (2), 472–479.

16. Li, L.; Hu, B.; Xia, L.; Jiang, Z., Determination of trace Cd and Pb in environmental and biological samples by ETV-ICP-MS after single-drop microextraction. *Talanta* 2006, *70* (2), 468–473.

17. Lin, M. Y.; Whang, C. W., Microwave-assisted derivatization and single-drop microextraction for gas chromatographic determination of chromium(III) in water. *J. Chromatogr. A* 2007, *1160* (1–2), 336–339.

18. Maltez, H. F.; Borges, D. L. G.; Carasek, E.; Welz, B.; Curtius, A. J., Single drop microextraction with *O,O*-diethyl dithiophosphate for the determination of lead by electrothermal atomic absorption spectrometry. *Talanta* 2008, *74* (4), 800–805.

19. Naseri, M. T.; Hemmatkhah, P.; Hosseini, M. R. M.; Assadi, Y., Combination of dispersive liquid–liquid microextraction with flame atomic absorption spectrometry using microsample introduction for determination of lead in water samples. *Anal. Chim. Acta* 2008, *610* (1), 135–141.

20. Naseri, M. T.; Hosseini, M. R. M.; Assadi, Y.; Kiani, A., Rapid determination of lead in water samples by dispersive liquid–liquid microextraction coupled with electrothermal atomic absorption spectrometry. *Talanta* 2008, *75* (1), 56–62.

21. Parthasarathy, N.; Pelletier, M.; Buffle, J., Hollow fiber based supported liquid membrane: a novel analytical system for trace metal analysis. *Anal. Chim. Acta* 1997, *350* (1–2), 183–195.

22. Romero, M.; Liu, J. F.; Mayer, P.; Jönsson, J. Å., Equilibrium sampling through membranes of freely dissolved copper concentrations with selective hollow fiber membranes and the spectrophotometric detection of a metal stripping agent. *Anal. Chem.* 2005, *77* (23), 7605–7611.

23. Shokoufi, N.; Shemirani, F.; Assadi, Y., Fiber optic–linear array detection spectrophotometry in combination with dispersive liquid–liquid microextraction for simultaneous preconcentration and determination of palladium and cobalt. *Anal. Chim. Acta* 2007, *597* (2), 349–356.

24. Takahashi, A.; Igarashi, S.; Ueki, Y.; Yamaguchi, H., X-ray fluorescence analysis of trace metal ions following a preconcentration of metal–diethyldithiocarbamate complexes by homogeneous liquid–liquid extraction. *Fresenius' J. Anal. Chem.* 2000, *368* (6), 607–610.

25. Takahashi, A.; Ueki, Y.; Igarashi, S., Homogeneous liquid–liquid extraction of uranium (VI) from acetate aqueous solution. *Anal. Chim. Acta* 1999, *387* (1), 71–75.

26. Xia, L.; Hu, B.; Jiang, Z.; Wu, Y.; Chen, R., 8-Hydroxyquinoline–chloroform single drop microextraction and electrothermal vaporization ICP-MS for the fractionation of aluminium in natural waters and drinks. *J. Anal. At. Spectrom.* 2005, *20* (5), 441–446.

27. Xia, L.; Hu, B.; Jiang, Z.; Wu, Y.; Liang, Y., Single-drop microextraction combined with low-temperature electrothermal vaporization ICPMS for the determination of trace Be, Co, Pd, and Cd in biological samples. *Anal. Chem.* 2004, *76* (10), 2910–2915.

28. Xia, L.; Wu, Y.; Hu, B., Hollow-fiber liquid-phase microextraction prior to low-temperature electrothermal vaporization ICP-MS for trace element analysis in environmental and biological samples. *J. Mass Spectrom.* 2007, *42* (6), 803–810.

29. Zeini Jahromi, E.; Bidari, A.; Assadi, Y.; Milani Hosseini, M. R.; Jamali, M. R., Dispersive liquid–liquid microextraction combined with graphite furnace atomic absorption spectrometry: ultra trace determination of cadmium in water samples. *Anal. Chim. Acta* 2007, *585* (2), 305–311.

30. Bidari, A.; Zeini Jahromi, E.; Assadi, Y.; Milani Hosseini, M. R., Monitoring of selenium in water samples using dispersive liquid–liquid microextraction followed by iridium-modified tube graphite furnace atomic absorption spectrometry. *Microchem. J.* 2007, *87* (1), 6–12.

31. Chamsaz, M.; Arbab-Zawar, M. H.; Nazari, S., Determination of arsenic by electrothermal atomic absorption spectrometry using headspace liquid phase microextraction after in situ hydride generation. *J. Anal. At. Spectrom.* 2003, *18* (10), 1279–1282.

32. Fan, Z., Determination of antimony(III) and total antimony by single-drop microextraction combined with electrothermal atomic absorption spectroscopy. *Anal. Chim. Acta* 2007, *585* (2), 300–304.

33. Fragueiro, S.; Lavilla, I.; Bendicho, C., Headspace sequestration of arsine onto a Pd(II)-containing aqueous drop as a preconcentration method for electrothermal atomic absorption spectrometry. *Spectrochim. Acta B* 2004, *59* (6), 851–855.

34. Fragueiro, S.; Lavilla, I.; Bendicho, C., Hydride generation–headspace single-drop microextraction–electrothermal atomic spectrometry method for determination of selenium in water after photoassisted prereduction. *Talanta* 2006, *68* (4), 1096–1101.

35. Xia, L.; Hu, B.; Jiang, Z.; Wu, Y.; Chen, R.; Li, L., Hollow fiber liquid phase microextraction combined with electrothermal vaporization ICP-MS for the speciation of inorganic selenium in natural waters. *J. Anal. At. Spectrom.* 2006, *21* (3), 362–365.

36. Jermak, S.; Pranaitytė, B.; Padarauskas, A., Headspace single-drop microextraction with in-drop derivatization and capillary electrophoretic determination for free cyanide analysis. *Electrophoresis* 2006, *27* (22), 4538–4544.

37. Jermak, S.; Pranaitytė, B.; Padarauskas, A., Ligand displacement, headspace single-drop microextraction, and capillary electrophoresis for the determination of weak acid dissociable cyanide. *J. Chromatogr. A* 2007, *1148* (1), 123–127.

38. Reddy-Noone, K.; Jain, A.; Verma, K. K., Liquid-phase microextraction–gas chromatography–mass spectrometry for the determination of bromate, iodate, bromide and iodide in high-chloride matrix. *J. Chromatogr. A* 2007, *1148* (2), 145–151.

39. Das, P.; Gupta, M.; Jain, A.; Verma, K. K., Single drop microextraction or solid phase microextraction–gas chromatography–mass spectrometry for the determination of iodine in pharmaceuticals, iodized salt, milk powder and vegetables involving conversion into 4-iodo-*N,N*-dimethylaniline. *J. Chromatogr. A* 2004, *1023* (1), 33–39.

40. Huang, K. J.; Wang, H.; Ma, M.; Sha, M. L.; Zhang, H. S., Ultrasound-assisted liquid-phase microextraction and high-performance liquid chromatographic determination of nitric oxide produced in PC12 cells using 1,3,5,7-tetramethyl-2,6-dicarbethoxy-8-(3′,4′-diaminophenyl)-difluoroboradiaza-*s*-indacene. *J. Chromatogr. A* 2006, *1103* (2), 193–201.

41. Birjandi, A. P.; Bidari, A.; Rezaei, F.; Hosseini, M. R. M.; Assadi, Y., Speciation of butyl and phenyltin compounds using dispersive liquid–liquid microextraction and gas chromatography–flame photometric detection. *J. Chromatogr. A* 2008, *1193* (1–2), 19–25.

42. Fan, Z.; Liu, X., Determination of methylmercury and phenylmercury in water samples by liquid–liquid–liquid microextraction coupled with capillary electrophoresis. *J. Chromatogr. A* 2008, *1180* (1–2), 187–192.

43. Gil, S.; Fragueiro, S.; Lavilla, I.; Bendicho, C., Determination of methylmercury by electrothermal atomic absorption spectrometry using headspace single-drop microextraction with in situ hydride generation. *Spectrochim. Acta B* 2005, *60* (1), 145–150.

44. Shioji, H.; Tsunoi, S.; Harino, H.; Tanaka, M., Liquid-phase microextraction of tributyltin and triphenyltin coupled with gas chromatography–tandem mass spectrometry: comparison between 4-fluorophenyl and ethyl derivatizations. *J. Chromatogr. A* 2004, *1048* (1), 81–88.

45. Chia, K. J.; Huang, S. D., Simultaneous derivatization and extraction of primary amines in river water with dynamic hollow fiber liquid-phase microextraction followed by gas chromatography–mass spectrometric detection. *J. Chromatogr. A* 2006, *1103* (1), 158–161.

46. Chiang, J. S.; Huang, S. D., Simultaneous derivatization and extraction of anilines in waste water with dispersive liquid–liquid microextraction followed by gas chromatography–mass spectrometric detection. *Talanta* 2008, *75* (1), 70–75.

47. Chiang, J.-S.; Huang, S.-D., Simultaneous derivatization and extraction of amphetamine and methylenedioxyamphetamine in urine with headspace liquid-phase microextraction followed by gas chromatography–mass spectrometry. *J. Chromatogr. A* 2008, *1185* (1), 19–22.

48. Deng, C.; Li, N.; Wang, L.; Zhang, X., Development of gas chromatography–mass spectrometry following headspace single-drop microextraction and simultaneous derivatization for fast determination of short-chain aliphatic amines in water samples. *J. Chromatogr. A* 2006, *1131* (1–2), 45–50.

49. Dziarkowska, K.; Jönsson, J. Å.; Wieczorek, P. P., Single hollow fiber SLM extraction of polyamines followed by tosyl chloride derivatization and HPLC determination. *Anal. Chim. Acta* 2008, *606* (2), 184–193.

50. Reddy-Noone, K.; Jain, A.; Verma, K. K., Liquid-phase microextraction and GC for the determination of primary, secondary and tertiary aromatic amines as their iodo-derivatives. *Talanta* 2007, *73* (4), 684–691.
51. Lee, H. S. N.; Sng, M. T.; Basheer, C.; Lee, H. K., Determination of basic degradation products of chemical warfare agents in water using hollow fibre–protected liquid-phase microextraction with in-situ derivatisation followed by gas chromatography–mass spectrometry. *J. Chromatogr. A* 2008, *1196–1197*, 125–132.
52. Chen, P. S.; Huang, S. D., Coupled two-step microextraction devices with derivatizations to identify hydroxycarbonyls in rain samples by gas chromatography–mass spectrometry. *J. Chromatogr. A* 2006, *1118* (2), 161–167.
53. Curyło, J.; Wardencki, W., Application of single drop extraction (SDE) gas chromatography method for the determination of carbonyl compounds in spirits and vodkas. *Anal. Lett.* 2006, *39* (13), 2629–2642.
54. Deng, C.; Li, N.; Wang, X.; Zhang, X.; Zeng, J., Rapid determination of acetone in human blood by derivatization with pentafluorobenzyl hydroxylamine followed by headspace liquid-phase microextraction and gas chromatography/mass spectrometry. *Rapid Commun. Mass Spectrom.* 2005, *19* (5), 647–653.
55. Deng, C.; Yao, N.; Li, N.; Zhang, X., Headspace single-drop microextraction with in-drop derivatization for aldehyde analysis. *J. Sep. Sci.* 2005, *28* (17), 2301–2305.
56. Dong, L.; Shen, X.; Deng, C., Development of gas chromatography–mass spectrometry following headspace single-drop microextraction and simultaneous derivatization for fast determination of the diabetes biomarker, acetone in human blood samples. *Anal. Chim. Acta* 2006, *569* (1–2), 91–96.
57. Fiamegos, Y. C.; Stalikas, C. D., Theoretical analysis and experimental evaluation of headspace in-drop derivatisation single-drop microextraction using aldehydes as model analytes. *Anal. Chim. Acta* 2007, *599* (1), 76–83.
58. Fiamegos, Y. C.; Stalikas, C. D., Gas chromatographic determination of carbonyl compounds in biological and oil samples by headspace single-drop microextraction with in-drop derivatisation. *Anal. Chim. Acta* 2008, *609* (2), 175–183.
59. Li, N.; Deng, C.; Yao, N.; Shen, X.; Zhang, X., Determination of acetone, hexanal and heptanal in blood samples by derivatization with pentafluorobenzyl hydroxylamine followed by headspace single-drop microextraction and gas chromatography–mass spectrometry. *Anal. Chim. Acta* 2005, *540* (2), 317–323.
60. Li, N.; Deng, C.; Yin, X.; Yao, N.; Shen, X.; Zhang, X., Gas chromatography–mass spectrometric analysis of hexanal and heptanal in human blood by headspace single-drop microextraction with droplet derivatization. *Anal. Biochem.* 2005, *342* (2), 318–326.
61. Liu, J.; Peng, J. F.; Chi, Y. G.; Jiang, G. B., Determination of formaldehyde in shiitake mushroom by ionic liquid-based liquid-phase microextraction coupled with liquid chromatography. *Talanta* 2005, *65* (3), 705–709.
62. Bagheri, H.; Saber, A.; Mousavi, S. R., Immersed solvent microextraction of phenol and chlorophenols from water samples followed by gas chromatography–mass spectrometry. *J. Chromatogr. A* 2004, *1046* (1–2), 27–33.
63. Basheer, C.; Lee, H. K., Analysis of endocrine disrupting alkylphenols, chlorophenols and bisphenol-A using hollow fiber–protected liquid-phase microextraction coupled with injection port–derivatization gas chromatography–mass spectrometry. *J. Chromatogr. A* 2004, *1057* (1–2), 163–169.
64. Basheer, C.; Parthiban, A.; Jayaraman, A.; Lee, H. K.; Valiyaveettil, S., Determination of alkylphenols and bisphenol-A: a comparative investigation of functional polymer-coated

membrane microextraction and solid-phase microextraction techniques. *J. Chromatogr. A* 2005, *1087* (1–2), 274–282.

65. Fattahi, N.; Assadi, Y.; Milani Hosseini, M. R.; Zeini Jahromi, E., Determination of chlorophenols in water samples using simultaneous dispersive liquid–liquid microextraction and derivatization followed by gas chromatography–electron-capture detection. *J. Chromatogr. A* 2007, *1157* (1–2), 23–29.

66. Fiamegos, Y. C.; Kefala, A. P.; Stalikas, C. D., Ion-pair single-drop microextraction versus phase-transfer catalytic extraction for the gas chromatographic determination of phenols as tosylated derivatives. *J. Chromatogr. A* 2008, *1190* (1–2), 44–51.

67. Fiamegos, Y. C.; Stalikas, C. D., In-drop derivatisation liquid-phase microextraction assisted by ion-pairing transfer for the gas chromatographic determination of phenolic endocrine disruptors. *Anal. Chim. Acta* 2007, *597* (1), 32–40.

68. Ito, R.; Kawaguchi, M.; Honda, H.; Koganei, Y.; Okanouchi, N.; Sakui, N.; Saito, K.; Nakazawa, H., Hollow-fiber-supported liquid phase microextraction with in situ derivatization and gas chromatography–mass spectrometry for determination of chlorophenols in human urine samples. *J. Chromatogr. B* 2008, *872* (1–2), 63–67.

69. Kawaguchi, M.; Ito, R.; Endo, N.; Okanouchi, N.; Sakui, N.; Saito, K.; Nakazawa, H., Liquid phase microextraction with in situ derivatization for measurement of bisphenol A in river water sample by gas chromatography–mass spectrometry. *J. Chromatogr. A* 2006, *1110* (1–2), 1–5.

70. Kawaguchi, M.; Ito, R.; Okanouchi, N.; Saito, K.; Nakazawa, H., Miniaturized hollow fiber assisted liquid-phase microextraction with in situ derivatization and gas chromatography–mass spectrometry for analysis of bisphenol A in human urine sample. *J. Chromatogr. B* 2008, *870* (1), 98–102.

71. Saraji, M.; Bakhshi, M., Determination of phenols in water samples by single-drop microextraction followed by in-syringe derivatization and gas chromatography–mass spectrometric detection. *J. Chromatogr. A* 2005, *1098* (1–2), 30–36.

72. Zhao, R. S.; Yuan, J. P.; Li, H. F.; Wang, X.; Jiang, T.; Lin, J. M., Nonequilibrium hollow-fiber liquid-phase microextraction with in situ derivatization for the measurement of triclosan in aqueous samples by gas chromatography–mass spectrometry. *Anal. Bioanal. Chem.* 2007, *387* (8), 2911–2915.

73. Kramer, K. E.; Andrews, A. R. J., Screening method for 11-nor-delta9-tetrahydrocannabinol-9-carboxylic acid in urine using hollow fiber membrane solvent microextraction with in-tube derivatization. *J. Chromatogr. B* 2001, *760* (1), 27–36.

74. Lee, H. S. N.; Sng, M. T.; Basheer, C.; Lee, H. K., Determination of degradation products of chemical warfare agents in water using hollow fibre–protected liquid-phase microextraction with in-situ derivatisation followed by gas chromatography–mass spectrometry. *J. Chromatogr. A* 2007, *1148* (1), 8–15.

75. Pardasani, D.; Kanaujia, P. K.; Gupta, A. K.; Tak, V.; Shrivastava, R. K.; Dubey, D. K., In situ derivatization hollow fiber mediated liquid phase microextraction of alkylphosphonic acids from water. *J. Chromatogr. A* 2007, *1141* (2), 151–157.

76. Saraji, M.; Farajmand, B., Application of single-drop microextraction combined with in-microvial derivatization for determination of acidic herbicides in water samples by gas chromatography–mass spectrometry. *J. Chromatogr. A* 2008, *1178* (1–2), 17–23.

77. Sun, S. H.; Xie, J. P.; Xie, F. W.; Zong, Y. L., Determination of volatile organic acids in oriental tobacco by needle-based derivatization headspace liquid-phase microextraction coupled to gas chromatography/mass spectrometry. *J. Chromatogr. A* 2008, *1179* (2), 89–95.

78. Varanusupakul, P.; Vora-adisak, N.; Pulpoka, B., In situ derivatization and hollow fiber membrane microextraction for gas chromatographic determination of haloacetic acids in water. *Anal. Chim. Acta* 2007, *598* (1), 82–86.

79. Wu, J.; Lee, H. K., Ion-pair dynamic liquid-phase microextraction combined with injection-port derivatization for the determination of long-chain fatty acids in water samples. *J. Chromatogr. A* 2006, *1133* (1–2), 13–20.

80. Wu, J.; Lee, H. K., Injection port derivatization following ion-pair hollow fiber–protected liquid-phase microextraction for determining acidic herbicides by gas chromatography/mass spectrometry. *Anal. Chem.* 2006, *78* (20), 7292–7301.

81. Fiamegos, Y. C.; Nanos, C. G.; Stalikas, C. D., Ultrasonic-assisted derivatization reaction of amino acids prior to their determination in urine by using single-drop microextraction in conjunction with gas chromatography. *J. Chromatogr. B* 2004, *813* (1–2), 89–94.

82. Oshite, S.; Furukawa, M.; Igarashi, S., Homogeneous liquid–liquid extraction method for the selective spectrofluorimetric determination of trace amounts of tryptophan. *Analyst* 2001, *126* (5), 703–706.

83. Reubsaet, J. L. E.; Loftheim, H.; Gjelstad, A., Ion-pair mediated transport of angiotensin, neurotensin, and their metabolites in liquid phase microextraction under acidic conditions. *J. Sep. Sci.* 2005, *28* (11), 1204–1210.

84. Batlle, R.; López, P.; Nerín, C.; Crescenzi, C., Active single-drop microextraction for the determination of gaseous diisocyanates. *J. Chromatogr. A* 2008, *1185* (2), 155–160.

85. Leinonen, A.; Vuorensola, K.; Lepola, L. M.; Kuuranne, T.; Kotiaho, T.; Ketola, R. A.; Kostiainen, R., Liquid-phase microextraction for sample preparation in analysis of unconjugated anabolic steroids in urine. *Anal. Chim. Acta* 2006, *559* (2), 166–172.

86. Stalikas, C. D.; Fiamegos, Y. C., Microextraction combined with derivatization. *Trends Anal. Chem.* 2008, *27* (6), 533–542.

87. Xu, L.; Basheer, C.; Lee, H. K., Chemical reactions in liquid-phase microextraction. *J. Chromatogr. A* 2009, *1216* (4), 701–707.

88. Gupta, M.; Jain, A.; Verma, K. K., Optimization of experimental parameters in single-drop microextraction–gas chromatography–mass spectrometry for the determination of periodate by the Malaprade reaction, and its application to ethylene glycol. *Talanta* 2007, *71* (3), 1039–1046.

89. Jiang, H.; Qin, Y.; Hu, B., Dispersive liquid phase microextraction (DLPME) combined with graphite furnace atomic absorption spectrometry (GFAAS) for determination of trace Co and Ni in environmental water and rice samples. *Talanta* 2008, *74* (5), 1160–1165.

90. Saraji, M.; Mousavinia, F., Single-drop microextraction followed by in-syringe derivatization and gas chromatography–mass spectrometric detection for determination of organic acids in fruits and fruit juices. *J. Sep. Sci.* 2006, *29* (9), 1223–1229.

91. Zhang, J.; Lee, H. K., Application of liquid-phase microextraction and on-column derivatization combined with gas chromatography–mass spectrometry to the determination of carbamate pesticides. *J. Chromatogr. A* 2006, *1117* (1), 31–37.

92. He, Y.; Lee, H. K., Liquid-phase microextraction in a single drop of organic solvent by using a conventional microsyringe. *Anal. Chem.* 1997, *69* (22), 4634–4640.

93. Ma, M.; Kang, S.; Zhao, Q.; Chen, B.; Yao, S., Liquid-phase microextraction combined with high-performance liquid chromatography for the determination of local anaesthetics in human urine. *J. Pharm. Biomed. Anal.* 2006, *40* (1), 128–135.

94. Liu, W.; Lee, H. K., Continuous-flow microextraction exceeding 1000-fold concentration of dilute analytes. *Anal. Chem.* 2000, *72* (18), 4462–4467.

95. Colombini, V.; Bancon-Montigny, C.; Yang, L.; Maxwell, P.; Sturgeon, R. E.; Mester, Z., Headspace single drop microextraction for the detection of organotin compounds. *Talanta* 2004, *63* (3), 555–560.

96. Kim, N. S.; Jung, M. J.; Yoo, Z. W.; Lee, S. N.; Lee, D. S., Headspace hanging drop liquid phase microextraction and GC-MS for the determination of linalool from evening primrose flowers. *Bull. Korean Chem. Soc.* 2005, *26* (12), 1996–2000.

97. Leong, M. I.; Huang, S. D., Dispersive liquid–liquid microextraction method based on solidification of floating organic drop combined with gas chromatography with electron-capture or mass spectrometry detection. *J. Chromatogr. A* 2008, *1211* (1–2), 8–12.

98. He, Y.; Lee, H. K., Continuous flow microextraction combined with high-performance liquid chromatography for the analysis of pesticides in natural waters. *J. Chromatogr. A* 2006, *1122* (1–2), 7–12.

99. Zhu, L.; Tay, C. B.; Lee, H. K., Liquid–liquid–liquid microextraction of aromatic amines from water samples combined with high-performance liquid chromatography. *J. Chromatogr. A* 2002, *963* (1–2), 231–237.

100. Batlle, R.; Nerin, C., Application of single-drop microextraction to the determination of dialkyl phthalate esters in food simulants. *J. Chromatogr. A* 2004, *1045* (1–2), 29–35.

101. Wu, J.; Ee, K. H.; Lee, H. K., Automated dynamic liquid–liquid–liquid microextraction followed by high-performance liquid chromatography–ultraviolet detection for the determination of phenoxy acid herbicides in environmental waters. *J. Chromatogr. A* 2005, *1082* (2), 121–127.

102. Barwick, V. J., Strategies for solvent selection: a literature review. *Trends Anal. Chem.* 1997, *16* (6), 293–309.

103. Aguilera-Herrador, E.; Lucena, R.; Cárdenas, S.; Valcárcel, M., Direct coupling of ionic liquid based single-drop microextraction and GC/MS. *Anal. Chem.* 2008, *80* (3), 793–800.

104. Aguilera-Herrador, E.; Lucena, R.; Cárdenas, S.; Valcárcel, M., Ionic liquid-based single-drop microextraction/gas chromatographic/mass spectrometric determination of benzene, toluene, ethylbenzene and xylene isomers in waters. *J. Chromatogr. A* 2008, *1201* (1), 106–111.

105. de Santana, F. J. M.; Bonato, P. S., Enantioselective analysis of mirtazapine and its two major metabolites in human plasma by liquid chromatography–mass spectrometry after three-phase liquid-phase microextraction. *Anal. Chim. Acta* 2007, *606* (1), 80–91.

106. El-Beqqali, A.; Kussak, A.; Abdel-Rehim, M., Determination of dopamine and serotonin in human urine samples utilizing microextraction online with liquid chromatography/electrospray tandem mass spectrometry. *J. Sep. Sci.* 2007, *30* (3), 421–424.

107. Ho, T. S.; Reubsaet, J. L. E.; Anthonsen, H. S.; Pedersen-Bjergaard, S.; Rasmussen, K. E., Liquid-phase microextraction based on carrier mediated transport combined with liquid chromatography–mass spectrometry: New concept for the determination of polar drugs in a single drop of human plasma. *J. Chromatogr. A* 2005, *1072* (1), 29–36.

108. Ho, T. S.; Vasskog, T.; Anderssen, T.; Jensen, E.; Rasmussen, K. E.; Pedersen-Bjergaard, S., 25,000-fold pre-concentration in a single step with liquid-phase micro-extraction. *Anal. Chim. Acta* 2007, *592* (1), 1–8.

109. Kuuranne, T.; Kotiaho, T.; Pedersen-Bjergaard, S.; Rasmussen, K. E.; Leinonen, A.; Westwood, S.; Kostiainen, R., Feasibility of a liquid-phase microextraction sample clean-up and liquid chromatographic/mass spectrometric screening method for selected anabolic steroid glucuronides in biological samples. *J. Mass Spectrom.* 2003, *38* (1), 16–26.

110. Raich-Montiu, J.; Krogh, K. A.; Granados, M.; Jönsson, J. Å.; Halling-Sørensen, B., Determination of ivermectin and transformation products in environmental waters using hollow fibre–supported liquid membrane extraction and liquid chromatography–mass spectrometry/mass spectrometry. *J. Chromatogr. A* 2008, *1187* (1–2), 275–280.

111. Romero-González, R.; Pastor-Montoro, E.; Martínez-Vidal, J. L.; Garrido-Frenich, A., Application of hollow fiber supported liquid membrane extraction to the simultaneous determination of pesticide residues in vegetables by liquid chromatography/mass spectrometry. *Rapid Commun. Mass Spectrom.* 2006, *20* (18), 2701–2708.

112. Tsai, T. F.; Lee, M. R., Liquid-phase microextration combined with liquid chromatography–electrospray tandem mass spectrometry for detecting diuretics in urine. *Talanta* 2008, *75* (3), 658–665.

113. Yamini, Y.; Reimann, C. T.; Vatanara, A.; Jönsson, J. Å., Extraction and pre-concentration of salbutamol and terbutaline from aqueous samples using hollow fiber supported liquid membrane containing anionic carrier. *J. Chromatogr. A* 2006, *1124* (1–2), 57–67.

114. de Santana, F. J. M.; de Oliveira, A. R. M.; Bonato, P. S., Chiral liquid chromatographic determination of mirtazapine in human plasma using two-phase liquid-phase micro-extraction for sample preparation. *Anal. Chim. Acta* 2005, *549* (1–2), 96–103.

115. González-Peñas, E.; Leache, C.; Viscarret, M.; Pérez de Obanos, A.; Araguás, C.; López de Cerain, A., Determination of ochratoxin A in wine using liquid-phase microextraction combined with liquid chromatography with fluorescence detection. *J. Chromatogr. A* 2004, *1025* (2), 163–168.

116. Sanagi, M. M.; See, H. H.; Ibrahim, W. A.; Naim, A. A., Determination of pesticides in water by cone-shaped membrane protected liquid phase microextraction prior to micro-liquid chromatography. *J. Chromatogr. A* 2007, *1152* (1–2), 215–219.

117. Keller, B. O.; Li, L., Nanoliter solvent extraction combined with microspot MALDI TOF mass spectrometry for the analysis of hydrophobic biomolecules. *Anal. Chem.* 2001, *73* (13), 2929–2936.

118. Sekar, R.; Wu, H. F., Quantitative method for analysis of monensin in soil, water, and urine by direct combination of single-drop microextraction with atmospheric pressure matrix-assisted laser desorption/ionization mass spectrometry. *Anal. Chem.* 2006, *78* (18), 6306–6313.

119. Shrivas, K.; Wu, H. F., Quantitative bioanalysis of quinine by atmospheric pressure-matrix assisted laser desorption/ionization mass spectrometry combined with dynamic drop-to-drop solvent microextraction. *Anal. Chim. Acta* 2007, *605* (2), 153–158.

120. Shrivas, K.; Wu, H. F., Single drop microextraction as a concentrating probe for rapid screening of low molecular weight drugs from human urine in atmospheric-pressure matrix-assisted laser desorption/ionization mass spectrometry. *Rapid Commun. Mass Spectrom.* 2007, *21* (18), 3103–3108.

121. Shrivas, K.; Wu, H. F., Modified silver nanoparticle as a hydrophobic affinity probe for analysis of peptides and proteins in biological samples by using liquid–liquid micro-extraction coupled to AP-MALDI-ion trap and MALDI-TOF mass spectrometry. *Anal. Chem.* 2008, *80* (7), 2583–2589.

122. Sudhir, P.-R.; Wu, H.-F.; Zhou, Z.-C., Identification of peptides using gold nano-particle-assisted single-drop microextraction coupled with AP-MALDI mass spectro-metry. *Anal. Chem.* 2005, *77* (22), 7380–7385.

123. Sun, S.; Cheng, Z.; Xie, J.; Zhang, J.; Liao, Y.; Wang, H.; Guo, Y., Identification of volatile basic components in tobacco by headspace liquid-phase microextraction

coupled to matrix-assisted laser desorption/ionization with Fourier transform mass spectrometry. *Rapid Commun. Mass Spectrom.* 2005, *19* (8), 1025–1030.

124. Wu, H. F.; Lin, C. H., Direct combination of immersed single-drop microextraction with atmospheric pressure matrix-assisted laser desorption/ionization tandem mass spectrometry for rapid analysis of a hydrophilic drug via hydrogen-bonding interaction and comparison with liquid–liquid extraction and liquid-phase microextraction using a dual gauge microsyringe with a hollow fiber. *Rapid Commun. Mass Spectrom.* 2006, *20* (16), 2511–2515.

125. Xie, J. P.; Sun, S. H.; Wang, H.-Y.; Zong, Y. L.; Nie, C.; Guo, Y. L., Determination of nicotine in mainstream smoke on the single puff level by liquid-phase microextraction coupled to matrix-assisted laser desorption/ionization Fourier transform mass spectrometry. *Rapid Commun. Mass Spectrom.* 2006, *20* (17), 2573–2578.

126. Sudo, T.; Igarashi, S., Homogeneous liquid–liquid extraction method for spectrofluorimetric determination of chlorophyll a. *Talanta* 1996, *43* (2), 233–237.

127. Donati, G. L.; Pharr, K. E.; Calloway Jr, C. P.; Nóbrega, J. A.; Jones, B. T., Determination of Cd in urine by cloud point extraction–tungsten coil atomic absorption spectrometry. *Talanta* 2008, *76* (5), 1252–1255.

128. Figueroa, R.; García, M.; Lavilla, I.; Bendicho, C., Photoassisted vapor generation in the presence of organic acids for ultrasensitive determination of Se by electrothermal–atomic absorption spectrometry following headspace single-drop microextraction. *Spectrochim. Acta B* 2005, *60* (12), 1556–1563.

129. Sigg, L.; Black, F.; Buffle, J.; Cao, J.; Cleven, R.; Davison, W.; Galceran, J.; Gunkel, P.; Kalis, E.; Kistler, D.; Martin, M.; Noël, S.; Nur, Y.; Odzak, N.; Puy, J.; van Riemsdijk, W.; Temminghoff, E.; Tercier-Waeber, M. L.; Toepperwien, S.; Town, R. M.; Unsworth, E.; Warnken, K. W.; Weng, L.; Xue, H.; Zhang, H., Comparison of analytical techniques for dynamic trace metal speciation in natural freshwaters. *Environ. Sci. Technol.* 2006, *40* (6), 1934–1941.

130. Xu, L.; Basheer, C.; Lee, H. K., Developments in single-drop microextraction. *J. Chromatogr. A* 2007, *1152* (1–2), 184–192.

131. Araujo, P.W.; Brereton, R. G., Experimental design: I. Screening. *Trends Anal. Chem.* 1996, *15* (1), 26–31.

132. Araujo, P. W.; Brereton, R. G., Experimental design: II. Optimization. *Trends Anal. Chem.* 1996, *15* (2), 63–70.

133. Araujo, P. W.; Brereton, R. G., Experimental design: III. Quantification. *Trends Anal. Chem.* 1996, *15* (3), 156–163.

134. Bezerra, M. A.; Santelli, R. E.; Oliveira, E. P.; Villar, L. S.; Escaleira, L. A., Response surface methodology (RSM) as a tool for optimization in analytical chemistry. *Talanta* 2008, *76* (5), 965–977.

135. Bianchi, F.; Careri, M., Experimental design techniques for optimization of analytical methods: I. Separation and sample preparation techniques. *Curr. Anal. Chem.* 2008, *4* (1), 55–74.

136. Ebrahimzadeh, H.; Yamini, Y.; Kamarei, F.; Shariati, S., Homogeneous liquid–liquid extraction of trace amounts of mononitrotoluenes from waste water samples. *Anal. Chim. Acta* 2007, *594* (1), 93–100.

137. Farajzadeh, M. A.; Bahram, M.; Mehr, B. G.; Jönsson, J. Å., Optimization of dispersive liquid–liquid microextraction of copper(II) by atomic absorption spectrometry as its oxinate chelate: application to determination of copper in different water samples. *Talanta* 2008, *75* (3), 832–840.

138. Lezamiz, J.; Jönsson, J. Å., Development of a simple hollow fibre supported liquid membrane extraction method to extract and preconcentrate dinitrophenols in environmental samples at ng L^{-1} level by liquid chromatography. *J. Chromatogr. A* 2007, *1152* (1–2), 226–233.

139. Martendal, E.; Budziak, D.; Carasek, E., Application of fractional factorial experimental and Box–Behnken designs for optimization of single-drop microextraction of 2,4,6-trichloroanisole and 2,4,6-tribromoanisole from wine samples. *J. Chromatogr. A* 2007, *1148* (2), 131–136.

140. Regueiro, J.; Llompart, M.; Garcia-Jares, C.; Garcia-Monteagudo, J. C.; Cela, R., Ultrasound-assisted emulsification–microextraction of emergent contaminants and pesticides in environmental waters. *J. Chromatogr. A* 2008, *1190* (1–2), 27–38.

141. Romero, J.; López, P.; Rubio, C.; Batlle, R.; Nerín, C., Strategies for single-drop microextraction optimisation and validation:. application to the detection of potential antimicrobial agents *J. Chromatogr. A* 2007, *1166* (1–2), 24–29.

142. Sobhi, H. R.; Yamini, Y.; Esrafili, A.; Abadi, R. H. H. B., Suitable conditions for liquid-phase microextraction using solidification of a floating drop for extraction of fat-soluble vitamins established using an orthogonal array experimental design. *J. Chromatogr. A* 2008, *1196–1197*, 28–32.

143. Tan, L.; Zhao, X. P.; Liu, X. Q.; Ju, H. X.; Li, J. S., Headspace liquid-phase microextraction of short-chain fatty acids in plasma, and gas chromatography with flame ionization detection. *Chromatographia* 2005, *62* (5–6), 305–309.

144. Tor, A.; Aydin, M. E., Application of liquid-phase microextraction to the analysis of trihalomethanes in water. *Anal. Chim. Acta* 2006, *575* (1), 138–143.

145. Vidal, L.; Canals, A.; Salvador, A., Sensitive determination of free benzophenone-3 in human urine samples based on an ionic liquid as extractant phase in single-drop microextraction prior to liquid chromatography analysis. *J. Chromatogr. A* 2007, *1174* (1–2), 95–103.

146. Vidal, L.; Domini, C. E.; Grané, N.; Psillakis, E.; Canals, A., Microwave-assisted headspace single-drop microextraction of chlorobenzenes from water samples. *Anal. Chim. Acta* 2007, *592* (1), 9–15.

147. Vidal, L.; Psillakis, E.; Domini, C. E.; Grané, N.; Marken, F.; Canals, A., An ionic liquid as a solvent for headspace single drop microextraction of chlorobenzenes from water samples. *Anal. Chim. Acta* 2007, *584* (1), 189–195.

148. Wu, J.; Lee, H. K., Orthogonal array designs for the optimization of liquid–liquid–liquid microextraction of nonsteroidal anti-inflammatory drugs combined with high-performance liquid chromatography–ultraviolet detection. *J. Chromatogr. A* 2005, *1092* (2), 182–190.

149. Xia, J.; Xiang, B.; Zhang, W., Determination of metacrate in water samples using dispersive liquid–liquid microextraction and HPLC with the aid of response surface methodology and experimental design. *Anal. Chim. Acta* 2002, *625* (1), 28–34.

150. Hauser, B.; Popp, P.; Kleine-Benne, E., Membrane-assisted solvent extraction of triazines and other semi-volatile contaminants directly coupled to large-volume injection-gas chromatography–mass spectrometric detection. *J. Chromatogr. A* 2002, *963* (1–2), 27–36.

151. Schellin, M.; Popp, P., Membrane-assisted solvent extraction of seven phenols combined with large volume injection-gas chromatography–mass spectrometric detection. *J. Chromatogr. A* 2005, *1072* (1), 37–43.

152. Grover, P. K.; Ryall, R. L., Critical appraisal of salting-out and its implications for chemical and biological sciences. *Chem. Rev.* 2005, *105* (1), 1–10.

153. López-Blanco, M. C.; Blanco-Cid, S.; Cancho-Grande, B.; Simal-Gándara, J., Application of single-drop microextraction and comparison with solid-phase microextraction and solid-phase extraction for the determination of α- and β-endosulfan in water samples by gas chromatography–electron-capture detection. *J. Chromatogr. A* 2003, *984* (2), 245–252.

154. Psillakis, E.; Kalogerakis, N., Application of solvent microextraction to the analysis of nitroaromatic explosives in water samples. *J. Chromatogr. A* 2001, *907* (1–2), 211–219.

155. Zhang, T.; Chen, X.; Liang, P.; Liu, C., Determination of phenolic compounds in wastewater by liquid-phase microextraction coupled with gas chromatography. *J. Chromatogr. Sci.* 2006, *44* (10), 619–624.

156. Zhao, R.; Chu, S.; Xu, X., Optimization of nonequilibrium liquid-phase microextraction for the determination of nitrobenzenes in aqueous samples by gas chromatography–electron capture detection. *Anal. Sci.* 2004, *20* (4), 663–666.

157. Schnobrich, C. R.; Jeannot, M. A., Steady-state kinetic model for headspace solvent microextraction. *J. Chromatogr. A* 2008, *1215* (1–2), 30–36.

158. Hyötyläinen, T.; Riekkola, M. L., Sorbent- and liquid-phase microextraction techniques and membrane-assisted extraction in combination with gas chromatographic analysis: a review. *Anal. Chim. Acta* 2008, *614* (1), 27–37.

159. de Jager, L.; Andrews, A. R. J., Preliminary studies of a fast screening method for cocaine and cocaine metabolites in urine using hollow fibre membrane solvent microextraction (HFMSME). *Analyst* 2001, *126* (8), 1298–1303.

160. Rasmussen, K. E.; Pedersen-Bjergaard, S., Developments in hollow fibre–based, liquid-phase microextraction. *Trends Anal. Chem.* 2004, *23* (1), 1–10.

161. Psillakis, E.; Kalogerakis, N., Developments in liquid-phase microextraction. *Trends Anal. Chem.* 2003, *22* (9), 565–574.

162. Pedersen-Bjergaard, S.; Rasmussen, K. E., Liquid-phase microextraction with porous hollow fibers, a miniaturized and highly flexible format for liquid–liquid extraction. *J. Chromatogr. A* 2008, *1184* (1–2), 132–142.

163. Jönsson, J. Å.; Mathiasson, L., Membrane extraction in analytical chemistry. *J. Sep. Sci.* 2001, *24* (7), 495–507.

164. Liang, P.; Xu, J.; Guo, L.; Song, F., Dynamic liquid-phase microextraction with HPLC for the determination of phoxim in water samples. *J. Sep. Sci.* 2006, *29* (3), 366–370.

165. Myung, S. W.; Yoon, S. H.; Kim, M., Analysis of benzene ethylamine derivatives in urine using the programmable dynamic liquid-phase microextraction (LPME) device. *Analyst* 2003, *128* (12), 1443–1446.

166. Wang, Y.; Kwok, Y. C.; He, Y.; Lee, H. K., Application of dynamic liquid-phase microextraction to the analysis of chlorobenzenes in water by using a conventional microsyringe. *Anal. Chem.* 1998, *70* (21), 4610–4614.

167. Xiao, Q.; Hu, B.; Yu, C.; Xia, L.; Jiang, Z., Optimization of a single-drop microextraction procedure for the determination of organophosphorus pesticides in water and fruit juice with gas chromatography–flame photometric detection. *Talanta* 2006, *69* (4), 848–855.

168. Xu, J.; Liang, P.; Zhang, T., Dynamic liquid-phase microextraction of three phthalate esters from water samples and determination by gas chromatography. *Anal. Chim. Acta* 2007, *597* (1), 1–5.

169. Chen, P. S.; Huang, S. D., Determination of ethoprop, diazinon, disulfoton and fenthion using dynamic hollow fiber–protected liquid-phase microextraction coupled with gas chromatography–mass spectrometry. *Talanta* 2006, *69* (3), 669–675.

170. Chiang, J. S.; Huang, S. D., Determination of haloethers in water with dynamic hollow fiber liquid-phase microextraction using GC-FID and GC-ECD. *Talanta* 2007, *71* (2), 882–886.

171. Hou, L.; Lee, H. K., Determination of pesticides in soil by liquid-phase microextraction and gas chromatography–mass spectrometry. *J. Chromatogr. A* 2004, *1038* (1–2), 37–42.

172. Hou, L.; Shen, G.; Lee, H. K., Automated hollow fiber–protected dynamic liquid-phase microextraction of pesticides for gas chromatography–mass spectrometric analysis. *J. Chromatogr. A* 2003, *985* (1–2), 107–116.

173. Huang, S. P.; Huang, S. D., Dynamic hollow fiber protected liquid phase microextraction and quantification using gas chromatography combined with electron capture detection of organochlorine pesticides in green tea leaves and ready-to-drink tea. *J. Chromatogr. A* 2006, *1135* (1), 6–11.

174. Huang, S. P.; Huang, S. D., Determination of organochlorine pesticides in water using solvent cooling assisted dynamic hollow-fiber-supported headspace liquid-phase microextraction. *J. Chromatogr. A* 2007, *1176* (1–2), 19–25.

175. Li, G.; Zhang, L.; Zhang, Z., Determination of polychlorinated biphenyls in water using dynamic hollow fiber liquid-phase microextraction and gas chromatography–mass spectrometry. *J. Chromatogr. A* 2008, *1204* (1), 119–122.

176. Pezo, D.; Salafranca, J.; Nerín, C., Development of an automatic multiple dynamic hollow fibre liquid-phase microextraction procedure for specific migration analysis of new active food packagings containing essential oils. *J. Chromatogr. A* 2007, *1174* (1–2), 85–94.

177. Zhao, L.; Lee, H. K., Liquid-phase microextraction combined with hollow fiber as a sample preparation technique prior to gas chromatography/mass spectrometry. *Anal. Chem.* 2002, *74* (11), 2486–2492.

178. Melwanki, M. B.; Fuh, M. R., Dispersive liquid–liquid microextraction combined with semi-automated in-syringe back extraction as a new approach for the sample preparation of ionizable organic compounds prior to liquid chromatography. *J. Chromatogr. A* 2008, *1198–1199*, 1–6.

179. Mohammadi, A.; Alizadeh, N., Automated dynamic headspace organic solvent film microextraction for benzene, toluene, ethylbenzene and xylene: renewable liquid film as a sampler by a programmable motor. *J. Chromatogr. A* 2006, *1107* (1–2), 19–28.

180. Saraji, M., Dynamic headspace liquid-phase microextraction of alcohols. *J. Chromatogr. A* 2005, *1062* (1), 15–21.

181. Shen, G.; Lee, H. K., Headspace liquid-phase microextraction of chlorobenzenes in soil with gas chromatography–electron capture detection. *Anal. Chem.* 2003, *75* (1), 98–103.

182. Jiang, X.; Basheer, C.; Zhang, J.; Lee, H. K., Dynamic hollow fiber–supported headspace liquid-phase microextraction. *J. Chromatogr. A* 2005, *1087* (1–2), 289–294.

183. Chen, C. C.; Melwanki, M. B.; Huang, S. D., Liquid–liquid–liquid microextraction with automated movement of the acceptor and the donor phase for the extraction of phenoxyacetic acids prior to liquid chromatography detection. *J. Chromatogr. A* 2006, *1104* (1–2), 33–39.

184. Hou, L.; Lee, H. K., Dynamic three-phase microextraction as a sample preparation technique prior to capillary electrophoresis. *Anal. Chem.* 2003, *75* (11), 2784–2789.

185. Jiang, X.; Oh, S. Y.; Lee, H. K., Dynamic liquid–liquid–liquid microextraction with automated movement of the acceptor phase. *Anal. Chem.* 2005, *77* (6), 1689–1695.

186. Lin, C. Y.; Huang, S. D., Application of liquid–liquid–liquid microextraction and high-performance liquid-chromatography for the determination of sulfonamides in water. *Anal. Chim. Acta* 2008, *612* (1), 37–43.

187. Keith, L. H.; Crummett, W.; Deegan, J.; Libby, R. A.; Taylor, J. K.; Wentler, G., Principles of environmental analysis. *Anal. Chem.* 1983, *55* (14), 2210–2218.

188. Loconto, P. R., *Trace Environmental Quantitative Analysis*: Principles, *Technique, and Applications*, Marcel Dekker, New York, 2001.

189. Xia, L.; Hu, B.; Wu, Y., Hollow fiber–based liquid–liquid–liquid microextraction combined with high-performance liquid chromatography for the speciation of organomercury *J. Chromatogr. A* 2007, *1173* (1–2), 44–51.

190. Basheer, C.; Alnedhary, A. A.; Rao, B. S. M.; Lee, H. K., Determination of organophosphorus pesticides in wastewater samples using binary-solvent liquid-phase microextraction and solid-phase microextraction: a comparative study. *Anal. Chim. Acta* 2007, *605* (2), 147–152.

191. Basheer, C.; Vetrichelvan, M.; Valiyaveettil, S.; Lee, H. K., On-site polymer-coated hollow fiber membrane microextraction and gas chromatography–mass spectrometry of polychlorinated biphenyls and polybrominated diphenyl ethers. *J. Chromatogr. A* 2007, *1139* (2), 157–164.

192. Buszewski, B.; Ligor, T., Single-drop extraction versus solid-phase microextraction for the analysis of VOCs in water. *LC·GC Eur.* 2002, *15* (2), 92–97.

193. Jiang, X.; Lee, H. K., Solvent bar microextraction. *Anal. Chem.* 2004, *76* (18), 5591–5596.

194. Lambropoulou, D. A.; Psillakis, E.; Albanis, T. A.; Kalogerakis, N., Single-drop microextraction for the analysis of organophosphorus insecticides in water. *Anal. Chim. Acta* 2004, *516* (1–2), 205–211.

195. Melwanki, M. B.; Hsu, W. H.; Huang, S. D., Determination of clenbuterol in urine using headspace solid phase microextraction or liquid–liquid–liquid microextraction. *Anal. Chim. Acta* 2005, *552* (1–2), 67–75.

196. Palit, M.; Pardasani, D.; Gupta, A. K.; Dubey, D. K., Application of single drop microextraction for analysis of chemical warfare agents and related compounds in water by gas chromatography/mass spectrometry. *Anal. Chem.* 2005, *77* (2), 711–717.

197. Psillakis, E.; Kalogerakis, N., Solid-phase microextraction versus single-drop microextraction for the analysis of nitroaromatic explosives in water samples. *J. Chromatogr. A* 2001, *938* (1–2), 113–120.

198. Psillakis, E.; Kalogerakis, N., Hollow-fibre liquid-phase microextraction of phthalate esters from water. *J. Chromatogr. A* 2003, *999* (1–2), 145–153.

199. Shen, G.; Lee, H. K., Hollow fiber–protected liquid-phase microextraction of triazine herbicides. *Anal. Chem.* 2002, *74* (3), 648–654.

200. Xiao, Q.; Yu, C.; Xing, J.; Hu, B., Comparison of headspace and direct single-drop microextraction and headspace solid-phase microextraction for the measurement of volatile sulfur compounds in beer and beverage by gas chromatography with flame photometric detection. *J. Chromatogr. A* 2006, *1125* (1), 133–137.

201. Ho, T. S.; Pedersen-Bjergaard, S.; Rasmussen, K. E., Recovery, enrichment and selectivity in liquid-phase microextraction: comparison with conventional liquid–liquid extraction *J. Chromatogr. A* 2002, *963* (1–2), 3–17.

202. López-Blanco, C.; Gómez-Álvarez, S.; Rey-Garrote, M.; Cancho-Grande, B.; Simal-Gándara, J., Determination of pesticides by solid phase extraction followed by gas chromatography with nitrogen–phosphorus detection in natural water and comparison with solvent drop microextraction. *Anal. Bioanal. Chem.* 2006, *384* (4), 1002–1006.

203. Zuin, V. G.; Schellin, M.; Montero, L.; Yariwake, J. H.; Augusto, F.; Popp, P., Comparison of stir bar sorptive extraction and membrane-assisted solvent extraction as enrichment techniques for the determination of pesticide and benzo[*a*]pyrene residues in Brazilian sugarcane juice. *J. Chromatogr. A* 2006, *1114* (2), 180–187.

CHAPTER 6

APPLICATIONS

6.1 INTRODUCTION

Ever since its inception in 1996,[1] applications of solvent microextraction (SME) have been growing dynamically. By the end of 2008 there were well over 600 relevant publications, which included not only the first direct immersion (DI) (single-drop) mode, but a number of important variants as well. Headspace (HS), dispersive (DLL), drop-to-drop (DD), hollow fiber–protected two-phase [HF(2)], membrane-assisted solvent extraction (MASE), hollow fiber–protected three-phase [HF(3)], solvent bar (SB), continuous flow (CF), directly suspended droplet (DSD), three-phase liquid–liquid–liquid (LLL), and electromembrane (EME) microextraction have been developed and are now in common use. This popularity is reflected by a large number of reviews which have been fully or partly devoted to this technique. They include general reviews on microextraction,[2–7] headspace solvent microextraction,[8–10] direct immersion,[4,10,11] and membrane-based microextraction.[2,12–20] In addition, several reviews have discussed applications of SME for the determination of various classes of analytes, such as pharmaceuticals and their residues,[21–26] traditional Chinese medicines,[27] volatile organic compounds,[28,29] pesticide residues,[30,31] and chlorophenols,[32] or for various matrices: air[28] and water,[28,29] foods,[30,33] vegetables,[34] marine environment,[35] and for analytical toxicology.[36,37]

Solvent Microextraction: Theory and Practice, By John M. Kokosa, Andrzej Przyjazny, and Michael A. Jeannot
Copyright © 2009 John Wiley & Sons, Inc.

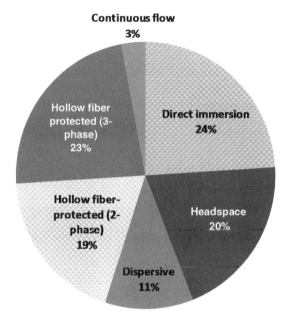

Continuous flow 3%

FIGURE 6.1. Frequency of use of various modes of solvent microextraction.

The frequency of use of various SME modes is almost evenly divided among direct immersion, headspace, and hollow fiber–protected two- and three-phase microextractions (see Figure 6.1 for details), but recently dispersive liquid–liquid microextraction and related techniques[38–70] (dispersive, temperature-controlled, ultrasound-assisted, cloud point, and homogeneous[71–91]) have experienced rapid growth as well, probably because of their speed and low detection limits.

So far, the majority of analytical applications of solvent microextraction involve environmental analysis (60%), followed by clinical and forensic, food and beverages, plant material, consumer products and pharmaceuticals, and physicochemical uses (Figure 6.2).

Solvent microextraction has been used for the isolation and enrichment of analytes from gaseous, liquid, and solid samples. By far the most common sample matrix analyzed is liquid: almost 90% of all SME applications described in the literature involve liquid samples, predominantly aqueous. Solid samples constitute about 10% and gaseous matrices less than 1% of developed SME procedures.

In the following sections, some representative examples of SME applications in analytical procedures for various matrices are given. The examples provided are not intended as a comprehensive list of all known applications of solvent microextraction. A complete compilation of all SME applications described in the literature as of December 2008, including matrix type, extraction conditions, final determination method, analytical characteristics, and references, is provided as Excel worksheets on the CD-ROM included with this book.

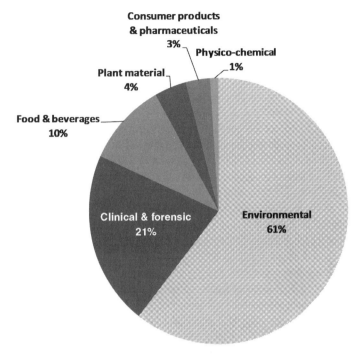

FIGURE 6.2. Applications of solvent microextraction.

6.2 GASEOUS SAMPLES

Despite the fact that gases have the simplest and cleanest matrices of all possible samples, only four papers have been published thus far using SME for gaseous matrices.[92–95] The analysis of air, the most common gaseous matrix investigated, presents unique problems. The most serious problem results from very low levels of analytes, sometimes present at concentrations down to $10^{-10}\%$. As a result, virtually all analytical procedures for the determination of trace contaminants in indoor or ambient air call for an analyte isolation/enrichment step. In some procedures, grab samples of air are collected in Tedlar gas sampling bags or Summa canisters, followed by either cryogenic concentration, or adsorption of the analytes on a solid sorbent bed, followed by solvent or thermal desorption and GC final determination. Alternatively, active or passive samplers can be used to isolate and enrich the analytes during field sampling, followed by thermal or solvent desorption and analysis proper. Theoretically, a solvent microextraction device, such as a microsyringe, can be used in place of an active or passive sampler. This approach would provide the same advantages that all solvent microextraction techniques offer: integration of the first several steps of the analytical process: sampling, analyte(s) isolation, and their enrichment. Of all solvent microextraction modes, only headspace SME is suitable for this operation, either in the single-drop mode or the hollow fiber–protected (two-phase) mode. In both cases, an organic solvent with a high boiling point and low vapor pressure should be selected to prevent solvent losses during the extraction process. If the single-drop mode is used, a microdrop of organic solvent is extruded from the tip of the syringe needle and exposed to a gaseous sample for a prescribed period of time. Next, the microdrop is retracted into the syringe barrel and injected into a gas chromatograph. This approach was used for the determination of nicotine in mainstream tobacco smoke[95] (see Figure 6.3). This mode suffers from three drawbacks: (1) the area of contact between the solvent and the sample is relatively small, thus lowering the amount of analyte transported across the phase boundary per unit time; (2) the volume of the microdrop is limited to no more than 5 μL; and (3) the drop can be dislodged easily, necessitating a repeat of the sampling process. The last drawback can be eliminated by using a supported-drop design,[92] in which a PTFE structure containing a stainless steel net is attached to the needle tip of a microsyringe, thus stabilizing the drop. The technique was used to determine diisocyanates in air.

The second mode, hollow fiber–protected headspace SME, overcomes the problems mentioned above, enables the use of larger solvent volumes (recall that the amount of analyte extracted is directly proportional to solvent volume!) and also protects the solvent from particulate matter that can be present in heavily polluted air. In this approach, a length of microporous polypropylene hollow fiber, typically Accurel Q 3/2, impregnated with an organic solvent is fitted onto the needle of a

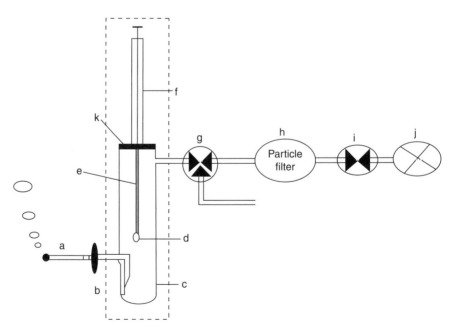

FIGURE 6.3. The new sampling system for cigarette smoke: (a) cigarette sample; (b) Cambridge filter holder; (c) gathering vial for cigarette smoke; (d) solvent drop for LPME; (e) needle; (f) microsyringe; (g) switch valve; (h) particle filter; (i) flow control valve; (j) sampling pump; (k) PTFE–silicone septum. (Reprinted with permission from Ref. 95; copyright © 2006 Wiley-Interscience.)

microsyringe filled with a volume of the same organic solvent. The solvent is extruded from the syringe barrel into the lumen of the fiber. The fiber with the solvent (still attached to the microsyringe) is then exposed to the gaseous sample for a definite time. Use of the fiber increases the interfacial area, resulting in faster transport of the analytes into the solvent. It also prevents the solvent from being dislodged from the tip of the syringe. At the end of the extraction period, the solvent with the extracted analytes is withdrawn from the fiber lumen into the microsyringe and analyzed. A modification of this approach was used to assess air quality inside vehicles and at filling stations by monitoring benzene, toluene, ethylbenzene, and xylenes by means of semipermeable devices.[94] In this procedure, a length of flat polyethylene tubing was filled with 100 μL of triolein and both ends of the tubing were then heat-sealed. In this way, a semipermeable membrane device (passive sampler) was formed. Next, the device was exposed to the air containing aromatic hydrocarbons. For the final determination, the sampler was placed in a headspace vial, heated to 150°C for 20 min, and 2 mL of the headspace from the vial analyzed by gas chromatography–mass spectrometry (GC-MS). Thus, the procedure combined two sample preparation techniques: headspace solvent microextraction and headspace analysis.

The determination of nitrophenols in air is another example of an approach combining two sampling and sample preparation techniques.[93] In the first step of this procedure, nitrophenols are isolated from air by impinger sampling and collected in a 0.01 M NaOH solution (active sampling). The pH of the sodium hydroxide solution is then lowered and the analytes extracted using HF(3)ME with a NaOH solution as the acceptor. The method detection limits range between 0.5 and 1.0 ng/L (3.1 to 46.7 ppb by volume in air) using HPLC-UV for the final determination. The procedure is inexpensive and does not require derivatization of nitrophenols. Since the first step involves exhaustive extraction, it eliminates the second major problem with trace component determination in gaseous samples using solvent microextraction (i.e., the effect of temperature and humidity on the air–organic solvent distribution constant). This dependence makes quantitative analysis quite challenging. All quantitative analysis methods are based on the assumption of direct proportionality between the analyte concentration in a sample and its concentration in the extract, where the proportionality constant is the distribution ratio. To obtain accurate results, the temperature and humidity conditions for standards must be the same as those for samples, which requires careful temperature and humidity measurements during solvent microextraction and reproduction of those conditions in the laboratory for standards.

The extraction of trace chemicals from air is rapid, owing to diffusion coefficients in gases typically being about four orders of magnitude larger than those in liquids. Consequently, extraction times are short, from 1 min to several minutes, which is a definite advantage. For example, in the determination of nicotine in mainstream smoke using HS-SDME, the extraction time is only 4 min.[95]

If a solvent microextraction device (microsyringe) is used for field sampling and needs to be transported to the laboratory to carry out the final determination, the extract has to be protected from analyte loss. The loss can be minimized by drawing the extract into the syringe barrel and plugging the needle with a GC septum. This approach will minimize analyte losses but will not eliminate them entirely. The major source of those losses is sorption of analytes in the septum, which is made of silicone rubber. For this reason, the analysis should be performed as soon as possible. Lowering the temperature of extract storage by keeping the sampling device in a container filled with dry ice is also helpful in minimizing analyte losses.

6.3 LIQUID SAMPLES

Almost all applications of solvent microextraction for liquid matrices described in the literature have been for aqueous samples. The only exceptions were vodkas and spirits,[96] edible oils,[97] personal care products (PCPs) (cologne, shaving lotion),[98] and engine oil.[99] Since all modes of solvent microextraction have been used for sample preparation of liquid matrices, the main problem the analyst faces is which technique should be used for a particular sample. The selection guidelines are discussed in detail in Chapters 4 and 5, and the following sections of Chapter 6 provide tables, which list SME modes that have been used for different classes of analytes in various types of matrices. Briefly, the selection criteria include matrix type, physical and chemical properties of analytes, amount of sample available, accuracy and precision required, concentration levels of analytes, number of steps in the sample preparation procedure, sample throughput, and whether the mode of microextraction lends itself to automation. Considering the fact that the main advantages of solvent microextraction are its simplicity and low cost, the preferred modes are single drop techniques, either headspace or direct immersion. Both require no additional equipment other than a microsyringe, a stirrer and (possibly) temperature control, and both can be readily automated.[100]

We recommend beginning with headspace solvent microextraction, since it can be used for any matrix, gaseous, liquid or solid, even complex and dirty, as long as the analytes are sufficiently volatile. For relatively clean aqueous matrices, free of suspended matter, and analytes with low to moderate polarity and volatility, direct immersion (single drop) is a good starting point. Analytes with low volatility or high polarity present in complex liquid samples, such as environmental water containing suspended solids, fruit juices, or physiological fluids, call for membrane-assisted microextraction. The HF(2)ME or membrane-assisted solvent extraction are best suited for analytes with medium and low polarity, while the use of HF(3) ME is appropriate for analytes of moderate or high polarity. For highly polar, hydrophilic analytes carrier-mediated HF(3)ME is recommended. In either case, some preliminary sample preparation steps, such as precipitation, filtration, or centrifugation, may be required prior to the microextraction.

Owing to a very high interfacial area between an organic solvent and a sample, rapid extraction of analytes from relatively clean aqueous matrices can be accomplished by dispersive liquid–liquid microextraction. In this mode, the extraction time can be as short as 20 s,[64] but typically it ranges from several minutes to a maximum of 30 min. However, to separate the donor (sample) and acceptor (extracting solvent), the mixture has to be centrifuged, which takes from 2 to 20 min.

In general, during solvent microextraction, liquid samples have to be agitated, either through vibration or magnetic stirring, to ensure efficient mass transfer to the extracting solvent. Two major advantages of hollow fiber–protected two-phase SEM over the direct-immersion (single-drop) mode are that in the former technique samples can be stirred or vibrated vigorously without any loss of the extracting solvent, and the interfacial area is expanded significantly compared to the drop.

In static solvent microextraction, the solvent is stagnant, whereas in a dynamic mode both a small volume of sample and the acceptor phase are in motion, the latter forming a thin film of organic solvent. The dynamic mode results in improvements in kinetics, reducing the extraction time (see Chapter 3 for details), but it does not improve the extraction efficiency. In both extraction modes, the amount of analytes extracted remains the same provided that the equilibrium between the two phases (donor and acceptor) has been reached. At the same time, the dynamic mode requires additional equipment, such a syringe pump[56,101–117] a variable-speed stirrer motor,[118–120] or an autosampler, which makes the technique more expensive, although some authors have performed dynamic SME manually.[121–125]

Aqueous matrices for which solvent microextraction has been applied vary greatly. Some are relatively clean and simple (e.g., tap, bottled, well, or pool water). They require little if any preliminary steps prior to extraction, and most often quantitative analysis can be performed by a calibration curve method (external calibration) for both direct and headspace modes. External calibration is useful if there is little variability in sample composition, or where matrix composition of samples is normalized prior to extraction. Other aqueous matrices, such as wastewater or biological samples, can be very complex or even multiphase (e.g., whole blood,[25,126–129] blood plasma,[25,50,65,123,127,129–165] blood serum,[151,159,166–172] or urine.[105,119,123,128,130,138–141,144,149,154,158,159,167,169,173–201]) These matrices may require a preliminary cleanup prior to the microextraction proper even if the headspace or hollow fiber–protected modes are to be used. Typically, these preliminary steps may include one or more of the following: filtration, protein precipitation, centrifugation, the release of bound analytes (e.g., drugs bound to proteins), and pH adjustment. For these matrices, especially those containing solids, multiple equilibria can exist during the analyte extraction (e.g., analyte–solid phase, analyte–aqueous phase, analyte–extracting solvent). Consequently, the best quantitative method in these cases is the standard addition method, either single-point or multipoint, which compensates for the matrix effect. However, even the standard addition method does not guarantee accurate results, because the added standard can be distributed among the phases in a different way than the analyte already present in the sample. Therefore, all newly developed procedures for complex matrices involving solvent microextraction have to be validated. The validation can be done by analyzing standard or certified reference materials[51,150,151,170,202–205] having matrices similar to those in intended applications, or by performing parallel analyses of the same sample using two methods: the one under development and a standard method known to be accurate. The results are then evaluated statistically (e.g., by Student's t-test) to see if the concentrations are significantly different from the values certified or from those obtained by the standard method.

In most solvent microextraction procedures, an internal standard is added to compensate for variability in extraction and/or analysis conditions. The standard is added either to the sample or to the acceptor phase. Preferably, the standard should be added to samples, because it can then compensate for the variability in both extraction conditions (stirring rate, temperature, ionic strength, pH, organic solvent volume, extraction time) and instrumental analysis (detector signal). If the standard is added to the organic solvent, it can only account for the variations in solvent volume and instrumental conditions.

If mass spectrometry is used as the final determination method, accurate results can be obtained by isotopic dilution,[202,206,207] in which the standard added is isotopically labeled, usually deuterated, or labeled with [13]C. Since both the analyte and its labeled analog are extracted at the same time and under identical conditions, many sources of error are eliminated. The drawbacks of this method are the limited availability and high cost of the labeled compounds.

6.4 SOLID SAMPLES

About 10% of all applications of solvent microextraction described in the literature involved solid samples. They ranged from environmental matrices, such as soil,[64,90,107,108,111,122,171,194,208–217] sediments,[202,212,218,219] dust,[124,171,220,221] or algae[222–224] to hair,[225] plant parts,[95,109,226–244] fruits and vegetables,[67,91,245–251] milk powder,[248] and polystyrene.[252]

Solid matrices cannot be treated directly by SME; thus, the analytes have to be removed from the matrix prior to the extraction proper. One approach commonly used for volatile analytes is to release them into the gaseous phase by heating solid samples to temperatures ranging from 40°C for soil[111] to 150°C for polystyrene,[252] followed by headspace solvent microextraction. Heating samples ensures quantitative release of volatile analytes and also faster kinetics of the process. On the other hand, higher sampling temperatures decrease the organic solvent–air distribution constant, resulting in lower sensitivity of the determination. This loss of sensitivity can be avoided if the extracting solvent is cooled while the sample is heated (Figure 6.4). In this approach, there are two independently operated temperature-controlling baths: one for heating the sample and the other for cooling the syringe with the solvent. In practice, solvents have been cooled to temperatures ranging from −6 to 4°C.[110,226,253–260] It should be kept in mind, however, that this approach complicates the extraction apparatus and may preclude full automation of the entire analytical procedure.

Another approach used to release analytes from solid matrices before they are isolated and enriched by SME involves extraction with a polar organic solvent, water, or a mixture of both. Some preliminary steps, including drying, homogenization, and/or screening, may have to be undertaken prior to the extraction. A novel technique, called *hydrodistillation–headspace solvent* microextraction (HD-HS-SDME), used primarily to isolate essential oils from aromatic plants and their parts, couples water extraction with solvent microextraction[233,235,236,241,242]

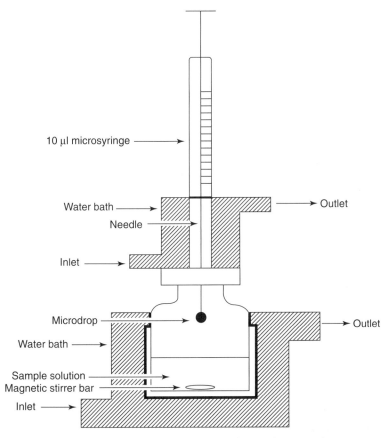

FIGURE 6.4. Schematic diagram of the headspace solvent microextraction apparatus with solvent cooling. (Reprinted with permission from Ref. 258; copyright © 2004 Elsevier.)

(Figure 6.5). In this technique, a small amount of plant material, typically from 0.7 to 4 g, is mixed with water and subjected to hydrodistillation. A microdrop of a high-boiling-point solvent is suspended in the headspace of a hydrodistilling sample. This arrangement results in a short extraction time, usually between 10 and 20 min (including refluxing), and consumes a small amount of plant material. In all other approaches, water or polar solvent extraction precedes SME.

Pure water is used to extract analytes from soil,[64,108,194,208,216] sediment,[218] and vegetables.[249] In some cases, the pH of a sample is adjusted by adding a buffer[225] or a NaOH solution,[216] or concentrated acetic acid is used as the extracting solvent.[202] The extraction time can be reduced significantly (to 5 min) by using pressurized hot water extraction (PHWE), a dynamic sample extraction technique, at temperatures between 100 and 374°C and pressures high enough to maintain the liquid state, followed by SME[231] (Figure 6.6). The polar organic solvents used for extraction of analytes from solid matrices include acetone,[90,251] methanol,[171] and a mixture of acetonitrile and methanol.[219] When polar solvents or their mixtures are used, the

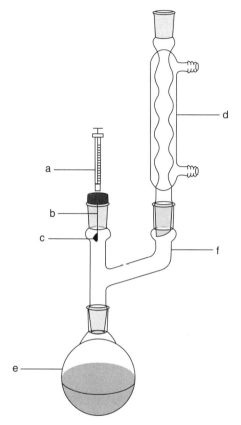

FIGURE 6.5. HD-HSME apparatus. a, syringe; b, needle; c, microdrop; d, condenser; e, round-bottomed flask; f, Claisen distillation head. (Reprinted with permission from Ref. 242; copyright © 2007 Wiley-Interscience.)

extracts have to be diluted with water prior to solvent microextraction. Finally, some procedures of analyte extraction from solid samples make use of mixtures of water and a polar solvent (e.g., acetone–water,[107,210,217] or methanol–water[250]).

Larger amounts of extracted analytes and faster kinetics of the process can be achieved by using sonication[64,171,216,251] or microwave-assisted extraction (MAE).[27,202,218] The use of microwave or ultrasound energy for heating the solution results in a significant reduction in the extraction time (usually to less than 30 min). In addition to having the advantage of a high extraction speed, MAE also enables a significant reduction in the consumption of organic solvent (typically less than 40 mL).

6.5 ENVIRONMENTAL APPLICATIONS OF SME

The majority (60%) of all published analytical applications of solvent microextraction have dealt with environmental samples, including air, water, soil and dust,

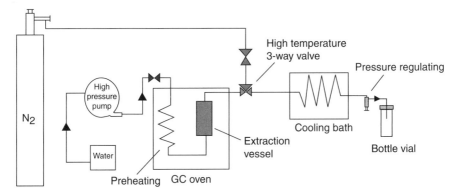

FIGURE 6.6. Typical equipment used for pressurized hot-water extraction. (Reprinted with permission from Ref. 231; copyright © 2005 Elsevier.)

sediments, and some biological materials, such as algae, fish, or pine needles. These applications are summarized in Table 6.1, in which the most commonly determined analytes and modes of their extraction are shown in bold face type. The table can be used as a quick guide to select the mode of microextraction for a specific environmental analysis. The fraction of environmental applications varies with the SME mode, ranging from 82% for dispersive liquid–liquid microextraction to 81% for continuous/cycle flow, 70% for hollow fiber–protected two-phase, 65% for direct immersion, 50% for headspace, and 48% for hollow fiber–protected three-phase. Table 6.1 reveals that the majority of SME applications have been developed for aqueous matrices. It is interesting to compare the detection limits for priority pollutants in aqueous samples obtained by optimized solvent microextraction procedures with the Environmental Protection Agency method requirements (method detection limits)[261] (Table 6.2). It is evident that the detection limits of the SME procedures published in the literature exceed the EPA method requirements, and hence the technique is well suited for environmental assays. The use of solvent microextraction for important groups of environmental pollutants is discussed below.

6.5.1 Volatile Hydrocarbons

As pollutants, alkanes are not considered as hazardous as aromatic hydrocarbons; therefore, very few papers describing their determination using SME have been published.[260,262] In contrast, the presence of aromatic hydrocarbons and BTEX (benzene, toluene, ethylbenzene, xylenes) compounds in particular is an indicator of environmental pollution by petroleum-based products. Consequently, the determination of BTEX in environmental samples ranks among the most common analytical tasks. Since BTEX compounds are volatile, the most frequent SME mode for their determination is headspace,[94,118,263–267] although direct immersion,[265] and hollow fiber–protected two-phase mode[268] have also been used. Validation of the procedures developed is carried out by determining relative recoveries from aqueous samples spiked with BTEX compounds. Since SME in most cases is a nonexhaustive extraction procedure, the relative recovery, defined as the ratio of the

TABLE 6.1 **Applications of SME in Environmental Analysis**

Matrix	Analytes[a]	SME Mode[a,b]
Drinking water (tap, bottled, well)	**Pesticides (organophosphorus**, organochlorine, triazine, pyrethroids, carbamates, chloroacetanilide, organosulfur)	CF, **DLL**, **SD**, HS, HF(2)
	Polycyclic aromatic hydrocarbons	CF, DLL, SD, HS, HF(2)
	Volatile halocarbons	CF, DLL, **HS**, **SD**, HF(2)
	Metal ions (derivatized)	CF, **DLL**, SD, HF(2)
	Phthalate esters	**DLL**, **SD**, HF(3)
	Phenols (alkyl, chloro)	SD, HS, HF(2), **HF(3)**
	Amines (**aromatic**, aliphatic, chlorinated anilines)	CF, HS, SD, HF(3)
	Chlorobenzenes	DLL, SD, **HS**
	Nitro compounds (nitrobenzenes, nitroaromatic explosives)	**SD**, HF(2)
	Hydrocarbons (alkanes, aromatic)	**IIS**
	Drugs (antidepressant, antipsychotic, NSAIDs)	SD, **HF(3)**
	Haloacetic acids	HS, HF(2), HF(3)
	Metalloids (selenium, arsenic)	DLL, HS
	Organometallic compounds	HS
	Polychlorinated biphenyls	HF(2)
	Polybrominated diphenyl ethers	DLL, SD, HF(2)
	Antimicrobial and antibacterial agent	HF(2)
	Antioxidants	DLL
	Antifouling agents	SD
	Methyl *tert*-butyl ether	HS
Fresh water (ground, surface, precipitation)	**Pesticides (organophosphorus, organochlorine, triazine,** pyrethroids, carbamates, chloroacetanilide, acidic herbicides,	CF, **DLL**, **SD**, HS, **HF(2)**, **HF(3)**
	Metal ions (derivatized)	CF, **DLL**, **SD**, HF(2), HF(3)
	Polycyclic aromatic hydrocarbons	DLL, **SD**, HS, **HF(2)**, HF(3)
	Phenols (alkyl, chloro)	DLL, HS, **SD**, HF(2), **HF(3)**
	Chlorobenzenes	DLL, SD, HS, HF(2)
	Aromatic compounds	SD, **HS**, HF(2), HF(3)
	Amines (aromatic, aliphatic)	SD, HS, HF(2), **HF(3)**
	Drugs (antidepressant, decongestant, antibiotic, **basic**, acidic, NSAIDs)	DLL, SD, HF(2), **HF(3)**
	Carbonyl compounds (aldehydes, ketones)	SD, HS
	Metalloids (selenium, arsenic, antimony)	CF, DLL, SD, HS, HF(2)
	Phthalate esters	DLL, SD

TABLE 6.1 *Continued*

Matrix	Analytes[a]	SME Mode[a,b]
	Polybrominated diphenyl ethers	DLL, SD, HF(2)
	Organohalogen compounds	SD, HS, HF(2)
	Cyanide	HS
	Nitroaromatic compounds (nitrobenzenes, explosives)	SD, HF(2)
	Hydrocarbons	SD, HS
	Surfactants	SD, HF(2)
	Chemical warfare agents	SD, HF(2)
	Esters	SD
	Alcohols	SD, HS
	Methyl *tert*-butyl ether	HS
	Organometallic compounds	DLL, SD, HF(3)
	Polychlorinated biphenyls	HF(2)
	Periodate	SD, HS
	Antifouling agents	SD
	Antimicrobial and antibacterial agent	HF(2), HF(3)
	Hydroxycarbonyls	HS
	Hydrophobic biomolecules	SD
	Alkylphosphonic acids	HF(2)
Seawater	**Pesticides** (organophosphorus, organochlorine)	DLL, SD, **HF(2)**
	Metal ions (derivatized)	CF, **DLL**, HF(2)
	Aromatic amines	SD, **HF(3)**
	Metalloids (selenium, arsenic, antimony)	CF, SD, HS
	Phthalate esters	DLL
	Polycyclic aromatic hydrocarbons	SD, HF(2)
	Polychlorinated biphenyls	HF(2)
	Polybrominated diphenyl ethers	HF(2)
	Aromatic compounds	CF, HS
	Organometallic compounds	DLL, SD, HF(3)
	Ammonia	HS
	Phenols	HF(2), HF(3)
	Halogen-containing anions	SD
	Antifouling agents	SD
Wastewater (sewage, drain, industrial effluent, landfill leachate)	**Pesticides** (organophosphorus, organochlorine)	DLL, SD, HS, **HF(2)**, HF(3)
	Phenols (alkyl, chloro)	CF, **SD**, **HF(3)**
	Amines (aromatic, aliphatic)	SD, HS, **HF(3)**
	Drugs (antidepressant, acidic, NSAIDs, antibacterial)	**HF(3)**
	Hydrocarbons (alkanes, aromatic)	**HS**, HF(2)

(Continued)

TABLE 6.1 *Continued*

Matrix	Analytes[a]	SME Mode[a,b]
	Metal ions (derivatized)	SD, HS, HF(3)
	Chlorobenzenes	SD, HS
	Phthalate esters	DLL, SD
	Polycyclic aromatic hydrocarbons	DLL, HS, HF(2)
	Polybrominated diphenyl ethers	DLL, HF(2)
	Aldehydes	HS
	Organometallic compounds	HS
	Cyanide	HS
	Long-chain fatty acids	HF(2)
	Degradation products of chemical warfare agents	HF(2)
Soil and dust	**Polycyclic aromatic hydrocarbons**	SD, **HF(2)**
	Pesticides	DLL, SD, HF(2)
	Chlorobenzenes	SD, HF(2)
	Polybrominated diphenyl ethers	HF(2)
	Antibiotic	SD
Sediments	Pesticides	HF(2)
	Polychlorinated biphenyls	HF(2)
	Polycyclic aromatic hydrocarbons	HF(2)
Biological material	Aromatic amines	CF, SD
(algae, fish, pine	Polycyclic aromatic hydrocarbons	HS, HF(2)
needles)	Organometallic compounds	HS, HF(3)
	Metal ions (derivatized)	SD
Air, tobacco smoke	Nicotine	HS
	Aromatic hydrocarbons	HS
	Diisocyanates	HS
	Nitrophenols	HF(3)

[a] Most common analytes and modes are shown in boldface type.
[b] SD, single drop; HS, headspace; CF, continuous flow; DLL, dispersive liquid–liquid; HF(2), hollow fiber–protected two-phase; HF(3), hollow fiber–protected three-phase.

concentrations found in a matrix to those in distilled or deionized water spiked with the same amounts of analytes, is generally used instead of absolute recovery. Recovery values close to 100% indicate a lack of matrix effect and the good accuracy of the technique. An example of BTEX determination in the aqueous matrix taking advantage of ionic liquid–based headspace SME[263] is shown in Figure 6.7. A paper by Ouyang et al.[100] compares the performance of fully automated static and dynamic HS-SDME, SDME, and HF(2) modes of solvent microextraction as applied to BTEX determination. The authors conclude that the exposed dynamic headspace mode provides the best performance. BTEX compounds were also determined in air inside vehicles and filling stations.[94] A detailed procedure for the determination of BTEX compounds in water by headspace and SDME is provided in Chapter 7.

TABLE 6.2   Comparison of Limits of Detection (LOD) for Optimized Solvent Micro-extraction Procedures with the U.S. EPA Method Detection Limits (MDL)

Analyte	EPA MDL (ng/L)	LOD for SME (ng/L)
BTEX (benzene, toluene, ethylbenzene, xylenes)	30–90	4.1–23.5
Polycyclic aromatic hydrocarbons	40	5–11
Phenols	1500–42,000	5–15
Polychlorinated biphenyls	60–100	2–10
Polychlorinated hydrocarbon solvents	10–100	5–40
Phthalate esters	22–640	2–130

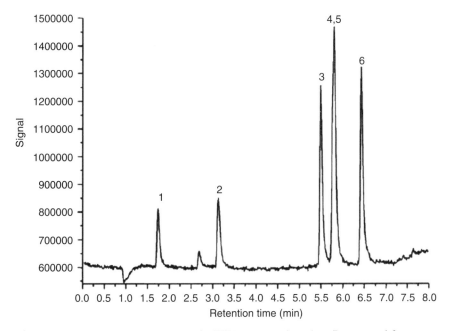

FIGURE 6.7. GC-MS chromatogram of BTEX compounds at 1 µg/L extracted from water using ionic liquid-based headspace solvent microextraction: (1) benzene; (2) toluene; (3) ethylbenzene; (4) *m*-xylene; (5) *p*-xylene; (6) *o*-xylene. (Reprinted with permission from Ref. 263; copyright © 2008 Elsevier.)

6.5.2   Volatile Halocarbons

Volatile halocarbons are ubiquitous in the water environment. Water disinfection by chlorination yields halogenated by-products, mostly trihalomethanes (THMs): chloroform, dichlorobromomethane, chlorodibromomethane, and bromoform, but also haloacetic acids and other compounds. Trichloroethane, trichloroethene, and tetrachloroethene are related to water pollution by industrial discharge. Since some of these compounds are considered to be carcinogens, they have been listed as priority pollutants in many countries. Owing to their volatility, the most common microextraction mode used for volatile halocarbons is headspace,[262,269,270] although the direct-immersion single-drop mode[271] and continuous flow[272] have also been employed. Finally, membrane-assisted solvent extraction (MASE), a membrane extraction technique somewhat similar to hollow fiber–protected SME, has been used successfully for the isolation and enrichment of volatile halocarbons.[273] In this approach, a headspace vial is filled with an aqueous sample. A nonporous polypropylene membrane bag is filled with 100 μL of an organic solvent and placed inside the vial. Since the MASE membranes are nonporous, MASE is essentially a three-phase system in which the membrane acts as the real interface.[274] To speed up the mass transfer kinetics, the membrane thickness is reduced to 30 μm and the extraction temperature is increased to about 40°C. The device is agitated to facilitate extraction of the analytes into the bag. Following the extraction, 1 μL of the extract is analyzed by gas chromatography with electron capture detection (GC-ECD) (Figure 6.8). The electron capture detector is used preferentially for halocarbon analyses, due to its selectivity and high sensitivity. The procedure was validated by comparing the results with those obtained by EPA method 524.2, which employs the purge-and-trap technique, and a good agreement between the two methods was found. The detection limits ranged between 5 ng/L for tetrachloroethene and 50 ng/L for chloroform, and the linear dynamic range extended from 5 ng/L to 150 μg/L. Trihalomethanes were determined by several authors by means of SME followed by GC-ECD determination. Some procedures made use of HS-SDME[259,275]; others employed SDME[271,276,277] or HF(2)ME[278] (Figure 6.9). A recent paper described the determination of THMs in drinking water by dispersive liquid–liquid microextraction, followed by GC-ECD determination.[52] The procedure developed has a number of advantages, including small sample volume (5 mL), a very short extraction time (less than 2 min), a wide linear dynamic range extending over three orders of magnitude (from 0.01 to 50 μg/L), very low detection limits (from 5 to 40 ng/L), and simplicity (only a centrifuge is required).

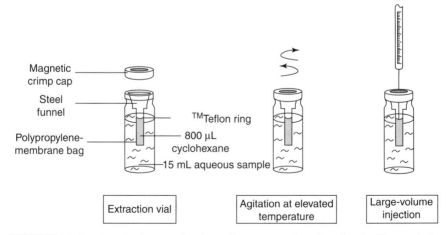

FIGURE 6.8. Device of solvent-assisted membrane extraction. (Reprinted with permission from Ref. 309; copyright © 2003 Elsevier.)

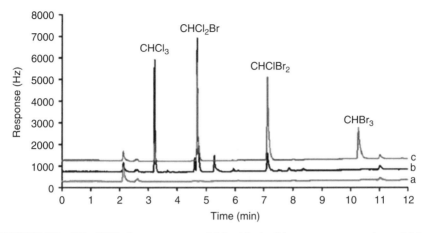

FIGURE 6.9. GC-μECD chromatograms of (a) a blank, (b) a tap water sample, and (c) a spiked water sample containing THMs at 10 μg/L using hollow fiber–supported two-phase microextraction. (Reprinted with permission from Ref. 278; copyright © 2006 Elsevier.)

A detailed procedure for the determination of THMs in drinking water by headspace SME is described in Chapter 7. Other water disinfection by-products, haloacetic acids, were extracted from water samples by using the HS-HF(2) mode following esterification with methanol.[279]

6.5.3 Volatile Polar Solvents

The volatile polar solvents include alcohols, aldehydes, ketones, and esters. Most of these compounds are at least somewhat soluble in water. Some are even miscible with water in every ratio. Classical liquid–liquid extraction provides poor extraction yields, owing to the unfavorable distribution constant of the analytes between the nonpolar organic solvent and water. Solvent microextraction can be used successfully to isolate and enrich such polar volatile compounds. Three strategies are possible and all have been used in practice. The first involves derivatization of the analytes into compounds with improved extractability due to their lower polarity. The derivatization is typically carried out either in the sample prior to the extraction proper (pre-extraction derivatization) or concurrently with it, or in the extracting solvent present in the drop or in the fiber concurrently with the extraction or following the extraction (postextraction derivatization) (see Chapter 5 or a comprehensive review of micro-extraction combined with derivatization[280]). It is important to realize that pre-extraction and concurrent extraction–derivatization can improve the sensitivity (enrichment factor) and selectivity of both the extraction step and the final determination method (detector response and resolution), while the postextraction derivatization can only improve the selectivity and sensitivity of the final determination.[261] Headspace single-drop microextraction with concurrent in-drop derivatization has been used to determine aldehydes in water. In one procedure,[281] five aldehydes, ranging from acetaldehyde to heptanal, were extracted from water samples into decane and simultaneously derivatized with O-2,3,4,5,6-(pentafluorobenzyl) hydroxylamine hydrochloride (PFBHA) dissolved in the extractant to yield oximes. The oximes were then determined by GC-MS. The limits of detection ranged between 0.08 and 0.32 µg/L, and the extraction time was only 6 min. In the second procedure,[282] aqueous solutions containing formaldehyde or hexanal were subjected to headspace SME using 2,4,6-trichlorophenylhydrazine as the derivatizing agent dissolved in 1-octanol (the extractant). The phenylhydrazones formed were determined by GC-MS. The detection limit for the two analytes was 3 µg/L. Both procedures were simple, highly sensitive, selective, rapid, convenient, and inexpensive. As we shall see later, derivatization of aldehydes coupled to microextraction is not limited to environmental samples, but is also common in clinical and forensic applications of SME. A detailed procedure for the determination of acetone using derivatization–microextraction is described in Chapter 7.

Hydroxyketones were determined in rainwater by a coupled two-step derivatization–microextraction procedure followed by GC-MS determination[102] (Figure 6.10). In the first step, hydroxyketones were derivatized in a sample with PFBHA and extracted by dynamic hollow fiber–protected SME using *o*-xylene as the extracting solvent. Next, the extract containing the oximes of the analytes was exposed in the headspace single-drop mode to the second derivatizing agent: bis(trimethylsilyl)trifluoroacetamide, which silylated the hydroxyl groups of the analytes. Although the complete procedure was relatively long (total extraction time for the two steps was over 60 min), it provided a wide linear dynamic range, typically three or more orders of magnitude, and low detection limits: from 0.023 to 4.75 µg/L.

The second strategy, which is more suitable for organic volatiles that are somewhat less polar (e.g., esters) involves the use of single-drop microextraction, either in headspace or direct-immersion mode, with a solvent that has some polar character and very low water solubility. Toluene is used most often in the direct-immersion mode, but other aromatic compounds have also been employed.

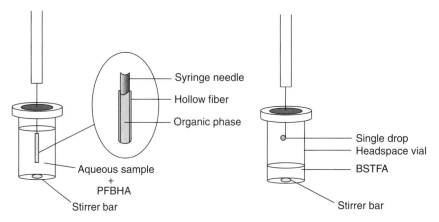

FIGURE 6.10. Devices of coupled two-step derivatization–microextraction of hydroxyketones. (Reprinted with permission from Ref. 102; copyright © 2006 Elsevier.)

An example of this approach is the determination of esters in water by HS-SDME with *p*-cymene as the extracting solvent, followed by gas chromatography with flame ionization detection (GC-FID).[283]

The third strategy makes use of a very polar solvent as the extractant in microextraction. In one approach, water is used as the extractant for HS-SDME. This approach, combined with derivatization, is quite common for the determination of some inorganic species and is discussed later in this section. However, it has also been used for the isolation of organic compounds (e.g., alcohols) from aqueous matrices.[284] It should be apparent that this procedure does not result in analyte enrichment, because the donor and acceptor solvent is the same (water), and hence the partition coefficient of the analytes between the donor and acceptor phase at equilibrium is close to 1. However, substantial sample cleanup can be achieved and the procedure completely eliminates the use of organic solvents. It can be used successfully in the determination of relatively high concentrations of polar volatile compounds in complex, dirty aqueous matrices, such as municipal wastewater or industrial effluent. In a modification of this approach termed *headspace water-based SME*,[285] some property of the aqueous donor and acceptor phase is modified to shift the equilibrium between the two phases to enhance the transfer of analytes from the sample to the extractant solution and thus improve the enrichment factor (partition coefficient). The *enrichment factor* is defined as the ratio of analyte concentration in the extracting solvent to that in the sample. For example, to extract analytes having an acidic character, such as phenols, the pH of the donor phase is lowered by acidification to suppress analyte ionization, while the pH of the acceptor phase is raised (NaOH solution is used for the extraction) to ionize phenols. In principle, this approach is analogous to the hollow fiber–protected three-phase microextraction except for the difference in the barrier between the donor and acceptor phases. In headspace water-based SME, the barrier is the headspace while in HF(3)ME the barrier is an organic solvent impregnated in the fiber pores. The headspace water-based SME was used for the isolation and enrichment of phenols from aqueous matrices,[285] and enrichment factors from 106 to 528 were obtained, but it could also be used for some volatile polar solvents (e.g., amines).

A novel approach to the use of polar organic solvents as the extractants in solvent microextraction involves the use of room-temperature ionic liquids (RTILs).[286] Ionic liquids have many unique properties, such as negligible vapor pressure, excellent thermal stability, and high viscosity, which allow the use of stable large drops, increasing the extraction yield. Their polarity is adjustable through selection of the appropriate cations and anions; thus, their viscosity, miscibility with water and organic solvents, and extractability of organic compounds can be tuned. Ionic liquids can be used in various modes of SME: HS-SDME,[263,265,287–292] SDME,[196,262,265,287,293–296] dispersive liquid–liquid,[38,68,69] or HF(3).[297] Until recently, the use of ionic liquids in solvent microextraction was limited by their incompatibility with gas chromatography. Because ionic liquids are nonvolatile, their use in analytical procedures has been limited to HPLC, capillary electrophoresis, and spectroscopic techniques as the final determination methods. However, two new papers[262,263] describe the design of a removable interface enabling direct introduction of the extracted analytes into a GC-MS system while preventing the ionic liquid from entering the column (Figure 6.11). The interface consists of three main integrated components: an injection zone, a removable unit, and a transfer line. The injection zone is equipped with a septum and connected to a carrier gas line. This zone is connected downstream to a removable unit made of a polymer tube packed with cotton. This tube can easily be removed for cleanup purposes. A transfer line, provided with a stainless steel needle, is used to connect the removable unit to the GC inlet. This design can significantly expand the scope of application of ionic liquids in solvent microextraction.

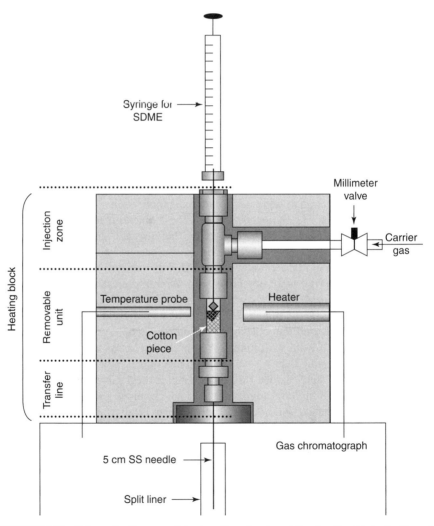

FIGURE 6.11. Schematic diagram of the interface developed for the direct introduction of ionic liquid containing the extracted analytes. SDME, single-drop microextraction; SS, stainless steel. (Reprinted with permission from Ref. 263; copyright © 2008 Elsevier.)

Methyl *tert*-butyl ether (MTBE), a volatile compound reasonably soluble in water, is used as a gasoline additive to improve the octane number. As a result, MTBE is a relatively common groundwater pollutant. Headspace SDME followed by GC-FID was used[254,298] to determine this analyte in aqueous matrices. By using solvent cooling, the detection limit obtained for MTBE was 60 ng/L, and the linear dynamic range extended over more than three orders of magnitude.[254] The procedure was used for a variety of environmental water samples.

6.5.4 Nonpolar Semivolatile Compounds

This group of analytes includes, for example, polycyclic aromatic hydrocarbons (PAHs), polychlorinated biphenyls (PCBs), polybrominated diphenyl ethers (PBDEs), and phthalates. Their origins in the environment are different: PAHs are formed as by-products of fuel burning (fossil or biomass), while PCBs and PBDEs enter the environment through their uses as flame retardants, coolants, insulating fluids, hydraulic fluids, and other uses. Phthalates are used as plasticizers. Along with some organochlorine pesticides, they are all considered to be persistent organic pollutants, although their hazards vary. Some PAHs have been identified as carcinogens, mutagens, and teratogens. PCBs are toxic and probable human carcinogens. PBDEs and phthalates are suspected endocrine disruptors and may pose health risks. Because these analytes are hydrophobic, in the environment they tend to accumulate in soil or sediment. As a result of this, it is often difficult to collect a representative sample and preserve it, since nonpolar semivolatile compounds have a strong tendency to adsorb on suspended solids, on the walls of sampling containers,[299] or even on the Teflon coating of the stir bar.[300] To reduce analyte losses taking place during sample collection, transport, storage, and preparation, the containers used for sampling should be silanized.[301] An alternative and better approach would be to use field sampling, which some SME modes, and hollow fiber–protected SME in particular, are capable of. As a result of the importance of nonpolar semivolatiles as environmental pollutants, a number of SME procedures have been developed for PAHs,[61,71,81–84,89,100,117,210–212,217,240,257,265,287,301–307] PCBs,[218,308,309] PBDEs,[54,171,308,310] and phthalates.[43,55,60,125,265,311,312] Most of these procedures have been developed for aqueous matrices,[61,81–84,89,100,117,257,306,307,309] and soil, dust, and sediments,[111,171,210–212,217,218] but few have dealt with plant material.[240,306] Owing to the low polarity of these analytes, their equilibrium distribution coefficient K values are very large, thus enabling high sensitivity and low detection limits when solvents that are nonpolar or of low polarity are used for their extraction. Hence, nonpolar semivolatiles have been extracted using alkanes and cycloalkanes, including hexane,[125,308] cyclohexane,[212,309] octane,[210] decane,[171] and undecane[100,307,310]; aromatic compounds: benzene[304] and toluene[211,218,240,301–303,311]; and chlorinated hydrocarbons: tetrachloroethane,[54] tetrachloroethene,[61] chloroform,[60,71,89] carbon tetrachloride,[55,312] and chlorobenzene.[43] Other extractants that have been used include alcohols, such as butanol[257] and 1-octanol,[117,217,305] ionic liquids,[265,287] perfluoro surfactants,[81–83] and even an aqueous solution of β-cyclodextrin.[306] Interestingly, some of these procedures have been fully automated,[100,211,212,309] thus enabling unattended analyses of a large number of environmental samples.

In addition to the most common SME modes: headspace,[257,265,287,306] hollow fiber–protected headspace,[111] direct immersion,[100,125,217,265,287,301,312,313] or hollow fiber–protected two-phase,[100,117,171,210–212,218,240,302,303,305,308,310,311] a surprisingly large number of analytical procedures for the determination of nonpolar semi-volatiles involve dispersive SME[43,54,55,60,61] or a related technique called homogeneous liquid–liquid extraction (HLLE).[71,81–83,89] In HLLE, the analyte in a homogeneous solution is extracted into a very small volume of sedimented phase formed from the solution by the phase separation phenomenon. The ternary component solvent system and the perfluorinated surfactant system are the two usual modes of homogeneous liquid–liquid extraction. In HLLE, the initial state is a homogeneous solution, so that the surface area of the interface between the aqueous phase and the water-miscible organic solvent phase is infinitely large. Consequently, vigorous agitation is not necessary. The difference between dispersive liquid–liquid microextraction and homogeneous liquid–liquid extraction is that in the former technique the extracting solvent is dispersed in the form of fine droplets throughout the sample and can be separated from the aqueous phase solely by centrifugation, whereas in the latter technique a true homogeneous solution is formed, which requires the addition of a reagent to increase the ionic strength or adjust the pH of solution prior to centrifugation in order to separate the sample from the acceptor solution.

A GC-FID chromatogram of PAHs determined in an aqueous sample following dispersive liquid–liquid microextraction is shown in Figure 6.12.[61] The advantages of this mode include short extraction time, typically no more than several minutes,

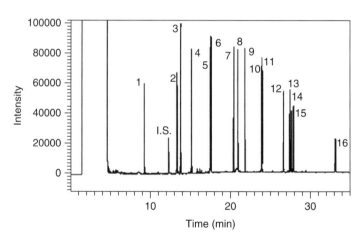

FIGURE 6.12. Chromatogram of surface water spiked at a concentration level of 5.0 μg/L of PAHs obtained by using DLLME combined with GC-FID. Peak identification: (1) naphthalene, (2) acenaphthylene, (3) acenaphthene, (4) fluorene, (5) phenanthrene, (6) anthracene, (7) fluoranthene, (8) pyrene, (9) benzofluorene, (10) benzo[a]anthracene, (11) chrysene, (12) benzo[e]acephenanthylene, (13) benzo[e]pyrene, (14) benzo[a]pyrene, (15) perylene, and (16) benzo[ghi]perylene, (I.S.) biphenyl. (Reprinted with permission from Ref. 61; copyright © 2006 Elsevier.)

simple equipment required, and low detection limits (e.g., 2 to 8 ng/L for phthalates when using GC-MS as the final determination technique[43]). A detailed procedure for the determination of PAHs in water by means of SDME, HS-SDME, and dispersive liquid–liquid microextraction is provided in Chapter 7.

Two other modes of solvent microextraction that have been used for the isolation and enrichment of nonpolar semivolatile compounds deserve mention, owing to the very low detection limits of the analytical procedures employing them. The first is membrane-assisted solvent extraction (MASE), which was used for the GC-MS determination of PCBs in river water[309] in a fully automated procedure with the detection limits in the range 2 to 10 ng/L. The second is continuous flow microextraction (CFME) combined with GC-MS, which was used in the determination of 16 PAHs in tap water samples with the detection limits ranging from 1 to 10 ng/L.[304] Continuous-flow SME differs from other modes of microextraction in that a drop of solvent fully and continuously makes contact with a fresh and flowing sample solution[10] (Figure 6.13). The presence of both diffusion and convection results in the high effectiveness of CFME, ensured by the fact that this mode always involves the equilibrium rather than exhaustive extraction. A shortcoming of this approach is the need for additional equipment, such as a microinfusion pump.

Finally, to obtain extremely high sensitivities and low detection limits in the determination of PAHs in aqueous matrices, a powerful approach involving combination of two or even three trace enrichment methods, including homogeneous liquid–liquid extraction, has been developed. In one procedure,[71] the analytes were first extracted from 1 L of an aqueous sample by solid-phase extraction (SPE), then eluted by a small volume (5 mL) of organic solvent, and the extract was further preconcentrated to just 20 µL by homogeneous liquid–liquid extraction. This last

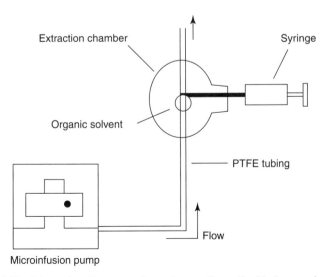

FIGURE 6.13. Schematic diagram of continuous-flow liquid-phase microextraction. (Reprinted with permission from Ref. 304; copyright © 2007 Elsevier.)

volume was analyzed by HPLC with fluorimetric detection. The overall enrichment factor was 50,000, resulting in detection limits at the sub-ng/L levels. In the second procedure,[82,83] PAHs in aqueous matrices were first extracted from 50 mL of a sample into 30 μL of the sedimented phase by homogeneous liquid–liquid extraction using a perfluoro surfactant, followed by an online preconcentration using the sweeping method, and capillary electrophoretic (CE) determination with UV detection. The detection limits were at the sub-μg/L level. The third procedure,[81] termed the *triplex concentration system*, combined the first two approaches and provided enrichment factors in excess of 1×10^7 and detection limits at the ng/L level by making use first of SPE, followed by homogeneous liquid–liquid extraction, and finally, online preconcentration by the sweeping method and CE-UV determination.

6.5.5   Polar Semivolatile Compounds

This group of environmental pollutants includes phenols (alkyl, chloro, and nitro), aromatic amines, and organic acids, such as long-chain fatty acids or alkylphosphonic acids. Phenolic compounds are ubiquitous in the environment, especially in environmental waters and soils. Phenols are important pollutants because of their wide use in many industrial processes, such as the manufacture of plastics, dyes, drugs, antioxidants, and pesticides. They are of great environmental concern, owing to their high toxicity. For this reason, a number of phenolic compounds are listed in the U.S. Environmental Protection Agency (EPA) list of priority pollutants. Aromatic amines have been used, among others, in the production of dyes, pharmaceuticals, explosives, pesticides, polymers, and polyurethanes, and consequently, they are commonly found in the environment. Aromatic amines are highly toxic to humans, and some of them have been classified as carcinogens. Fatty acids are not hazardous, but if present in water, they can inhibit a variety of microbial populations and affect the wastewater treatment process. Alkylphosphonic acids are the degradation products of nerve agents. The fact that all these analytes are polar suggests that most of the approaches used for the extraction of volatile polar solvents should also be applicable to polar semivolatiles, although in some cases the microextraction step may have to be modified. For example, polar semivolatile compounds are amenable to headspace SDME, but as a result of the lower volatility of phenols and aromatic amines, to improve Henry's law constants, sample temperatures used in this mode are raised to 50 to 80°C for phenols[285,291,314] and to 90°C for chlorinated anilines.[288] At the same time, the extracting solvents in all these cases are highly polar, to increase the air–organic solvent distribution constant and hence the amount of analyte extracted; they include ionic liquids[288,291] (as we recall, these solvents are nonvolatile and have excellent thermal stability), aqueous NaOH solution,[285] an acetonitrile–water (1:1) mixture,[314] and an aqueous acid solution for the extraction of amines.[182]

Since all polar semivolatiles have either acidic (phenols, fatty acids, alkylphosphonic acids) or basic (amines) properties, they lend themselves to three-phase SME. In this mode, analytes are extracted from an aqueous sample to an organic solvent and simultaneously back-extracted from the organic solvent to the acceptor solution, usually a few microliters of water at the appropriate pH (see Figure 2.6). The organic solvent is therefore an interface between the two aqueous solutions.[274]

To achieve analyte isolation and enrichment, the acid–base properties of the analytes are exploited. For acidic compounds, the pH of the sample (donor) solution is adjusted to a value below the pK_a of the analytes to form neutral species, which can be extracted by the organic solvent, while the pH of the acceptor solution is adjusted to a value above their pK_a values so that the analytes are converted into hydrophilic ionic species, which are effectively excluded from the organic liquid membrane and therefore accumulate in the aqueous acceptor solution (see Chapter 3 for details). For example, in the HF(3)ME procedure for the determination of dinitrophenols in water, the pH values of the sample and acceptor phase were 2 and 10, respectively. Conversely, if the analytes are basic in nature, the donor solution has to be made basic and the acceptor solution acidic. For example, the optimized donor and acceptor solutions for the microextraction of aromatic amines were 0.1 M NaOH and 0.5 M HCl, respectively.[315] In addition to analyte enrichment, three-phase SME provides efficient sample cleanup, eliminating nonionic interferences and reducing matrix effects, which is of particular importance in environmental and clinical analyses. The three-phase SME mode is realized in two fundamental designs: one employing a hollow fiber [HF(3)ME] and the other exploiting a thin layer of organic solvent placed between the sample and the acceptor phase. The latter approach is usually called *liquid–liquid–liquid microextraction* (LLLME) (Figure 6.14). A small

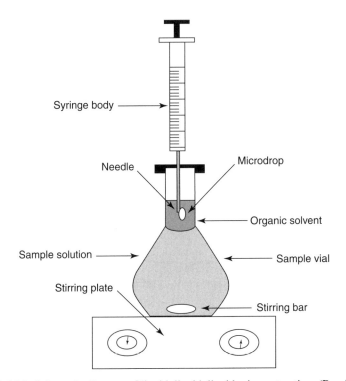

FIGURE 6.14. Schematic diagram of liquid–liquid–liquid microextraction. (Reprinted with permission from Ref. 452; copyright © 2008 Elsevier.)

volumetric flask[316–318] or a Teflon ring[319] can be used for LLLME, which makes the experimental setup very simple. There are two requirements for an organic solvent to be used in LLLME. First, it must be immiscible with water, and second, its density must be lower than that of water. The hollow fiber–protected three-phase mode makes use of a short length of porous hollow fiber which has been impregnated with an organic solvent to separate the sample solution on the outside of the fiber from the acceptor solution in the lumen of the fiber. The most commonly used solvents are 1-octanol and dihexyl ether (see Chapter 5), because both exhibit a certain degree of polarity while having low volatility and solubility in water, which ensures a long-term stability of the solvent film in the fiber pores and compatibility with the fiber material (polypropylene).

HF(3)ME has been used in the determination of phenolic compounds[112,297,320–326] and aromatic amines.[106,315,327] LLLME was also employed for this purpose (phenols[316,325] and aromatic amines[317,319,328,329]). The two modes were compared side by side in the procedure for determination of aniline derivatives in water and found to be almost the same in terms of their analytical characteristics, except for enrichment factors, which were somewhat higher for HF(3)ME.[330]

Owing to the fact that the final extract has to be compatible with the final determination technique, the analytical procedures using three-phase SME call for either high-performance liquid chromatography or capillary electrophoresis. The detection limits are typically at the µg/L or sub-µg/L level, but they can be lowered substantially using a two-step microextraction procedure in which the aqueous acceptor solution from the first step becomes (after pH readjustment) the donor solution for the second step.[327] When this two-step three-phase SME procedure was used for the determination of aromatic amines in water, the detection limits were in the range 10 to 250 ng/L, and the linear dynamic range covered from over two to over three orders of magnitude.

The third strategy, adopted from the microextraction of volatile polar compounds, is derivatization of the analytes. Several different approaches have been used for microextraction combined with derivatization of polar semivolatiles. The first approach involved preextraction (in-sample) derivatization and was used in the analytical procedures for the determination of phenols, amines, and alkylphosphonic acids. Phenolic compounds were derivatized in aqueous samples through acetylation with acetic anhydride,[331,332] amines were derivatized with pentafluorobenzaldehyde (PFBAY)[333] or iodine,[334] and alkylphosphonic acids were converted into esters by alkylation with propyl bromide.[214] Derivatization was followed by the extraction using SDME[331,332,334] or HF(2)ME.[214,333]

In the second approach, concurrent in-drop microextraction–derivatization, HS-SDME in the single-drop mode was used for the isolation/enrichment of amines from water samples.[335] The derivatizing agent dissolved in the organic solvent was PFBAY, and the derivatives were determined by GC-MS. Phenolic endocrine disruptors were determined using in-drop derivatization combined with concurrent SDME assisted by ion-pairing transfer, followed by GC-MS or GC-FID final determination.[180] In this procedure, an ion-pairing agent tetrabutylammonium bromide (TBAB) was added to the sample solution, whose pH was preadjusted to 10.5.

The phenolate ions formed ion pairs with the tetrabutylammonium cation (Q^+), which could then cross the liquid–liquid interface due to the lipophilicity of Q^+ (the phase transfer step). The drop of organic solvent contained the derivatizing agent (ethyl chloroformate), which reacted with the phenolate ions, converting them into O-ethoxycarbonyl derivatives and releasing the free ion Q^+, which was then transferred back to the aqueous phase. The entire extraction–derivatization step took just 12 min, although the detection limits (on the order of μg/L) were not as low as those of the other SME procedures employing GC-MS. An analogous procedure for the isolation and enrichment of 15 phenols employed TBAB for ion pairing and p-toluenesulfonyl chloride to convert the analytes into the tosylated derivatives prior to their GC-MS determination.[336]

This approach, applicable for highly polar or ionic analytes with poor extractability in hollow fiber–protected SME, in which a relatively hydrophobic ion-pairing reagent (called a carrier) with acceptable water solubility forms ion pairs with the analytes, followed by extraction of the ion-pair complexes into the organic phase (and into an aqueous acceptor phase in the three-phase mode), is called a *carrier-mediated SME*[25] (see Chapter 3 for more details). Carrier can be added to either the sample solution or the organic liquid membrane.

We noted earlier that dispersive liquid–liquid microextraction (DLLME) enables rapid extractions as a result of very large interfacial area between the donor and the acceptor solutions. In a DLLME extraction–derivatization procedure for the determination of chlorophenols in water, the disperser solvent (acetone), added to 5 mL of the sample, contained both the extraction solvent (chlorobenzene) and the derivatizing agent (acetic anhydride). The acetylation reaction was complete within a few seconds, and the entire SME step lasted only about 2 min prior to final determination by GC-ECD.[47] Similarly, DLLME combined with derivatization was employed for simultaneous derivatization with PFBAY and extraction of anilines from wastewater, followed by GC-MS determination.[42]

The third approach to derivatization of polar semivolatiles involves postextraction derivatization, which is carried out in two different ways: either in-syringe or in-port. In the former strategy,[337] the analytes (phenols) are first extracted from water samples into a microdrop of organic solvent using SDME. After extraction, derivatization is carried out in the syringe barrel by drawing 0.5 μL of a powerful silylating agent N,O-bis(trimethylsilyl)acetamide (BSA), into the syringe. The silylation reaction is complete in 5 min at 50°C. The procedure yields low detection limits, in the range 4 to 61 ng/L, when combined with GC-MS. In the latter strategy, derivatization of the extracted analytes is carried out in the injection port of gas chromatograph. In one approach,[338,339] phenols are extracted from water samples by HF(2)ME, and 2 μL of the extract are injected into the GC-MS injection port, followed by 2 μL of the silylating agent, BSTFA. The injector is operated in the stop-flow mode after a holding time of 2 min. In this way, the sample and the derivatizing agent are retained in the injection port for 2 min, in which time the analytes are completely derivatized. The procedure yields very low detection limits: between 5 and 15 ng/L when Accurel Q 3/2 polypropylene hollow fiber is used for the extraction,[338] and between 0.07 and 2.34 ng/L when the fiber is replaced with a functional polymer-coated fiber.[339] In another approach,[116] long-chain fatty acids in aqueous samples are converted into ion pairs by the addition of tetra-butylammonium hydrogen sulfate. The ion pairs are extracted into the organic solvent (acceptor phase) in the hollow fiber. In the injection port of the GC-MS, the ion pairs are converted quantitatively into butyl esters. The detection limits for the procedure are in the range 9.3 to 15 ng/L.

Other modes of SME have been used for the direct extraction (without derivatization) of phenolic compounds and amines from aqueous matrices. In the case of phenols, the pH of the samples had to be adjusted to a low value to suppress ionization of the analytes. Continuous-flow microextraction was used for the isolation and enrichment of aromatic amines from green algae[222–224] and phenolic compounds from wastewater samples[340]; membrane-assisted solvent extraction was employed in the procedure for the determination of seven phenols in contaminated groundwater,[341] and HF(2)ME was part of the GC-MS procedure for the determination of phenols in water,[188,208] but the detection limits of all these procedures were inferior to the limits attainable when using microextraction–derivatization, and typically were at the μg/L level.

Finally, a combination of two sample preparation techniques; solid-phase extraction and dispersive liquid–liquid microextraction, was used for the preconcentration of chlorophenols from aqueous samples prior to their determination by GC-ECD.[48] The linear dynamic range of the procedure extended over four orders of magnitude, and the detection limits were in the range 0.5 to 100 ng/L.

6.5.6 Metal Ions, Metalloid Ions, and Organometallic Compounds

A variety of metal ions have been extracted from environmental and biological samples by solvent microextraction. Among the most commonly determined and hazardous heavy metals were mercury,[38,78,170] lead,[58,59,151,170,204,342–345] cadmium,[66,170,342,344,346,347] copper,[45,76,170,344,348] cobalt,[51,62,347] uranium,[73,88] silver,[74] palladium,[62,76,170,347] nickel,[51] and chromium.[221] In addition to metals, SME was used for the isolation and enrichment of metalloids: arsenic,[349,350] antimony,[351] and selenium,[40,162,352,353] as well as organometallic compounds: organomercury,[203,205,265,318] organotin,[41,202,265,354,355] and organomanganese.[356] Three strategies are employed for solvent microextraction of these species. In the first, the analytes are converted in situ into volatile derivatives, which are then extracted by means of direct immersion or dispersive SME and determined by gas chromatography. This strategy is used for the determination of chromium[221] (see Table 6.3) and organotin species[41,355] in water. In one procedure, tributyltin (TBT) and triphenyltin (TPT) are derivatized with sodium tetrakis(4-fluorophenyl)borate, extracted into 3 μL of α,α,α-trifluorotoluene using SDME, and determined by gas chromatography–tandem mass spectrometry. The detection limit for TBT is 0.36 ng/L.[355] In the other procedure, butyl and phenyltin compounds are ethylated with sodium tetraethylborate, extracted using dispersive liquid–liquid microextraction, and determined by gas chromatography with flame photometric detection (GC-FPD) with the detection limits at the 0.21-ng/L level.[41] It should be mentioned here that some organometallic compounds, such as methylcyclopentadienylmanganese tricarbonyl, are sufficiently volatile to enable their direct extraction (without derivatization) by HS-SDME, followed by GC-MS determination.[356]

In the second strategy, the analytes are derivatized in-sample (preextraction derivatization) to form volatile derivatives, which are then sequestered by HS-SDME, and determined by electrothermal atomic absorption spectrometry (ET AAS). Because the analytes are extracted from the gas phase, rapid analysis of complex samples is possible. The derivatives are either hydrides (arsine,[349,350] selenium hydride,[353] tin hydride,[357] or methylmercury hydride[203]), formed in the reaction of analytes with sodium tetrahydroborate, or methyl derivatives formed by photo-assisted reaction with organic acids.[352] In both cases, the volatile derivatives are sequestered in the microdrop of an aqueous solution of Pd(II), except for one procedure for the determination of arsenic,[349] where silver diethyldithiocarbamate is used for sequestration.

In the third strategy, metal or semimetal ions are chelated in situ, forming neutral complexes, which can be isolated and enriched by various SME modes, and determined by spectroscopy or gas chromatography. The most sensitive procedures for the determination of selected heavy metal ions involving SME are compiled and compared to EPA method detection limits in Table 6.3. An inspection of the Table reveals that except for silver and nickel, the analytical procedures incorporating solvent microextraction provide detection limits substantially lower than the EPA method requirements and are therefore well suited for the determination of hazardous heavy metals in the environment.

TABLE 6.3 Detection Limits of Selected Metals for SME Coupled with Various Final Determination Techniques[a]

Analyte	Chelating Agent	SME Mode	Final Determination Technique	Detection Limit (ng/L)	
				SME	EPA
Hg(II)	DDTC	HF(2)	ET-ICP-MS	3.3	5
Pb(II)	8-HQ	CF	ET-ICP-MS	2.9	1000
Cd(II)	APDC	DLL	ETAAS	0.6	100
Co(II)	BZA	CF	ETAAS	0.99	2
U(VI)	CH$_3$COO$^-$/TBA$^+$	HLLE	VIS	140	1000
Ag(I)	TSCC4P	HLLE	ETAAS	5	5
Pd(II)	BZA	CF	ETAAS	1.5	5000
Ni(II)	PAN	DLL	ETAAS	33	10
Cr(III)	Htfa	SD	GC-FPD	500	1000

[a] DDTC, diethyldithiocarbamate; 8-HQ, 8-hydroxyquinoline; APDC, ammonium pyrrolidine dithiocarbamate; BZA, benzoylacetone; TBA$^+$, tetrabutylammonium cation; TSCC4P, tetraspirocyclohexylcalix[4]pyrrole; PAN, 1-(2-pyridylazo)2-naphthol; Htfa, 1,1,1-trifluoroacetylacetone; ET-ICP-MS, electrothermal inductively coupled plasma mass spectrometry; ETAAS, electrothermal atomic absorption spectrometry; VIS, visible spectrophotometry; GC-FPD, gas chromatography with flame photometric detection; SD, single drop; CF, continuous flow; DLL, dispersive liquid–liquid; HF(2), hollow fiber–protected two-phase; HLLE, homogeneous liquid–liquid extraction.

6.5.7 Other Inorganic Analytes

Solvent microextraction has been used for the isolation and preconcentration of free cyanide, weak acid dissociable cyanide, ammonia, and bromide, iodide, iodate bromate, and ions in aqueous matrices. Free cyanide[184] and weak acid–dissociable cyanide[220] are extracted by HS-SDME using an aqueous Ni(II)–NH$_3$ solution as the extractant [the product formed is Ni(CN)$_4^{2-}$ ion] and determined by capillary electrophoresis with UV detection (CE-UV). The detection limit is around 2 μg/L. Ammonia is extracted from seawater with the pH adjusted to 12 by HS-SDME with an aqueous solution of phosphoric acid as the extracting solvent, followed by CE-UV determination. The LOD of this procedure is 27 μg/L. The halogen-containing anions BrO$_3^-$, IO$_3^-$, Br$^-$, are determined by derivatization–microextraction, followed by GC-MS.[358] The oxyhalides are reduced with ascorbic acid to halides and converted to 4-bromo-2,6-dimethylaniline and 4-iodo-2,6-dimethylaniline by their reaction with 2-iodosobenzoate in the presence of 2,6-dimethylaniline, and the derivatives extracted by SDME using toluene as the organic solvent. The detection limits for the procedure are 20, 15, 20, and 10 ng/L for bromate, iodate, bromide, and iodide, respectively.

6.5.8 Pesticides

The wide application of pesticides in agricultural and nonagricultural areas has resulted in extensive environmental pollution with these analytes, which presents considerable hazards to human health through either direct exposure to pesticides or through their residues in food and drinking water. The analysis of pesticides poses special problems, owing to their diverse chemical properties, such as volatility and polarity. Some classes of pesticides [e.g., organochlorine pesticides (OCPs)], are nonpolar, whereas others (e.g., acidic herbicides) are very polar semivolatiles. Most pesticides are sufficiently volatile and thermally stable to be amenable to gas chromatography, although their volatility is not high enough to warrant the use of headspace SDME, and the preferred mode is direct immersion. Other pesticides have to be determined by high-performance liquid chromatography or capillary electrophoresis. Alternatively, they have to be derivatized prior to determination by gas chromatography.

As a result of high versatility of solvent microextraction, it has been widely used in analytical procedures for the determination of pesticides.[31] Most of the procedures have been developed for water samples,[39,57,61,63,68,69,101,103,108,110,115,121,215,219,302,359–393] but some have been applied to soil[64,90,107,108,215,216] and sediments,[218,219] fruits and vegetables,[67,249,251,394] and beverages.[109,368,382,384,395–399]

6.5.8.1 Organochlorine Pesticides
Organochlorine insecticides were historically the first class of synthetic pesticides, used on a large scale in the 1950s and 1960s. Their use has largely been discontinued as a result of high persistence, bioaccumulation, and toxicity. Organochlorine pesticides (OCPs) are considered to be priority pollutants that are still monitored in a variety of environmental matrices, and a number of analytical procedures involving SME have been developed for their

determination. Since OCPs are nonpolar semivolatile compounds, the preferred modes of SME and solvents used for their extraction are similar to those discussed in Section 6.5.4 (Nonpolar Semivolatile Compounds). Hence, the SME modes used for OCPs included HF(2),[107–110,218,302,362,363,385,395] membrane-assisted solvent extraction,[367,368,399] SDME,[251,364,365,377,382,389] headspace,[292] and dispersive SME.[61] It is apparent that the preferred modes for the extraction of OCPs are HF(2) and SDME. The extracting solvents are predominantly nonpolar hydrocarbons: toluene, hexane, isooctane, and tetradecane, although more polar 1-octanol, 1-nonanol, or even a polar ionic liquid have also been used. Since OCPs are thermally stable and have chlorine atoms in their structures, they have been determined mostly by GC-ECD[109,364,365,377,382] and GC-MS.[107,108,110,218,251,302,362,363,367,368,385,394,395,399] An example of HF(2) extraction of organochlorine pesticides from spiked artificial seawater and their GC-MS determination is shown in Figure 6.15,[362] and a detailed SDME procedure for the determination of pesticides in soil is provided in Chapter 7.

6.5.8.2 Organophosphorus Pesticides

Organophosphorus insecticides have been intended as a replacement for OCPs due to their better degradability. However, many of them are highly toxic and suspected carcinogens, mutagens, and endocrine disruptors. Consequently, the extraction of organophosphorus pesticides (OPPs) by SME has received considerable attention. By far the most common SME mode used in the analytical procedures for the determination of OPPs is direct immersion.[121,359,370,373,374,376,378,397] The other modes used include dispersive,[39,67,69] hollow fiber–protected two-phase,[103,361,372] membrane-assisted solvent extraction,[368,396,399] and continuous-flow microextraction.[366,369] In the majority of these procedures, toluene is the acceptor solvent, although in dispersive liquid–liquid microextraction, chlorobenzene is used most often, owing to the

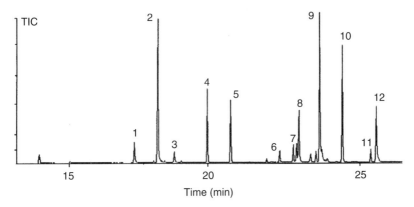

FIGURE 6.15. Chromatogram of 12 OCPs in an OCP-spiked (40 μg/L) artificial seawater sample after LPME using a hollow fiber membrane. Peaks: (1) α-BHC; (2) lindane; (3) β-BHC; (4) heptachlor; (5) aldrin; (6) dieldrin; (7) endrin; (8) endosulfan; (9) p,p'-DDT; (10) p,p -DDD; (11) endrin aldehyde; (12) methoxychlor. (Reprinted with permission from Ref. 362; copyright © 2002 Elsevier.)

requirement of DLLME that the extracting solvent have a density higher than that of water. Other extractant solvents ranged from hexane and cyclohexane to carbon tetrachloride and undecanol. Many of the final determination methods exploit highly sensitive and phosphorus-selective gas chromatographic detectors, such as the flame photometric (FPD)[39,67,359,370,384,397] or nitrogen phosphorus (NPD) [372,376,378] detectors. Using GC-FPD, detection limits as low as 1 to 5 ng/L were obtained for 13 organophosphorus pesticides.[359] Alternatively, universal detection by GC-MS[103,361,368,373,379,396,399] or HPLC-UV[69,121,366,369,374] can be employed.

6.5.8.3 Carbamates

Carbamates are also widely used as insecticides instead of OCPs and OPPs, due to their better degradability, although they may pose potential hazards to humans, being suspected mutagens and carcinogens. Hollow fiber–protected two-phase is the most common SME procedure in the determination of carbamates.[107,206,372,387] In addition, SDME,[378] dispersive liquid–liquid micro-extraction,[63] and HF(3)[249] have been used. The organic solvents used for micro-extraction included 1-octanol, toluene, isooctane, tetrachloroethane, and dihexyl ether doped with trioctylphosphine oxide (TOPO). TOPO is added into the membrane liquid to enhance the extraction efficiency of weak organic acids with low pK_a and K_{ow}. The addition of TOPO can significantly improve the behavior of the membrane liquid, due to the hydrogen bonding between TOPO and the analytes. The final determination methods employed GC-NPD,[372,378] which is selective for nitrogen-containing compounds (all carbamates contain nitrogen atoms), GC-MS,[107,206,387] HPLC-UV,[63] or HPLC-MS.[249]

6.5.8.4 Pyrethroids

Pyrethroids are synthetic insecticides similar to the natural pesticide pyrethrum, which is produced by chrysanthemum flowers. Pyrethroids are easily degradable in the environment and only moderately toxic; therefore, they are increasingly applied in lieu of organochlorine and organophosphorus pesticides. Two procedures for the determination of pyrethroids using solvent microextraction have been developed: one involving direct immersion followed by GC-ECD detection,[382] the other using dispersive SME with HPLC-UV detection.[68]

6.5.8.5 Triazine Herbicides

Triazine herbicides are used to control some grasses and broadleaf weeds. Although they are of low toxicity to humans, atrazine, a member of this class, has been classified as a possible human carcinogen. A variety of direct SME modes have been used for the extraction of triazines, including membrane-assisted solvent extraction,[367,368,399] HF(3),[249,381] SD,[360,386] dispersive,[57] continuous flow,[369] and HF(2)ME.[215] In addition, another micro-extraction technique, called *microporous membrane liquid–liquid extraction* (MMLLE), was used for the determination of triazines.[379] This technique can be considered a hybrid of continuous flow and HF(2)ME. In MMLLE, the membrane, either a flat sheet or a hollow fiber, is used as a barrier between two phases, one of them organic (acceptor), filling the membrane pores and one side of the membrane, and the other the aqueous donor solution (sample) on the other side of the membrane. The donor phase is pumped at a constant rate, while the acceptor phase usually remains stagnant, although in some designs it is mobile as well. Following the extraction, the extract is eluted to a sample loop in a large-volume GC injection valve and injected online into the gas chromatograph. The main advantage of MMLLE is that it can be coupled online to gas chromatography. The solvents used for the microextraction of triazines have covered a range of polarity index, from hexane and cyclohexane, through carbon tetrachloride and toluene to 1% TOPO in dihexyl ether [for HF(3) microextraction], chlorobenzene (in dispersive liquid–liquid microextraction), 1-octanol, and butyl acetate. The enriched analytes were determined mostly by GC-MS,[57,215,360,367,368,399] HPLC-UV,[369,381,386] GC-FID,[379] or HPLC-MS.[249]

6.5.8.6 *Phenoxy Acids*

Phenoxy acid herbicides have been used heavily in agriculture and forestry for over 50 years. As a result, they or their metabolites have been detected in soil as well as surface water and groundwater. As we discussed earlier in this chapter, the best strategy for microextraction of analytes with acidic or basic properties involves either HF(3)ME or derivatization: in-sample, in-fiber or in-drop, or in-injection port. In the former approach, the detection method must be compatible with the extraction procedure, which implies the use of either high-performance liquid chromatography, or capillary electrophoresis. If the analytes are derivatized prior to microextraction, they can subsequently be determined by gas chromatography. In accordance with these considerations, almost all microextraction procedures for phenoxy acid herbicides are based on hollow fiber–protected three-phase mode with HPLC-UV detection.[101,115,375,380,398] One procedure involves HF(2)ME with ion-pair formation, injection-port derivatization, and GC-MS determination.[383] As we mentioned before, in the analytical procedures employing HF(3)ME, the pH of the donor and acceptor phases must be adjusted to maximize the enrichment factor. Hence, in the extraction of phenoxy acids the donor phase is made 0.05 to 0.5 M in HCl, while the acceptor phase is from 0.01 M to 1 M NaOH. Also, the organic solvents used have some capability of hydrogen bonding and include 1-octanol, 2% TOPO in dihexyl ether, and a 1:1 mixture of 1-octanol and dihexyl ether.

6.5.8.7 *Substituted Amides, Anilides, and Thiocarbamates*

These analytes are used extensively to control weeds in agricultural crops. Consequently, natural waters and soil often contain the residues of this class of herbicides. So far, SME has been a part of just a few analytical procedures for the determination of amide, anilide, and thiocarbamate pesticides. Most of these make use of SDME,[216,376,388] while the rest involve hollow fiber–protected two-phase,[107] continuous flow,[369] and microporous membrane liquid–liquid extraction.[379] The detection methods include GC-MS,[107] GC-FID,[379] GC-NPD,[376] GC-ECD,[388] and HPLC-UV,[216,369] and the acceptor phase solvents range from toluene through dichloromethane and carbon tetrachloride to ethyl acetate.

6.5.8.8 *Fungicides*

Fungicides are chemical compounds that are used to prevent the spread of fungi in crops and gardens. They have very diverse chemical structures. As a result of their widespread use, crops and environmental water have been found to be contaminated with fungicide residues. Most microextraction procedures for fungicides employ membrane extraction: either HF(2)[219,391,392] or microporous membrane liquid–liquid extraction,[379,393] although the direct-immersion mode[390] is also used. The methods of final determination of fungicides include GC-ECD,[219,390,391] HPLC-UV,[392,393] and GC-FID.[379] In most cases, toluene is used as the extraction solvent, but hexane and 1-octanol are also employed.

6.5.8.9 *Organosulfur Pesticides*   Organosulfur pesticides are extracted from environmental samples using two SME modes: hollow fiber–protected two-phase and dispersive, followed by determination by GC-FPD.[64] The two modes were compared; dispersive SME was more rapid and offered better enrichment factors, while the HF(2) mode was more robust for complex samples such as soil.

Due to highly diverse structures as well as physical and chemical properties of different classes of pesticides, it is unlikely that an SME procedure allowing simultaneous determination of all pesticides could ever be developed. Among the microextraction techniques discussed above, most have been applied to just one class of pesticides. However, several procedures were more comprehensive and enabled solvent microextraction and determination of more than one group of these analytes. For example, a membrane-assisted solvent extraction procedure followed by large-volume injection (LVI) GC-MS allows the determination of 47 organochlorine, organophosphorus, and triazine pesticides at concentrations down to 5 to 50 ng/L.[368] Fifteen pesticides, including organophosphorus, organochlorine, triazine, phenylurea, and chloroacetanilide, have been determined by microporous membrane liquid–liquid extraction and GC-FID.[379] Twenty-three organonitrogen pesticides: carbamates, triazines, and phenylureas, were extracted from vegetables by hollow fiber–protected three-phase SME and determined by HPLC-MS with the detection limits at the level 0.1 to 2.9 μg/kg.[249] Organophosphorus, organochlorine, and triazine pesticides were determined successfully in sugarcane juice by membrane-assisted solvent extraction, followed by large-volume injection GC-MS.[399] The examples above indicate that so far, the most versatile SME modes were those exploiting membrane extraction.

The SME detection limits obtained for selected groups of pesticides extracted from aqueous samples are compiled in Table 6.4. The values listed represent the most sensitive procedure and the SME mode used is shown below or next to the detection limit. An inspection of Table 6.4 reveals that the final determination techniques employing gas chromatography are generally more sensitive than procedures making use of high-performance liquid chromatography with UV detection. Gas chromatography is therefore the preferred detection method for all analytes that are sufficiently volatile and thermally stable. SME-GC detection limits compare favorably with the EPA method requirements. For example, the EPA method detection limits for organophosphorus, organochlorine, carbamate, and triazine pesticides are 4 to 2000, 2.5 to 75, 100 to 200, and 100 to 800 ng/L, respectively. The analytical procedures involving SME in the best combination with the detection method offer detection limits well below these values.

TABLE 6.4 Detection Limits (ng/L) for SME of Pesticides in Aqueous Samples Coupled with Various Detection Methods[a]

Pesticide Class	GC-MS	GC-ECD	GC-FID	GC-NPD	HPLC-UV
Organophosphorus	0.3–11.4 HF(2)	—	—	1–72 HF(2)	170–290 DLL
Organochlorine	0.3–17.3 HF(2)	1–12 SD	1.6–6.2[b] MASE	—	50–80 HS
Pyrethroid	—	10–11 SD	—	—	280–600 DLL
Carbamates	200–800 HF(2)	—	—	72 HF(2)	1000 DLL
Triazines	1–5[b] MASE	—	3–7.8[b] MMLLE	—	30–60 SD
Phenoxy acids	0.51–13.7[c] HF(2)	—	—	—	100–400 HF(3)
Amides, anilides	—	0.2–114 SD	2.4 MMLLE	1000 SD	
Fungicides	—	6–10 SD	2–6.4[b] HF(2)	—	1100–1900 HF(2)

[a] SD, single drop; HS, headspace; DLL, dispersive liquid–liquid; HF(2), hollow fiber–protected two-phase; HF(3), hollow fiber–protected three-phase; MASE, membrane-assisted solvent extraction; MMLLE, microporous membrane liquid–liquid extraction.
[b] With large-volume injection (100 μL).
[c] Following derivatization.

The discussion above related mostly to aqueous matrices. The determination of pesticides in solid samples introduces unique problems associated with the release and transfer of the analytes from the original matrix to the acceptor phase. As discussed earlier in the chapter, some preextraction steps are typically required in the analytical procedures for the determination of pesticides in solid matrices. Furthermore, the detection limits in such cases are usually higher than those for water samples. For example, to improve extraction efficiency of organochlorine pesticides from marine sediments, microwave-assisted (MAE) HF(2) extraction followed by GC-MS is used, and the detection limits range from 70 to 700 ng/kg [parts per trillion (ppt)].[218] The data in Table 6.4 indicate that for the same analytes, extraction mode, and detection method, the detection limits for aqueous samples range from 0.3 to 17.3 ng/L (ppt), which is substantially less. In another study, organochlorine and organonitrogen pesticides in soil were determined by HF(2)-GC-MS, and the detection limits were at the 50 to 100 μg/kg (ppb) level.[107] Pressurized hot water extraction (PHWE)–microporous membrane liquid–liquid extraction coupled online with GC-MS was used to determine pesticides in grapes[394] with the limits of quantification of 0.3 to 1.8 μg/kg. A fungicide, vinclozolin, was extracted from sediments by ultrasound-assisted extraction (USE), subjected to HF(2) and determined by GC-ECD with the LOD of 0.5 μg/kg.[219] In the examples above, when a solvent microextraction technique is coupled to a more conventional extraction method (MAE, PHWE, USE), the resulting combination provides more efficiency, shortens sample preparation time, improves detection limits, and reduces solvent use.[31]

Among emerging pollutants, pharmaceuticals at trace levels are of particular concern, owing to their health effects and ubiquity in the aquatic environment. However, as the SME procedures for extraction of pharmaceuticals from environmental samples are virtually identical to those for bioanalytical samples, except for the omission of some preliminary sample treatment steps, such as filtration, they are discussed in the next section.

6.6   CLINICAL AND FORENSIC APPLICATIONS OF SME

Clinical and forensic analysis is the second most common area of analytical applications of solvent microextraction. Samples for which SME has been used include whole blood and blood components (plasma, serum), urine, saliva, hair, and milk. These applications are summarized in Table 6.5, in which the most commonly determined analytes and modes of their extraction are shown in boldface type. The table is intended as a guide to quickly select the mode of microextraction for a specific clinical or forensic analysis. The fraction of clinical and forensic applications varies with the SME mode and equals to 50% for hollow fiber–protected three-phase, 18% for SDME, 16% for hollow fiber–protected two-phase, 12% for headspace, 6% for continuous/cycle flow, and 6% for dispersive liquid–liquid microextraction. The SME mode most often used in clinical and forensic applications for drugs is hollow fiber–protected three-phase, since the analytes that are most commonly determined have basic or acidic functional groups. With a large difference in volume between the sample (typically, 1 to 4 mL) and the acceptor solution (typically 1 to 30 μL), analytes can be enriched substantially. In addition to analyte enrichment, HF(3)ME provides high selectivity and significant sample cleanup. The majority of SME applications have been developed for urine samples, followed by blood and its components. By far the most commonly determined analytes in clinical and forensic samples are drugs in body fluids. Microextraction and bioanalysis of drugs were reviewed recently,[23,25] and so was sample preparation for pharmaceutical analysis.[22] Also, two general reviews of solvent microextraction techniques had sections on biomedical applications.[15,20] As a result of their widespread use, drugs are also commonly found in the environment, especially in water and wastewater. Two recent reviews discuss sample preparation in analysis of pharmaceuticals in environmental samples[24] and analytical methods for tracing pharmaceutical residues in water and wastewater.[21]

The second important class of analytes frequently determined in biological samples includes the indicators of human exposure to environmental pollution, such as heavy metal ions, polychlorinated biphenyls, polybrominated diphenyl ethers, and chlorophenols. The technique used to sample and analyze blood, urine, breast milk, and other tissues to assess human exposure to environmental pollution is called biomonitoring. Finally, the third group of analytes determined in body fluids is represented by the biomarkers of certain diseases or disorders. For example, acetone is regarded as the biomarker of diabetes and its blood concentration could be used as the accessorial tool for diagnosis of diabetes. Similarly, a high level of certain aldehydes,

TABLE 6.5 Applications of SME in Clinical and Forensic Analysis

Matrix	Analytes[a]	SME Mode[a,b]
Blood and blood components (whole blood, plasma, serum)	**Drugs** (antidepressant, antiepileptic, psychoactive, polar, active herbal ingredients, **basic**, polar, hydrophilic, illicit, antipsychotic, NSAIDs, antimalarial, local anaesthetic)	**HF(3)**, DLL, SD, HS, HF(2)
	Alkaloids	SD, HF(3)
	Carbonyl compounds (aldehydes, ketones)	**HS**, SD
	Organic acids	HS, HF(3)
	Peptides and proteins	SD
	Ammonia	HS, SD
	Polychlorinated biphenyls	HF(2)
	Polybrominated diphenyl ethers	HF(2)
	Polyamines	HF(3)
Urine	**Drugs** (**basic**, antidepressant, psychoactive, NSAIDs, hydrophilic, antimalarial, active herbal ingredients, diuretics, cough suppressant, β_2-agonist)	**SD, HF(2), HF(3)**
	Metal ions (derivatized)	CF
	Drugs of abuse (and metabolites) (cocaine, ephedrine, amphetamines, tetrahydrocannabinol, phencyclidine, anabolic steroids)	**SD**, HS, **HF(2), HF(3)**
	Carbonyl compounds (aldehydes, ketones)	SD
	Alkaloids	SD, HF(3)
	Amino acids, peptides, and proteins	SD
	Phenolic endocrine disruptors	SD, HF(2)
	Organic acids	HF(3)
	Cyanide	HS
	Chlorophenols	HF(2)
	Polyamines	HF(3)
	Heterocyclic aromatic amines	HF(3)
	Phosphate esters	HF(3)
Saliva	Drugs of abuse	HF(2)
	Cyanide	HS
Hair	Metal ions (derivatized)	CF
	Drugs of abuse	HF(3)
Milk	Drugs (antidepressant, basic)	HF(3)

[a] Most common analytes and modes are shown in boldface type.
[b] SD, single drop; HS, headspace; CF, continuous flow; DLL, dispersive liquid–liquid; HF(2), hollow fiber–protected two-phase; HF(3), hollow fiber–protected three-phase.

such as formaldehyde, acetaldehyde, hexanal, heptanal, or malondialdehyde, can be indicative of cancer. An elevated ammonia blood level is considered a strong indicator of an abnormality in nitrogen homeostasis, most commonly related to liver dysfunction, while elevated concentrations of polyamines are tumor markers.

As mentioned in Section 6.3, biological samples generally represent complex matrices, requiring some preliminary operations prior to solvent microextraction. *Blood plasma* is the liquid component of blood, in which the blood cells are suspended. It makes up about 55% of total blood volume and contains mostly water, but also dissolved proteins, mineral ions, glucose, hormones, and clotting factors. Plasma is the second most commonly used biological sample (after urine). *Blood serum* is blood plasma without fibrinogen or other clotting factors. Drugs will bind to plasma and serum proteins to various degrees, depending on their physical and chemical properties. In general, acid and neutral drugs bind primarily to albumin, and basic drugs mainly to α-acid glycoprotein.[400] The interactions between drug and protein are a combination of ionic, hydrophobic, and polar. Prior to microextraction of the analytes of acidic or basic character, the pH of the sample has to be adjusted to a high value for basic analytes and to a low value for acidic analytes to suppress their ionization and facilitate extraction into the organic solvent. This pH adjustment normally suppresses any ionic interactions. Hydrophobic drug–protein interactions can be suppressed by addition of an organic solvent, generally methanol. Methanol is capable of disrupting both hydrophobic and polar interactions. The drug–protein binding can also be suppressed by diluting the samples with water, a polar solvent that can compete with the drug for the protein binding sites. This common sample preparation step has one more advantage: Sample dilution improves the rate of mass transfer of analytes into the acceptor phase by increasing diffusion coefficients of analytes in the sample. Owing to the fact that most analytes determined in samples of blood components are hydrophilic, the majority of analytical procedures for blood plasma and serum make use of three-phase modes: either hollow fiber–protected[129,131,132,135,138,140,143–149,152,153,155,156,158,163,164,169,172] or liquid–liquid–liquid microextraction.[133,139,154,159,165] Microextraction of very hydrophilic analytes may require carrier-mediated HF(3)ME.[157,201]

Whole blood has the most complex matrix of all body fluids. Consequently, the two solvent microextraction techniques most commonly used for whole blood are HS[401–405] and HF(3)ME,[126–129] which require least sample pretreatment. Typically, only pH adjustment and/or dilution with water are needed in the analytical procedures, unless analyte derivatization takes place.

Urine is an aqueous solution of metabolic wastes (such as urea), dissolved salts, and organic materials. Urine is the most commonly studied biological matrix, owing to its ease of noninvasive collection and because it is a universal means of excretion of both analytes and their metabolites. It has a longer detection time window for many analytes than other body fluids, and it contains high concentrations of metabolites. The SME mode used for extraction of urine samples depends on properties of analytes. For hydrophilic compounds, the majority of solvent microextraction procedures involve either hollow fiber–protected three-phase[127,128,138,140,144,145,149,153–156,158,169,177,183,187,190–192,197,199,200,406–408] or liquid–liquid–liquid microextraction[139,159,181,201] with HPLC or CE detection. Hydrophobic compounds are typically extracted by SDME[119,167,174,175,180,193,194,196,198,296,409–415] or hollow fiber–protected two-phase SME[105,142,145,158,176,185,186,188,189,195] and determined by gas chromatography. Preliminary sample preparation steps before

SME usually include pH adjustment and, in some cases, dilution with water. However, if urine samples have to be made alkaline, addition of base to the samples results in formation of a fine precipitate which upon stirring dislodges the microdrop in direct immersion SME.[175] Consequently, samples to be analyzed by SDME-GC require filtration prior to extraction.

For some types of body fluids, such as blood, the volume of sample available is limited, which can be an obstacle in certain investigations, for example in pharmacokinetic studies, where blood samples have to be collected repeatedly over a period of time. In such cases, solvent microextraction procedures have to be modified to enable extraction from small volumes of sample solutions. One approach suitable for handling small sample volumes is a variant of direct-immersion SME, called *drop-to-drop solvent microextraction* (DD-SME). In this mode, both the sample and organic solvent volumes are in the order of microliters. In one procedure, trimeprazine was extracted from 8 µL of urine and blood of rats using 0.6 µL of toluene.[130] In another procedure, 30 µL-samples of urine and plasma were used to extract quinine with 2 µL of *m*-xylene.[123] The second approach to small-volume sampling exploits miniaturized hollow fiber–protected three-phase SME to extract drugs from 50 µL of blood or plasma samples.[129,148] Owing to a small sample/extractant volume ratio, the enrichment factors in the two approaches are poor and the main advantage of these procedures, other than the small sample volume, is selectivity, which is provided by an extensive sample cleanup.

Human *saliva* is composed mostly of water but also includes electrolytes, mucus, antibacterial compounds, and various enzymes. Similarly to urine, saliva is attractive as a biological sample because of its noninvasive collection. The main disadvantage related to saliva sampling is limited sample size compared to urine. Furthermore, the window of detection for analytes in saliva is short, and analyte concentrations are often low. Accordingly, only a few solvent microextraction procedures have used saliva for sampling.[184,416,417] Owing to a high viscosity of saliva samples, they have to be diluted with water to improve mass transfer kinetics and reduce the analyte–protein interactions. In addition, the pH of saliva samples may have to be adjusted to convert analytes into nonionized forms. The procedures were developed to screen for cocaine and its metabolites in saliva by HF(2)ME followed by gas chromatography with pulsed discharge helium ionization detector (GC-PDHID),[416] to determine nicotine in saliva by HF(3)ME-HPLC-UV,[417] and to determine free cyanide by headspace SDME with in-drop derivatization followed by CE-UV.[184]

Hair can also be sampled noninvasively and it provides a record of exposure to analytes. One inconvenience of using hair as a biological sample is that it requires a preliminary preparation prior to solvent microextraction: hydrolysis or digestion in order to transfer analytes into solution. Human hair samples, digested by a mixture of nitric and perchloric acid in the presence of hydrogen peroxide, were used to determine the concentration of trace metal ions (Be, Co, Cd, and Pd) by low-temperature electrothermal ICP-MS, following continuous-flow microextraction with benzoylacetone as both extractant and chelating agent.[347] Basic drugs of abuse were determined in hair by HPLC-UV, following digestion with methanol at 50°C for 5 h and surfactant-enhanced hollow fiber–protected three-phase SME.[225]

The main constituents of *milk* include water, proteins, lipids, and carbohydrate (lactose). The lipids form an emulsion suspended in water. The analysis of milk can serve as a means of investigating the exposure of the population to a variety of chemicals. Hollow fiber–protected three-phase SME was used for the extraction of four basic antidepressant drugs from human breast milk.[418] Direct SME from breast milk samples provided low recoveries (18 to 38%), because the drugs were partially bound to the sample matrix (fat). Therefore, prior to extraction, the breast milk had to be acidified to release drugs interacting with the matrix and the majority of fat removed by centrifugation. After the pH was adjusted to alkaline with NaOH to deionize the analytes, the drugs were extracted from the supernatant through a thin layer of organic solvent present in the pores of a porous hollow fiber and into the acidic acceptor solution. The acceptor solution was analyzed directly by capillary electrophoresis with a limit of quantification (LOQ) of about 50 µg/L.

Although hollow fiber SME is a simple and effective sample preparation technique, it is relatively time consuming. One extraction takes typically between 10 and 120 minutes. The kinetics of mass transfer can be improved significantly by using so-called *electromembrane extraction* (EME), in which the driving force for extraction is an electrical potential difference applied across the supported liquid membrane.[149,156,419,420] The equipment for EME is the same as for HF(3)ME, except for the addition of two electrodes and a dc power supply (see Figure 3.8). The analytes in the EME system are ionized in both the sample and acceptor solutions and this promotes electrokinetic migration across the organic liquid membrane. For basic analytes, a positive electrode is placed in the sample and the negative electrode is located in the acceptor solution, while the potential is reversed for acidic compounds. The use of an electrical potential difference as the driving force typically shortens the extraction time to 5 min per extraction while retaining large enrichment factors and high extraction efficiency[149] (see Chapter 3 for more details). Thus far, all applications of electromembrane extraction involved the determination of drugs in biological matrices, including basic drugs in human breast milk.[149] Two other solvent microextraction procedures have been developed for bovine milk samples. In one of them, ammonia was determined by HS-SDME-CE-UV,[405] and the other was used for the determination of phenoxy acid herbicides by HF(3)ME-HPLC-UV.[398] In the former procedure, no sample pretreatment was necessary other than dilution with water. The latter procedure required sample acidification and centrifugation prior to solvent microextraction.

The selection of an appropriate mode of solvent microextraction in the analytical procedures used for the analysis of body fluids should be based on the following guidelines:

1. Headspace SME should be applied whenever possible. As a rule of thumb, analytes with a molar mass below 200 g/mol and/or without groups forming hydrogen bonds i.e., NRH groups, NH_2 groups, and OH groups are suitable for headspace SME because they are likely to have a sufficiently high vapor pressure[400]. In most cases, when headspace SME is used for the extraction of analytes from body fluids, no preliminary sample preparation steps are

required unless in situ derivatization of the analytes is carried out. Typically, polar volatile analytes such as aldehydes and ketones will require in-sample or in-drop/fiber derivatization. Alternatively, the extractant will have to be a polar solvent (e.g., ionic liquid).

2. For semivolatile hydrophobic analytes, either hollow fiber–protected two-phase or direct-immersion mode is preferred. The former can be used for more complex samples, such as plasma or whole blood, while the latter is suitable for less complex samples, such as urine, particularly if they are diluted and/or filtered.

3. Hydrophobic analytes with acid or base functional groups can be extracted effectively by one of the three-phase modes, either hollow fiber–protected three-phase or liquid–liquid–liquid microextraction. Highly hydrophilic analytes, which are very soluble in water, call for carrier-mediated HF(3)ME.

Typical detection limits in the analytical procedures for the determination of drugs in body fluids are 1 to 100, 5 to 500, and 2 to 50 µg/L for HF(3)ME, SDME, and HF(2)ME, respectively. The detection limits for the same analytes determined in environmental waters, which represent much simpler matrices, are usually less than 1 µg/L.

The following groups of drugs have been extracted from biological samples using solvent microextraction:

1. *Gastrointestinal tract/metabolism:* antacids,[157] antiemetics and anti-nauseants,[157,412] antidiarrheal agents,[149,155–157,419–421] H$_2$ antagonists[147,148,421]

2. *Blood and blood-forming organs:* anticoagulants[422]

3. *Cardiovascular system:* β-blockers,[119,144,147,148,157,199,421] vasoconstrictors,[119,147,148,328] vasodilators,[421] antiarrhytmics,[147,148,421] diuretics,[195] coronary therapeutics[157]

4. *Dermatologicals:* antihistamines[130,412,421]

5. *Genitourinary system:* uricosurics,[422] anabolic steroids[187,189]

6. *Infections and infestations:* antibiotics[194]

7. *Brain and nervous system:* local anesthetics,[165,411] analgesics,[50,139,144–146,149,153,155–158,178,192,296,407,419–424] antiepileptics,[153,158,168] psycholeptics,[126,145,146,149,153,155,156,158,169,412] psychoanaleptics[65,119,127,129–131,133,135,136,140,143,149,155–157,161,193,412,418]

8. *Antiparasitic products:* antimalarial agents[152,177,412,425]

9. *Respiratory system:* bronchodilators,[190,191,199] decongestants,[119,147,148,190,191] H$_1$ antagonists,[145] antihistamines,[153,157] antitussives[145,149,153,155,156]

10. *Illicit drugs*[105,119,127,128,144,146–148,153–155,158,174,181,182,186,187,201,225,409,421,426]

An example of determination of three antidepressant drugs in urine and plasma samples using hollow fiber–protected three-phase SME followed by HPLC-UV is shown in Figure 6.16.[140]

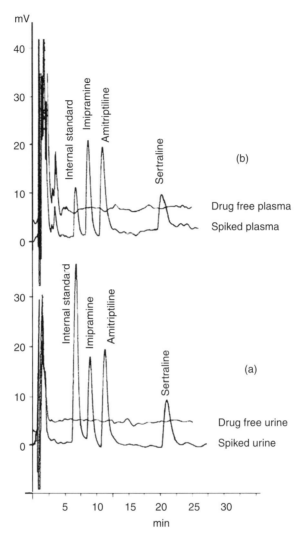

FIGURE 6.16. HPLC chromatogram of antidepressants in (a) urine spiked at 15.0 µg/L and (b) plasma spiked at 30.0 µg/L after extraction by hollow fiber–protected three-phase SME. Conditions: sample: 11 mL 0.01 M in NaOH (pH 12); organic solvent: *n*-dodecane; acceptor: 24 µL of 0.1 M phosphate buffer; stirring speed: 700 rpm; extraction time: 45 min. (Reprinted with permission from Ref. 140; copyright © 2007 Elsevier.)

Environmental pollutants present in biological samples can serve as indicators of human exposure to hazardous chemicals. Several procedures involving SME have been developed for the determination of nonpolar semivolatiles, such as poly-chlorinated biphenyls in whole blood by HF(2)ME-GC-MS,[134] polybrominated diphenyl ethers in blood serum by HF(2)ME-GC-ICP-MS,[171] and polar semi-volatiles: phenolic endocrine disruptors in urine by HF(2)ME-GC-MS[185,188] and

SDME-GC-MS,[180] and diphenyl phosphate ester in urine by HF(3)ME with CE-UV or HPLC-MS detection. Thus, the extractions of these pollutants (except for diphenyl phosphate ester) were based on hollow fiber–protected two-phase or direct-immersion mode, followed by GC-MS. The detection limits were at the level 15 to 40 ng/L for PBDEs, 70 to 940 ng/L for PCBs, 0.02 to 20 μg/L for phenolic endocrine disruptors, and 14 μg/L for diphenyl phosphate ester. The procedure for determination of trace metals in hair and urine samples by continuous flow–ICP-MS[170] was described earlier in this section.

Solvent microextraction has also found use for the isolation and enrichment of biomarkers of some diseases and disorders from biological samples. Thus, acetone, which is regarded as the biomarker of diabetes, was determined by headspace SDME-GC-MS in whole blood,[401–403] plasma,[141] and urine[141] samples, following derivatization with O-2,3,4,5,6-(pentafluorobenzyl)hydroxylamine hydrochloride (PFBHA) in-sample[401,403] or in-drop,[402] or in-drop derivatization with 2,4,6-trichlorophenylhydrazine (TCPH).[141] Depending on the analytical procedure used, the detection limits were between 0.036 and 1 μg/L. Aldehydes, whose elevated level can be indicative of cancer, were determined by HS-SDME-GC-MS in whole blood[403,404] as well as urine and plasma,[141] following in-sample derivatization with PFBHA[403] or in-drop derivatization with PFBHA[404] and TCPH.[141] The detection limits varied from 0.014 to 10 μg/L.

An elevated ammonia blood level is considered a strong indicator of liver dysfunction. Ammonia was extracted from whole blood by means of headspace SME and determined by CE-UV.[405] The extractant was an aqueous solution of phosphoric acid, and sample pH had to be adjusted to 12. The detection limit of the procedure was 27 μg/L of NH_4^+. Polyamines, whose elevated levels can be a tumor marker, were determined in urine and plasma by hollow fiber–protected three-phase SME, followed by precolumn derivatization with tosyl chloride and HPLC-UV analysis with the detection limits in the range 1.4 to 2.6 μg/L.[138]

Finally, since headspace SDME is well suited for the extraction of nonpolar volatile analytes, it should be applicable to forensic investigations of suspected arson cases. A detailed procedure for the determination of fire accelerants in arson investigations using HS-SDME is provided in Chapter 7.

6.7   APPLICATION OF SME IN FOOD AND BEVERAGE ANALYSIS

Food and beverage applications of solvent microextraction focus on three classes of analytes. The first one includes natural constituents of food and beverages, such as alcohols, volatile phenols, esters, carbonyl compounds, volatile flavor components, or organic acids. Chemical compounds deliberately added to food or beverages, such as caffeine added to soft drinks, iodide added to salt, or food dyes, are the second class. Undesirable natural or anthropogenic contaminants present in food and beverage samples (e.g., mycotoxins, chloroanisoles, pesticides, or residual solvents) constitute the third class. Samples for which SME has been used include alcoholic and nonalcoholic beverages, milk, edible oils and sauces, fruits and vegetables, salt, potato chips, honey, and salted eggs. These applications are summarized in Table 6.6, in which the most commonly determined analytes and modes of their extraction are shown in boldface type. The table can serve as a guide to select the mode of microextraction for a specific food or beverage analysis. The most common SME modes employed in the analytical procedures for food and beverage analysis are SD, HS, HF(2), and dispersive liquid–liquid microextraction. A quick glance at Table 6.6 reveals that for many of the analytes listed there, including pesticides, phenols, carbonyl compounds, polychlorinated biphenyls, aromatic hydrocarbons, drugs, and ammonia, SME procedures have already been discussed; therefore, some of them will be mentioned only briefly here.

Flavor and aroma are among the most important quality criteria of fresh and processed foods. Headspace sampling of the volatile fraction of vegetable matrices was reviewed recently.[34] Most flavor and aroma compounds are volatile, and procedures for their isolation from food and beverage samples take advantage of this volatility by incorporating headspace or direct-immersion solvent microextraction, followed by GC determination. The detection limits for liquid samples typically range from 1 to 50 μg/L. SDME-GC-MS was used to extract and determine 49 flavor components from a number of fruit and vegetable juices,[245] to characterize postharvest changes in volatile constituents of apricots,[246] and to elucidate the distribution of volatile compounds in the skin and pulp of Queen Anne's pocket melon.[247] Headspace-SME-GC-MS procedures have been developed for the analysis of flavor components of clove buds[237] and orange juice[427] and for investigating antimicrobial vapor-phase activities of the essential oil constituents of cinnamon, thyme, and oregano.[239] A procedure for the extraction and determination of volatile sulfur compounds in beer and beverage involving HS-SDME-GC-FPD was developed with detection limits in the range 0.2 to 1.9 μg/L.[428] Volatile phenols were extracted from red wines using dispersive SME and determined by GC-MS with detection limits from 28 to 44 μg/L.[46] Alcohols and esters are important components of beer and wine flavor. Several solvent microextraction procedures were developed for the extraction and determination of these analytes in beer[120,429–431] and wine[283,431] samples. Since both alcohol and ester flavor components are volatile, headspace SME is the preferred mode of extraction,[120,429,430] although direct immersion was also used.[431] A comparison of the two extraction modes, headspace and direct immersion, for the extraction of esters from wine and beer samples revealed that HS-SDME yielded lower detection limits (0.87 to 2.33 μg/L), except for ethyl acetate (523 μg/L).[283] The extracting solvents are typically alcohols (1-octanol, ethylene glycol) or aromatic compounds (o-xylene, p-cymene). The detection limits for alcohols using GC-MS and GC-FID as the final determination technique are on the order of 1 to 97 μg/L[120] and 1 to 170 mg/L,[429–431] respectively. These relatively high LOD values reflect the fact that alcohols are polar volatiles, and their direct extraction (without derivatization) does not provide high enrichment factors. However, since the levels of alcohols in beer and wine are on the order of mg/L or more, the procedures developed are adequate for this type of analysis.

Phenolic acids were determined in fruits and fruit juices by a procedure combining SDME extraction using hexyl acetate, followed by in-syringe derivatization

TABLE 6.6   Applications of SME in Food and Beverage Analysis

Matrix	Analytes[a]	SME Mode[a,b]
Alcoholic beverages (wine, spirits, vodka, beer)	**Pesticides** (organophosphorus, organochlorine, fungicides, carbamates, pesticide residue)	**SD, HF(2)**
	Alcohols	SD
	Volatile phenols	DLL
	Esters	SD, HS
	Carbonyl compounds	SD
	Haloanisoles	HS
	Polychlorinated biphenyls	**MASE**
	Volatile sulfur compounds	SD
	Mycotoxin (ochratoxin A)	HF(2)
Nonalcoholic beverages (tea, coffee, fruit juices)	**Pesticides** (organophosphorus, organochlorine, organosulfur, pyrethroids)	CF, DLL, SD, **HF(2)**
	Volatile flavor components	SD
	Volatile sulfur compounds	SD
	Volatile organic compounds	HS
	Aromatic hydrocarbons	HS
	Polycyclic aromatic hydrocarbons	MASE
	Organic acids	SD
	Alkaloid (caffeine)	SD
	Polychlorinated biphenyls	MASE
Milk (human, bovine, powdered)	**Drugs** (antidepressant, basic)	**HF(3)**
	Iodine	SD
	Ammonia	HS
	Phenoxy herbicides	HF(3)
Edible oils and sauces	Residual solvents	SD
	Active constituents of essential oils	HS
	Carbonyl compounds	HS
	Food dyes	HF(2)
Fruits and vegetables	**Pesticides** (organophosphorus, organochlorine, residue)	DLL, SD, HF(2), HF(3)
	Formaldehyde	SD
	Organic acids	SD
	Volatile flavor components	SD
Salt	Iodine	SD
	Halogen-containing anions	SD
Potato chips	Aldehyde (hexanal)	HS
Honey	Insecticide	HS
Salted eggs	Food dyes	HF(2)

[a] Most common analytes and modes are shown in boldface type.
[b] SD, single drop; HS, headspace; CF, continuous flow; DLL, dispersive liquid–liquid; HF(2), hollow fiber–protected two-phase; HF(3), hollow fiber–protected three-phase; MASE: membrane-assisted solvent extraction.

with BSA, and GC-MS.[250] With an extraction time of 20 min, detection limits were found to be between 0.6 and 164 µg/L. GC-MS chromatograms of the extract of unspiked and spiked pomegranate juice are shown in Figure 6.17.

Caffeine is naturally present in some beverages, such as tea or coffee, but it is also added to other beverages (e.g., soft drinks). A rapid drop-to-drop SME procedure using just 7-µL samples, a 0.5-µL drop of chloroform, and an extraction time of 5 min, followed by GC-MS, was developed to determine caffeine in beverages with a detection limit of 4 µg/L.[432] Relative recoveries close to 100% indicated freedom of the procedure from matrix effects.

Sudan dyes were extracted from and determined in edible oils, sauces, and salted eggs.[433] Two microextraction procedures: solvent bar and HF(2)ME and two detection techniques: HPLC-UV and HPLC-MS were compared, and both were found to be similar in terms of extraction and analytical performance, yielding detection limits of 0.09 to 0.95 and 2.5 to 6.2 µg/L for HPLC-UV and HPLC-MS, respectively.

SME has been used in the determination of contaminants of foods and beverages. Solvent microextraction procedures were developed to isolate and enrich pesticides from wine,[206,390,395,396,434] fruit juices,[384,396,397] fruits and vegetables,[67,249,251] and bovine milk.[398] SDME in combination with GC-FID or GC-ECD was used to determine residual solvents in edible oils.[97] Polychlorinated biphenyls were extracted by membrane-assisted solvent extraction from wine and fruit juice and determined by GC-MS.[309] Benzo[*a*]pyrene, OCPs, OPPs, and triazine herbicides were determined in sugarcane juice by membrane-assisted solvent extraction–GC-MS with detection limits from 0.004 to 0.56 µg/L.[399] Ochratoxin A, a mycotoxin produced by several fungal species from *Aspergillus* and *Penicillium* genera and a known

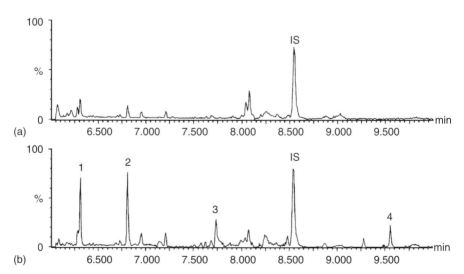

FIGURE 6.17. GC-MS chromatograms obtained by SDME of (a) pomegranate juice and (b) pomegranate juice spiked with a standard solution of phenolic acids at 90 to 740 µg/L. I.S., internal standard; 1, cinnamic acid; 2, *p*-hydroxybenzoic acid; 3, *o*-coumaric acid; 4, caffeic acid. (Reprinted with permission from Ref. 451; copyright © 2006 Wiley-VCH Verlag GmbH & Co.)

nephrotoxin, teratogen, and carcinogen, was extracted from wine by HF(2)ME using 1-octanol as the extractant and determined by HPLC with fluorimetric detection.[435] The method was found to be adequate for quantitative determination of ochratoxin A in wines, having the detection limit of 0.2 µg/L.

6.8 APPLICATION OF SME IN THE ANALYSIS OF PLANT MATERIAL

Over 20 papers on solvent microextraction have dealt with the analysis of plant parts, such as seeds,[241] flowers,[227,238,244] roots,[228,230,232] leaves,[95,243,306] fruits,[231] aerial parts,[229,233,235,236] buds,[237] and needles.[240,306] These applications are summarized in Table 6.7. It is evident that almost all procedures targeted volatile organic compounds, both polar and nonpolar, which called for headspace SDME as the extraction mode, followed by GC-MS for identification and quantitative analysis. The main purpose of these procedures was to obtain the chromatographic profile to serve as a fingerprint of the sample analyzed (i.e., to attempt to identify individual analytes and to determine their relative abundance, although in a few cases specific chemical compounds, such as panaxynol[230] or paeonol,[232] were determined. The profile represents all the detectable chemical compounds present in the extract, and it is used to authenticate and identify the sample even if the constituents are unknown. This approach is considered an efficient way to evaluate the quality of complex matrixes.[436] A major field in which this strategy is employed is the study of medicinal herbs, since fingerprint chromatography is accepted by the World Health Organization as a technique suitable for identification and evaluation of the quality of medicinal plants. A number of SME procedures have been developed for the chromatographic analysis of active components of traditional Chinese medicines (TCMs),[228–232,234,244] and they were reviewed recently.[27] Other procedures aimed at the determination of essential oil components of other plants,[227,233,235–238,241,242] nicotine[95] and volatile basic components in tobacco,[243] or environmental pollutants in foliage and pine needle samples.[109,240,306] As a result of the convenience of headspace sampling, in many cases very few preliminary steps have to be undertaken prior to microextraction. Plant samples are usually air

TABLE 6.7 Applications of SME in the Analysis of Plant Material[a]

Matrix	Analytes	SME Mode[b]
Plant parts: seeds, flowers, roots, leaves, fruits, aerial parts, buds, needles	**Volatile organic compounds, essential oil components**	**HS**
	Bioactive alcohols	HS
	Volatile basic tobacco components	HS
	Volatile organic acids	HS
	Organochlorine pesticides	HF(2)
	Polycyclic aromatic hydrocarbons	HF(2)

[a] Most common analytes and modes are shown in boldface type.
[b] HS, headspace; HF(2), hollow fiber–protected two-phase SME.

dried, ground, sometimes screened, and placed in the extraction vial.[227,228,234,237,244] Other plant samples require more elaborate preliminary steps, such as microwave distillation (MD),[229] microwave-assisted extraction (MAE),[229,232] ultrasound-assisted extraction (USE),[109,240] hydrodistillation,[233,235,236,241,242] and pressurized hot water extraction (PHWE)[230,231] before microextraction proper, especially if quantitative analysis of the analytes is to be carried out. These steps were already discussed in Section 6.4 except for microwave distillation (MD), a combination of microwave heating and dry distillation at atmospheric pressure. Plant material is placed in a microwave reactor without any water or organic solvent added. The heating of the water present in plant cells leads to their disruption, releasing essential oils that are evaporated by the water of the plant material. A cooling system outside the microwave oven condenses the distillate. Compared to traditional alternatives, MD offers some advantages, such as reduced extraction times and energy savings.[27]

We should keep in mind that the application of headspace solvent microextraction alone, without any preextraction steps, to plant material is limited in two ways: (1) only volatile components are extracted, and (2) only semiquantitative analysis can be carried out as a result of problems with the calibration procedure due to solid matrices. However, it is sufficient to obtain the chromatographic profile of a sample of plant material. Coupling an extraction technique such as PHWE, MAE, MD, or supercritical fluid extraction (SFE) to solvent microextraction: headspace for volatile components or hollow fiber–protected two-phase for semivolatiles, followed by GC or HPLC, enables the determination of absolute concentrations of the analytes. If the concentrations are sufficiently high, the analytes can also be determined directly in the extract from PHWE, MAE, MD, or SFE without using SME as an intermediate step in the procedure. Solvent microextraction should be incorporated in the analytical procedure if the preliminary extract needs to be cleaned up and/or the analytes enriched. An example of a chromatographic profile of the essential oil from *Fructus amomi*, a traditional Chinese medicine, obtained by pressurized hot water extraction followed by headspace SME and GC-MS is shown in Figure 6.18.

The only other applications of SME to the analysis of plant material involved the determination of nonpolar semivolatiles (OCPs, PAHs), for which HF(2)ME was most appropriate. Six organochlorine pesticides were determined in green tea leaves by ultrasound-assisted extraction with water, followed by HF(2)ME using 1-octanol as the extractant, and GC-ECD.[109] The detection limits of the procedure ranged from 11 to 28 µg/kg. Thirteen polycyclic aromatic hydrocarbons were extracted from pine needles by ultrasound-assisted extraction with a water–acetone (4:1 v/v) mixture, isolated and enriched by HF(2)ME using toluene as the organic solvent, and determined by GC-MS with the detection limits between 0.01 and 0.95 µg/kg (ppb).[240]

6.9 APPLICATION OF SME IN THE ANALYSIS OF CONSUMER PRODUCTS AND PHARMACEUTICALS

Applications of solvent microextraction in the analysis of consumer products and pharmaceuticals are given in Table 6.8. Clearly, at this point these applications are

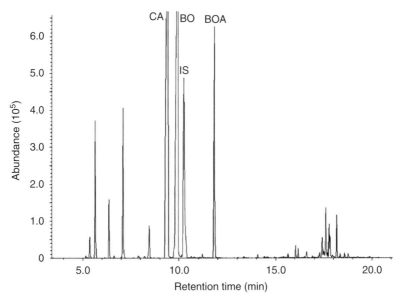

FIGURE 6.18. Total ion chromatogram of the essential oil in *Fructus amomi*, a traditional Chinese medicine from the Guangdong area, China, by PHWE–HS-SDME–GC–MS (CA, camphor, BO, borneol; BOA, borneol acetate; I.S., internal standard—menthol). (Reprinted with permission from Ref. 231; copyright © 2005 Elsevier.)

fairly limited, and most analytes of interest, such as phenols, carboxylic acids, and essential oil components, have been discussed in preceding sections. Accordingly, in this section we focus on phthalate esters and residual solvents. Phthalates have

TABLE 6.8 Applications of SME in the Analysis of Consumer Products and Pharmaceuticals

Matrix	Analytes	SME Mode[a]
Cosmetics (cologne, aftershave)	Phthalate esters	SD
Liquid food simulants	Phthalate esters	SD
	Phenols	HF(3)
	Carboxylic acids	HF(3)
	Antimicrobial agents	SD, HS
Active food packaging	Essential oil components	HF(2)
Polymer (polystyrene)	Styrene	HS
Engine oil	Gasoline and BTEX components	HS
Pharmaceuticals (preparations, oils, active ingredients)	Residual solvents	SD, HS
	Antiepileptic drug	HS
	Triphenylphosphine oxide	HF(2)

[a] SD, single drop; HS, headspace; HF(2), hollow fiber–protected two-phase; HF(3), hollow fiber–protected three-phase SME.

been used as plasticizers in the manufacture of PVC resins, cellulose, clothes, adhesives, medical products, and food packaging of indispensable value for the industry. Because of their widespread use and poor degradability, they are also ubiquitous in the environment. They are being phased out of numerous products in the European Union and United States over health concerns regarding their endo-crine-disrupting properties and suspected carcinogenicity. The most commonly used phthalates include bis-2-ethylhexyl phthalate (DEHP), di-*n*-butyl phthalate (DBP), and butylbenzyl phthalate (BBP). Due to their potential risk to human health and the environment, the U.S. Environmental Protection Agency has set the max-imum contamination level (MCL) for DEHP in water systems at 6 μg/L, and recommended that concentrations above 0.6 μg/L be closely monitored.

Since phthalate esters are slightly polar or nonpolar semivolatiles, the best modes for their microextraction are SDME,[98,125,265,312,313,437] dispersive,[43,55,60] or hollow fiber–protected two-phase[311] SME. One strategy, a modification of dispersive liquid–liquid microextraction called *ultrasound-assisted emulsification–microextraction* (USAEME), takes advantage of acoustically emulsified media formed by 10 mL of sample and 100 μL of chloroform (enrichment factor = 100) to extract phthalates in 10 min. The dispersed droplets of chloroform act as efficient liquid–liquid microextractors in the continuous phase, and later they are readily separated by centrifugation.[60] GC-MS detection limits for the procedure are between 28 and 133 ng/L. USAEME is an environmentally friendly procedure, because it does not use any disperser solvent. Typically, final determination in the analysis of phthalates is performed by GC-MS,[43,60,311,313] GC-FID,[98,125,437] or HPLC-UV[55,265,312] with detection limits from 2 to 133 ng/L, 0.02 to 4.3 μg/L, and 0.64 to 2.2 μg/L, respectively.

Although most of the procedures were used for water samples,[43,55,60,125,265,311–313] one involved food simulants[437] and another consumer products (cologne, after-shave).[98] The analysis of real samples revealed a widespread presence of phthalates. For example, the analysis of mineral water in PET bottles with a push-pull closure revealed the presence of diethyl phthalate, dibutyl phthalate, and DEHP as principal contaminants at the level 0.05 to 0.65 μg/L.[311] GC-MS chromatograms of six phthalate esters extracted by dispersive liquid–liquid microextraction from mineral water prior to and after being spiked at the concentration level of 0.8 μg/L of each analyte are shown in Figure 6.19.[43]

Pharmaceuticals are often contaminated by volatile organic solvents, which are used in manufacturing and processing of these products. Many of them are toxic and/or carcinogenic. Therefore, human exposure to these contaminants should be minimized. The International Conference on Harmonization (ICH) has re-commended acceptable amounts for residual solvents in pharmaceuticals for the safety of the patient. For example, ICH standards for xylene, pyridine, hexane, and chloroform in pharmaceutical products are 2170, 200, 290, and 60 μg/g (ppm), respectively. So far, only three SME procedures for the determination of volatile residual solvents in pharmaceuticals have been developed.[97,438,439] Owing to vola-tility of the analytes, all three procedures take advantage of the headspace mode. In the first one, volatile hydrocarbons and chlorinated hydrocarbons are determined in

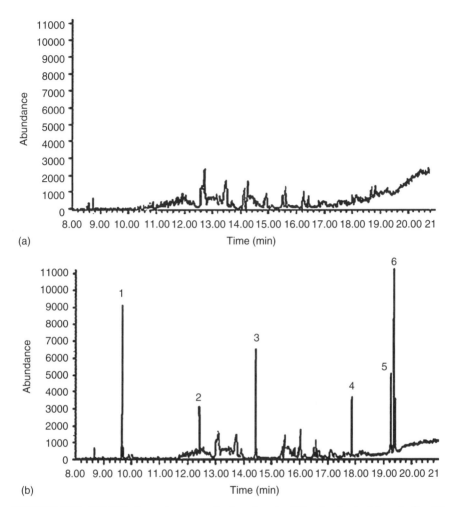

FIGURE 6.19. GC-MS chromatograms of mineral water (Ploor) prior to (a) and after (b) spiking with phthalate esters at the concentration level of 0.8 µg/L of each analyte and dispersive liquid–liquid microextraction using chlorobenzene as the extraction solvent. 1, Dimethyl phthalate; 2, diallyl phthalate; 3, di-*n*-butyl phthalate; 4, benzyl butyl phthalate; 5, dicyclohexyl phthalate; 6, bis-2-ethylhexyl phthalate. (Reprinted with permission from Ref. 43; copyright © 2007 Elsevier.)

edible oils and vitamins using benzyl alcohol as the extractant and GC-FID or GC-ECD as the detection method for hydrocarbons and chlorocarbons, respectively.[97] The detection limits are between 0.11 and 0.37 µg/g for GC-FID and between 0.001 and 0.049 µg/g for GC-ECD. Four chlorocarbons were found in a sample of vitamin A: chloroform (0.70 µg/g), 1,1,1-trichloroethane (0.26 µg/g), carbon tetrachloride (0.52 µg/g), and tetrachloroethene (> 1 µg/g).

An HS-SDME-GC-FID procedure was developed to determine four volatile solvents: tetrahydrofuran, methanol, ethanol, and dichloromethane in pharmaceutical products. The procedure made use of 1-octanol as the extracting solvent. The detection limits ranged from 2 to 60 µg/g.[438]

N-Methylpyrrolidone was used to extract 12 volatile organic solvents from pharmaceuticals by HS-SDME, followed by GC-MS.[439] The procedure was carried out both manually and in an automated mode. The detection limits were at the sub-ppm (µg/g) level. Ethanol, isopropanol, and triethylamine were found in commercial pharmaceutical products: multivitamin tablets, ginseng, and dietary supplement tablets.

Gasoline diluent and BTEX hydrocarbons were determined in used engine oil by a procedure combining headspace SME and GC-MS.[99] The procedure yielded results comparable to those obtained by a standard ASTM method, but avoided the need for cryogenic cooling and expensive sampling interfaces.

6.10 OUTLOOK FOR FUTURE ANALYTICAL APPLICATIONS OF SME

A rapid growth in the scope of applications of solvent microextraction observed over the last 20 or so years can be attributed to a number of advantages of this technique. Although SME will never replace other, alternative sample preparation methods, such as solid-phase extraction (SPE), solid-phase microextraction (SPME), pressurized liquid extraction (PLE), purge and trap (PT), and others, it has earned a prominent position among them as a sensitive, versatile, and inexpensive alternative. When the analytical chemist considers a particular task, which involves (among other factors) the selection of a sample preparation method, solvent microextraction certainly deserves serious consideration, owing to its features:

1. *Versatility.* The developed SME procedures include inorganic, semimetallic, and organic analytes; volatile and semivolatile compounds; and polar and nonpolar analytes. The technique is compatible with a broad variety of samples, and covers environmental, clinical and forensic, foods, beverages, pharmaceuticals, vegetables and fruits, plant material, air, soil, water, body fluids, and others.

2. *Simplicity and low cost.* In its simplest implementation, direct immersion (single drop) and headspace SME require only a sampling vial, a micro-syringe, a magnetic stirrer and stir bar, and a small volume of extracting solvent. Hollow fiber–protected modes add to the equipment above a length of inexpensive, disposable hollow fiber. No instrument modification is required in most cases.

3. *Reduction in the number of steps of analytical procedures.* Solvent microextraction combines sampling, sample cleanup, and analyte enrichment into one step, thus reducing the chance of contamination and/or analyte losses.

4. *Compatibility with a large number of final determination methods.* Solvent microextraction has been used with a number of separation and spectroscopic

techniques. The former include gas chromatography, high-performance liquid chromatography, and capillary electrophoresis. The latter are UV/VIS spectrophotometry, electrothermal atomic absorption spectrometry, electrothermal inductively coupled plasma mass spectrometry, flow injection analysis tandem mass spectrometry, and matrix-assisted laser desorption/ionization mass spectrometry.

5. *High sample throughput.* Some SME modes enable rapid analysis due to the short extraction time required. In this respect, three modes stand out: headspace, electromembrane, and dispersive SME. In headspace SME, typical extraction times are between 4 and 15 min and less commonly up to 60 min. For EME, extraction times of several minutes are usual, while for dispersive liquid–liquid microextraction, extraction times are commonly 2 to 15 min with a maximum of 30 min. Hence, HS-SDME can be used for a rapid extraction of volatiles, EME is suitable for analytes of high or moderate polarity, and dispersive liquid–liquid microextraction is well suited for rapid extraction of nonpolar or slightly polar semivolatiles. For example, an automated HS-SDME-GC-ECD procedure for the determination of trihalomethanes in tap water developed in the authors' laboratory had a total analysis time, counted from loading samples into the autosampler until the end of chromatographic run, of 25 min per sample.

6. *High selectivity.* A wide range of extracting solvents have been used in SME, volatile and nonvolatile, both nonpolar and polar. In addition to traditional organic solvents immiscible with or slightly soluble in water, such as 1-octanol, hexadecane, toluene, or hexane, typically used in HS, SD, HF(2), or HF (3) modes, the list includes ionic liquids [for HS, SD, and HF(3) SME], water, and aqueous solutions for HS and HF(3) solvent microextraction. Mixtures of solvents are also used in some applications. This offers a tunability of solvent properties, such as polarity, hydrophobicity, volatility, and chromatographic behavior. It is also important to notice that all solvents extract analytes by absorption rather than adsorption, resulting in higher upper detection limits (wider linear dynamic range) and minimal competition among analytes, which can be a problem with adsorptive SPME coatings.

7. *Good quantitation and low detection limits.* Many SME procedures have been validated against standard methods and found to be accurate and reproducible. They also offer low detection limits adequate for many applications (e.g., see Table 6.2).

8. *Renewable extracting agent.* Fresh solvent and fresh hollow fiber for each sample eliminate analyte carryover. There is no need for conditioning, as is the case in SPME.

9. *Field sampling.* Most devices used in SME are small and portable. Therefore, they can be employed for on-site sampling. A microsyringe with a hollow fiber attached to the needle can be used for headspace sampling of gaseous, liquid, or solid samples or for direct-immersion sampling of liquid samples. A short length of hollow fiber or flat membrane filled with an extracting solvent

and sealed at both ends can be used as passive samplers for monitoring the workplace atmosphere. With portable gas chromatographs available, the analyses could be performed at the site of sample collection, thus reducing the errors associated with sample storage and handling.

10. *Automation with commercially available equipment.* Several fully automated SME techniques taking advantage of a commercial CTC CombiPal autosampler have been developed and characterized in a seminal paper by Pawliszyn's group.[100] The automated modes included static headspace SDME (a drop of solvent is suspended at the tip of a microsyringe needle and exposed to the headspace of the sample solution), exposed dynamic HS-SDME (the solvent is exposed in the headspace of sample vial for a specific period of time and then withdrawn into the barrel of the syringe, a procedure that is repeated a number of times), unexposed dynamic HS-SDME (the solvent is moved inside the needle and the barrel of a syringe, and the gaseous sample is withdrawn into the barrel and then ejected), static single-drop SME (SDME) (a drop of solvent is suspended at the tip of a microsyringe needle and immersed directly into the sample solution), dynamic SDME (the solvent is moved inside the needle and the barrel of a syringe, and the sample solution is withdrawn and ejected), and hollow fiber–protected two-phase SME [HF(2)ME] (see Figure 2.4). Among the three HS-SDME techniques evaluated, the exposed dynamic HS-LPME technique provided the best performance. For SDME, the dynamic process enhanced the extraction efficiency and the method precision achieved was comparable with the static SDME technique. The precision of the fully automated HF(2)ME was quite acceptable (RSD values below 6.8%) and the concentration enrichment factors were better than the SDME approaches. The authors of this book also used the CombiPal autosampler with the Cycle Composer software to perform static and dynamic headspace and direct-immersion solvent microextraction for a variety of analytes, including trihalomethanes, polycyclic aromatic hydrocarbons, BTEX hydrocarbons, and pesticides and generally found the automated procedures to be reliable and robust (see Chapter 4 for examples of chromatograms).

The trend toward miniaturization of all extraction methods has resulted in the development of improved techniques, collectively called solvent microextraction, that use much smaller amounts of organic solvents, permit automation and higher sample throughput, online coupling to final determination methods, and provide much higher extraction efficiencies. Several areas of growth in the number of applications of SME can be predicted, with environmental, clinical, and forensic applications in the vanguard. The two areas that need more development, but could be very useful and would extend the scope of application of SME significantly are field analysis and industrial hygiene, where membrane-based microextraction samplers could play a primary role.

More SME applications can be expected for quick screening of certain classes of analytes. Headspace hollow fiber–protected SME is well suited for the rapid determination of volatile organic compounds (e.g., residual solvents) in gaseous, liquid, and solid matrices. Electromembrane extraction provides rapid extraction of moderately and strongly polar analytes, such as drugs, from complex biological and environmental matrices while dispersive liquid–liquid microextraction is the preferred rapid extraction mode for nonpolar and slightly polar semivolatiles, such as persistent organic pollutants, from relatively clean matrices such as water. The extraction of polar analytes presents a challenge for many sample preparation methods. SME is compatible with both HPLC and CE and can be used for the extraction of both polar volatiles and polar semivolatiles. For the former analytes, headspace SME with polar extracting solvents, such as aqueous solutions or ionic liquids, is applicable. The latter analytes are well handled by either HF(3)ME or carrier-mediated HF(3)ME.

Further commercialization and automation of SME equipment and procedures is anticipated. The existing automated online membrane extraction procedures were reviewed recently.[440] Several automated procedures involving direct immersion, headspace, and membrane-assisted SME have been developed.[100,115,368,393,441,442] The availability of commercial equipment, particularly for HF(2) and HF(3) techniques, would greatly extend the use of solvent microextraction, especially in environmental, clinical, and forensic areas.

A more expanded use of ionic liquids in SME is expected, owing to their unique properties and compatibility with a variety of final determination methods. A recent review discussed the application of ionic liquids in analytical chemistry.[286] Seventeen SME procedures taking advantage of ionic liquids have been developed so far.[38,68,69,196,262,263,265,287,288,290–297] Ionic liquids can prove particularly valuable for the extraction of strongly polar analytes.

Solvent microextraction can find use in most sample preparation procedures, with analytes ranging from metal ions, through volatile organic compounds, to nonpolar and polar semivolatiles as long as suitable solvent and equipment are available. In its simplest implementation, manual direct immersion or headspace mode, no equipment is needed other than that already present in any analytical laboratory. Analytical chemists are therefore encouraged to give it a try. To this end, Chapters 4 and Chapter 7 should be a good starting point and provide practical guidelines on the development of analytical procedures involving solvent microextraction.

6.11 PHYSICOCHEMICAL APPLICATIONS OF SME

The use of solvent microextraction is not limited to analytical applications, but can include the measurement of properties that characterize the extraction system. The measurements can help investigate multiphase equilibria in the matrix: for example, drug–protein[161,166,443] or hormone–protein[444] binding, kinetics of the partitioning process,[267,445] mechanistic aspects of in-drop derivatization,[282] and pharmacokinetic studies,[50,130,137,446] or determine octanol–water partition coefficients.[447,448] These applications are discussed briefly below.

6.11.1 Study of Drug–Protein Binding

Drug–protein binding is the reversible interaction of drugs with proteins in plasma. The extent to which binding occurs varies and depends on the physicochemical properties of the drug, the affinity between the drug and protein, the drug and protein concentrations, and the presence of other substances, which either compete with the drug for binding sites or displace it through the allosteric effects.[161] The degree of drug–protein binding has a significant effect on the pharmacokinetic and pharmacodynamic outcomes in vivo. Many critical pharmacokinetic parameters are a function of the unbound drug fraction. Consequently, the quantitative determination of drug–protein binding is important in clinical drug development.[166] Although several different methods have been used to determine free drug concentrations in plasma as well as drug–protein binding constants, including affinity chromatography, ultrafiltration, ultracentrifugation, equilibrium dialysis, microdialysis, and capillary electrophoresis, solvent microextraction has an advantage over these methods. Under appropriate conditions (very large sample/solvent volume ratio, very low enrichment factor, or continuous-flow mode), SME extracts an insignificant fraction of the total analyte, depletion is negligible (no more than 5%) and the equilibrium between a drug and a protein in plasma is not perturbed. SME measures only the concentration of free (unbound) analyte; the bound drug and the free protein are not extracted.

One approach to studying drug–protein interactions takes advantage of three-phase membrane-assisted microextraction in two different configurations. In one configuration,[443] a flat microporous PTFE membrane impregnated with dihexyl ether with 5% TOPO separates a flowing donor solution (a drug in buffer solution or in plasma) and a stagnant acceptor solution (a phosphate or citrate buffer). When the entire sample passed through the membrane contactor, acceptor solution is pumped into the injection loop of a high-performance liquid chromatograph. In this case, the donor phase is replenished continuously, the equilibrium conditions in the donor solution represent the situation in the entire sample, and any equilibrium in which the analyte participates is undisturbed. In the second configuration,[161] a batch microextraction, a 1.5-cm microporous polypropylene hollow fiber impregnated with the same extracting solvent as in the first approach is filled with the acceptor solution and placed inside a vial containing the donor solution. Following extraction, the hollow fiber is removed from the donor solution and the acceptor phase analyzed by HPLC. In this case, in order not to disturb drug–protein equilibrium, the volume of sample has to be large enough for the analyte concentration not to be influenced by the extraction. The pH of the donor solutions (buffer solution of the drug or plasma spiked with drug) is kept constant and equal to 7.5 in both configurations. The pH of the acceptor phase has to be adjusted individually for each drug to ensure non-exhaustive (equilibrium) extraction with low enrichment factors. The same amount of a drug is extracted from both the buffer solution and spiked plasma. The ionic strength of the donor and acceptor buffer solutions is equal. The concentration of the drug in the acceptor solution is determined by HPLC, after extraction from both buffer and plasma solutions. The fraction of the free drug in plasma, α_P, is calculated from the experimentally obtained values of equilibrium concentration of analyte in the acceptor phase, C_A, after extraction with the same drug concentrations from both plasma (C_A^P) and buffer (C_A), in the following manner:[443]

$$\alpha_P = \frac{C_A^P \alpha_D K_D}{C_A \alpha_{Dd} K_D^P} \tag{6.1}$$

where α_{Dd} and α_D are the fractions of analyte in extractable form in the plasma and buffer donor solutions, respectively, and K_D and K_D^P are the partition coefficients between the donor phase and the organic phase for the buffer and plasma solutions, respectively.

If the donor buffer solution has the same pH and ionic strength as the plasma, the following assumptions are justified: $\alpha_{Dd} = \alpha_D$ and $K_D = K_D^P$, and equation (6.1) is reduced to

$$\alpha_P = \frac{C_A^P}{C_A} \tag{6.2}$$

In practice, a series of standard drug solutions are prepared in buffer and extracted using SME. After equilibrium is reached, the extracts are analyzed and calibration curves prepared. The calibration curves are plotted as the peak area of the drug in the extractant against drug concentration in the buffer solution. Next, known amounts of protein are added to standard solutions, which are then subjected to SME and the extract analyzed. The free drug concentration is calculated from the calibration curve based on the measured peak area.

The drug–protein binding constant can be determined from a Scatchard plot:

$$\frac{r}{[D]} = -rK_a + NK_a \tag{6.3}$$

where r is the moles of drug bound per mole of protein and [D] is the concentration of free drug, K_a is the equilibrium constant of drug binding to protein, and N is the number of binding sites per protein molecule. The concentration of free drug, [D], is determined by SME for various total concentration values, and bound drug is calculated by subtracting free drug from the total spiked into the sample. Knowing the concentration of bound drug and total protein concentration, it is possible to calculate r. When $r/[D]$ is plotted on the y-axis and r is plotted on the x-axis, the resulting slope of the linear regression line is $-K_a$ and the x-intercept is NK_a.

The second strategy for studying drug–protein interactions employs the same approach as above but uses hollow fiber–protected two-phase microextraction with 1-octanol as the extracting solvent rather than HF(3)ME to determine equilibrium concentration of a free drug.[166] To ensure negligible depletion of the free drug (<5%) from the sample, the volumes of donor and acceptor solutions are 15 mL and 25 μL, respectively.

The third approach, employed for the determination of a free hormone, progesterone, in a protein solution, takes advantage of single-drop (direct-immersion) microextraction.[444] A 1-μL drop of octane suspended from the tip of a microsyringe needle in 500 mL of stirred 1% aqueous solution of bovine serum albumin (BSA) is used to extract unbound progesterone. The very small phase ratio employed, 2×10^{-6} mL of organic per milliliter of water, avoids perturbation of the aqueous solution equilibria. The binding constant was calculated from the measured free steroid concentration and calculated bound steroid concentration and total protein concentration and found to be in good agreement with literature values. The binding constant could be measured at both equilibrium and nonequilibrium extraction times. In addition, extraction kinetics was investigated and found to obey a convective–diffusive mass transfer model which accounted for diffusion of both free and bound analyte in the aqueous Nernst diffusion film adjacent to the interface.

6.11.2 Study of Kinetics of the Partitioning Process

Kinetics of solvent microextraction is considered in detail in Section 3.3. In this section we discuss some of the experimental results on mass transfer characteristics of SME. Single-drop microextraction in direct-immersion mode was investigated using a 1-μL drop of octane suspended in 1 mL of a stirred aqueous solution from the tip of a microsyringe needle.[445] Four analytes were used as model compounds: malathion, 4-methylacetophenone, 4-nitrotoluene, and progesterone. The amounts of analytes extracted were measured as a function of time by gas chromatography and found to follow a first-order extraction rate curve, which yielded the overall mass transfer coefficient for the analytes, $\overline{\beta}_o$. For a given analyte, $\overline{\beta}_o$ varied linearly with stirring rate. At a fixed stirring rate, $\overline{\beta}_o$ was directly proportional to the diffusion coefficient of the analyte D_{aq}. These findings supported the film theory of convective–diffusive mass transfer, as opposed to penetration theory.

Detailed kinetic studies of headspace SDME were carried out in a system consisting of 1 μL of 1-octanol suspended over 0.5 mL of aqueous solution in a 1.0-mL conical vial.[267] The analytes were benzene, toluene, ethylbenzene, and xylene (BTEX). In this case, mass transfer in the headspace was assumed to be a fast process, since diffusion coefficients in the gas phase are typically about 10^4 times larger than corresponding diffusion coefficients in liquids. Therefore, the rate of extraction should be limited either by slow mass transfer in aqueous phase or slow mass transfer in organic phase (1-octanol), or both. Plots of the peak area against the stirring time were generated at various stirring rates to understand the rate-determining step(s) in the overall extraction process. Extraction rate data were found to obey a simple first-order kinetic model, from which extraction-rate constants were calculated. The rate of extraction increased with stirring rate, indicating that aqueous-phase mass transfer was a limiting step in the overall extraction process. However, slow mass transfer in the aqueous phase alone did not account for the overall extraction rate. The resistance to mass transfer from the headspace into the organic drop, resulting from slow diffusion of analytes into 1-octanol, was the second rate-determining step.

6.11.3 Study of Mechanistic Aspects of In-Drop Derivatization

The objective of derivatization is to convert the analytes into derivatives that are extracted more efficiently and to improve the sensitivity and selectivity of their final determination, usually by one of chromatographic techniques. As discussed previously in this chapter, one implementation of this approach is simultaneous extraction–derivatization. In this case, the derivatizing agent dissolved in the extraction solvent reacts with the analyte to form a chemical product that remains in the solvent. The advantage of this strategy is that two steps are combined, resulting in shorter sample preparation time. The theory assumes that the derivatizing agent is in excess so that its concentration remains constant throughout the extraction process. Therefore, reaction rate is proportional only to analyte concentration in the organic solvent (i.e., the reaction is assumed to be pseudo-first order).

The combination of extraction and derivatization is usually described by two limiting cases. The first takes place when mass transfer to the organic solvent with the derivatizing agent is fast compared with the reaction rate. In other words, at any time during the extraction, the extracting phase is at equilibrium with the analyte in a well-stirred sample, resulting in a uniform reaction rate throughout the solvent (in drop or in fiber). This is a typical case, since the equilibration time for well-agitated samples is usually very short compared to a typical rate constant.[5] The second limiting case occurs when mass transfer to the extracting solvent is slow compared with the reaction rate. This condition can take place when the reaction is very fast. As a result, the product is formed close to the interface between the extracting phase and the sample matrix.

Two aldehydes, formaldehyde and hexanal, were used as model analytes in theoretical analysis and experimental evaluation of kinetics and mechanism of headspace single-drop microextraction with in-drop derivatization.[282] Their selection was based on different reaction rates with the derivatizing agent. The derivatizing agent was 2,4,6-trichlorophenylhydrazine (TCPH) dissolved in 1-octanol, which served as the extracting solvent. The products—hydrazones—were determined by GC-MS. TCPH and the hydrazones were proved not to desorb from 1-octanol; therefore, the reaction took place exclusively in the organic phase and not in headspace. In other words, during extraction–derivatization the analytes present in the gaseous phase diffused from that phase across the interface into the drop that contained TCPH. It was also assumed that equilibration of the analytes between the aqueous sample and headspace in a well-agitated system occurred quickly compared to diffusion into the organic drop and chemical reaction. For hexanal, mass transfer to the 1-octanol drop was found to be fast compared to the reaction rate. In this case, the rate of the overall extraction process was controlled by the reaction rate. Consequently, the steady-state concentration profile of hexanal is expected to be uniform throughout the 1-octanol drop, since diffusion is fast enough to replenish the aldehyde reacting within the drop. For formaldehyde, on the other hand, the reaction rate constant was larger by about an order of magnitude compared to that of hexanal, so the rates of mass transfer and chemical reaction were of the same order of magnitude. Therefore, both processes contributed to the overall extraction rate. As a result, the steady-state concentration of formaldehyde should be nonuniform, decreasing as a function of distance from the drop surface to the center due to the inability of diffusion to keep up with the rate of reaction. The simultaneous diffusion and chemical reaction result in enhanced mass transfer of the aldehydes into 1-octanol. Detection limits for aqueous matrices were 0.1 and 0.03 μM for formaldehyde and hexanal, respectively, and repeatability expressed as relative standard deviations was 3.4% and 4.9% for 1 μM hexanal and 0.3 μM formaldehyde, respectively. This demonstrates the applicability of the extraction–derivatization approach in SME.

6.11.4   Pharmacokinetic Studies Using SME

Pharmacokinetic studies involve the determination of time–concentration profiles for drugs in body fluids. In practice, samples of urine, blood, or blood components have to be collected before administration of a drug and then at specific time intervals. Next, the drug has to be extracted from the samples and its concentration determined. Since this is clearly an analytical task, the sample preparation method selected should meet certain conditions to be used in pharmacokinetics. First, it should use small sample volumes, as the amount of sample can be limited, particularly in the collection of blood. Second, it should be capable of extensive sample cleanup and analyte enrichment, owing to matrix complexity of body fluids and low analyte concentrations. Third, extraction time should be short to avoid further changes in drug concentration as a result of its interaction with matrix components. Drop-to-drop,[130] headspace,[137] and cloud–point[50,446] solvent microextraction procedures, which meet the foregoing criteria, have been used to determine concentration–time profiles for drugs in rat,[50,130] rabbit,[137] and human plasma.[446] All the procedures used small sample volumes: 0.5 mL,[50,137] 1 mL,[446] or as little as 8 μL in the case of drop-to-drop microextraction.[130] This last technique seems to be particularly well suited for pharmacokinetic studies, not only because it required very small sample volumes, but also no sample pretreatment or dilution was needed. An example of concentration–time profile for venlafaxine, a second–generation antidepressant drug, extracted from human plasma using cloud-point microextraction, is shown in Figure 6.20.[446]

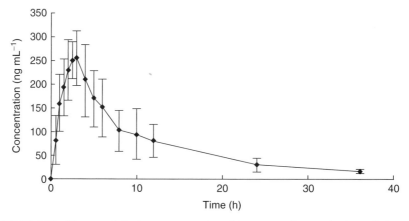

FIGURE 6.20. Mean venlafaxine concentration–time curve after a single oral dose of 150 mg venlafaxine in nine healthy volunteers. (Reprinted with permission from Ref. 446; copyright © 2008 Elsevier.)

6.11.5 Determination of Octanol–Water Partition Coefficients by SME

The octanol–water partition coefficient, K_{ow}, is the ratio of the concentration of a chemical in 1-octanol and in water at equilibrium and at a specified temperature. The logarithm of the octanol–water partition coefficient, log K_{ow} (or log P in some sources), is a quantitative thermodynamic measure of the lipophilic/hydrophilic balance of an organic compound. It is used in medicinal chemistry and rational drug design as a measure of molecular hydrophobicity. Hydrophobicity affects drug absorption, bioavailability, hydrophobic drug–receptor interactions, metabolism of molecules, as well as their toxicity. In addition, log K_{ow} has become a key parameter in studies of the environmental fate of chemicals, their bioaccumulation, soil–water partition coefficients, and sediment–water partition coefficients. The log K_{ow} values can be predicted using computer software, (e.g., ACD/LogP DB by Advanced Chemistry Development Inc., Toronto, Canada) or determined experimentally. The experimental methods fall into two categories, direct and indirect. Direct methods include the shake-flask and slow-stirring methods,[449] which are time-consuming, tedious, and require large amounts of chemicals. Indirect methods of estimating log K_{ow} take advantage of chromatographic techniques. They relate log K_{ow} values to the chromatographic capacity factor k'. The capacity factors are calculated from chromatographic retention data for standards with known log K_{ow} value and plotted against the known log K_{ow} to yield a standard curve. The capacity factors for compounds with unknown log K_{ow} are then used to determine log K_{ow} from these standard curves. Indirect methods are faster and easier to automate, but give good estimates only for compounds with similar functional groups or belonging to homologous series.

Hollow fiber–protected two-phase SME is a rapid, simple, and inexpensive alternative that can be used to determine log K_{ow} values of organic compounds.[447] The use of hollow fiber ensures the stability of organic solvent under vigorous stirring. The principle of the method is based on partitioning of the analytes between the bulk aqueous solution and 1-octanol in the pores and lumen of the porous fiber (see Figure 2.3). After equilibrium between the two phases has been reached, the acceptor and donor solutions are analyzed separately by HPLC and the equilibrium concentrations of the analytes in the two phases determined. The octanol–water partition coefficient can then be calculated directly from its definition:

$$K_{ow} = \frac{C_o}{C_w} \tag{6.4}$$

where C_o and C_w are the equilibrium concentrations of analyte in the organic (1-octanol in this case) and aqueous phases, respectively. It was determined experimentally that the time needed to reach equilibrium does not exceed 60 min. The log K_{ow} values determined by this procedure for 10 organic compounds are in good agreement with the literature data, the difference ranging from 0.6 to 6.6%.

As discussed in Chapter 3, when considering the distribution equilibria of acids and bases, the partition coefficient has to be replaced with the distribution ratio D, which is the ratio of the sum of the concentrations of all species in the organic phase to the sum of the concentrations of all species in the aqueous phase. For example, for a weak monoprotic acid HA, $D = K_{ow}\alpha_{HA}$, where α_{HA} is the fraction of the acid present in its neutral (protonated) form. Environmentally important weak acids, alkylphenols and chlorophenols, were investigated by means of negligible depletion techniques: HF(2)ME[448] and HF(3)ME,[450] respectively, followed by HPLC determination. The two procedures provided important environmental parameters, including freely dissolved concentration, a key parameter for environmental fate, exposure, and the effects of a chemical, distribution coefficients D, and binding to dissolved organic matter D_{DOC}. The experimental results agreed well with reported values for these parameters.

REFERENCES

1. Liu, H.; Dasgupta, P. K., Analytical chemistry in a drop: solvent extraction in a microdrop. *Anal. Chem.* 1996, *68* (11), 1817–1821.
2. Hyötyläinen, T.; Rickkola, M. L., Sorbent- and liquid-phase microextraction techniques and membrane-assisted extraction in combination with gas chromatographic analysis: a review. *Anal. Chim. Acta.* 2008, *614* (1), 27–37.
3. Majors, R. E., Miniaturized approaches to conventional liquid–liquid extraction. *LC•GC N. Am.* 2006, *24* (2), 118–130.
4. Miller, K. E.; Synovec, R. E., Review of analytical measurements facilitated by drop formation technology. *Talanta* 2000, *51* (5), 921–933.
5. Pawliszyn, J., Sample preparation: quo vadis? *Anal. Chem.* 2003, *75* (11), 2543–2558.
6. Pawliszyn, J.; Pedersen-Bjergaard, S., Analytical microextraction: current status and future trends. *J. Chromatogr. Sci.* 2006, *44* (6), 291–307.
7. Smith, R. M., Before the injection: modern methods of sample preparation for separation techniques. *J. Chromatogr. A* 2003, *1000* (1–2), 3–27.
8. Hakkarainen, M., Developments in multiple headspace extraction. *J. Biochem. Biophys. Methods* 2007, *70* (2), 229–233.
9. Lambropoulou, D. A.; Konstantinou, I. K.; Albanis, T. A., Recent developments in headspace microextraction techniques for the analysis of environmental contaminants in different matrices. *J. Chromatogr. A* 2007, *1152* (1–2), 70–96.
10. Xu, L.; Basheer, C.; Lee, H. K., Developments in single-drop microextraction. *J. Chromatogr. A* 2007, *1152* (1–2), 184–192.
11. Psillakis, E.; Kalogerakis, N., Developments in single-drop microextraction. *Trends Anal. Chem.* 2002, *21* (1), 53–63.
12. Barri, T.; Jönsson, J. Å., Advances and developments in membrane extraction for gas chromatography: techniques and applications. *J. Chromatogr. A* 2008, *1186* (1–2), 16–38.
13. Cordero, B. M.; Pavón, J. L. P.; Pinto, C. G.; Laespada, E. F. L.; Martinez, R. C.; Gonzalo, E. R., Analytical applications of membrane extraction in chromatography and electrophoresis. *J. Chromatogr. A* 2000, *902* (1), 195–204.
14. Jönsson, J. Å., Membrane extraction for sample preparation: a practical guide. *Chromatographia* 2003, *57* (Suppl. 1), S317–S324.
15. Jönsson, J. Å.; Mathiasson, L., Membrane-based techniques for sample enrichment. *J. Chromatogr. A* 2000, *902* (1), 205–225.
16. Jönsson, J. Å.; Mathiasson, L., Membrane extraction in analytical chemistry. *J. Sep. Sci.* 2001, *24* (7), 495–507.
17. Jönsson, J. Å.; Mathiasson, L., Sample preparation perspectives: membrane extraction for sample preparation. *LC•GC Eur.* 2003, *16* (10), 683–690.

18. Lee, J.; Lee, H. K.; Rasmussen, K. E.; Pedersen-Bjergaard, S., Environmental and bioanalytical applications of hollow fiber membrane liquid-phase microextraction: a review. *Anal. Chim. Acta* 2008, *624* (2), 253–268.
19. Majors, R. E.; Jönsson, J. Å.; Mathiasson, L., Membrane extraction for sample preparation. *LC•GC N. Am.* 2003, *21* (5), 424–438.
20. Psillakis, E.; Kalogerakis, N., Developments in liquid-phase microextraction. *Trends Anal. Chem.* 2003, *22* (9), 565–574.
21. Fatta, D.; Nikolaou, A.; Achilleos, A.; Meriç, S., Analytical methods for tracing pharmaceutical residues in water and wastewater. *Trends Anal. Chem.* 2007, *26* (6), 515–533.
22. Fu, X.; Liao, Y.; Liu, H., Sample preparation for pharmaceutical analysis. *Anal. Bioanal. Chem.* 2005, *381* (1), 75–77.
23. Lord, H.; Pawliszyn, J., Microextraction of drugs. *J. Chromatogr. A* 2000, *902* (1), 17–63.
24. Mutavdžić Pavlović, D.; Babić, S.; Horvat, A. J. M.; Kaštelan-Macan, M., Sample preparation in analysis of pharmaceuticals. *Trends Anal. Chem.* 2007, *26* (11), 1062–1075.
25. Pedersen-Bjergaard, S.; Rasmussen, K. E., Bioanalysis of drugs by liquid-phase microextraction coupled to separation techniques. *J. Chromatogr. B* 2005, *817* (1), 3–12.
26. Rasmussen, K. E.; Pedersen-Bjergaard, S., Developments in hollow fibre–based, liquid-phase microextraction. *Trends Anal. Chem.* 2004, *23* (1), 1–10.
27. Deng, C.; Liu, N.; Gao, M.; Zhang, X., Recent developments in sample preparation techniques for chromatography analysis of traditional Chinese medicines. *J. Chromatogr. A* 2007, *1153* (1–2), 90–96.
28. Demeestere, K.; Dewulf, J.; De Witte, B.; Van Langenhove, H., Sample preparation for the analysis of volatile organic compounds in air and water matrices. *J. Chromatogr. A* 2007, *1153* (1–2), 130–144.
29. Marczak, M.; Wolska, L.; Chrzanowski, W.; Namieśnik, J., Microanalysis of volatile organic compounds (VOCs) in water samples: methods and instruments. *Microchim. Acta* 2006, *155* (3–4), 331–348.
30. Hercegová, A.; Dömötörová, M.; Matisová, E., Sample preparation methods in the analysis of pesticide residues in baby food with subsequent chromatographic determination. *J. Chromatogr. A* 2007, *1153* (1–2), 54–73.
31. Lambropoulou, D. A.; Albanis, T. A., Liquid-phase microextraction techniques in pesticide residue analysis. *J. Biochem. Biophys. Methods* 2007, *70* (2), 195–228.
32. Quintana, M. C.; Ramos, L., Sample preparation for the determination of chlorophenols. *Trends Anal. Chem.* 2008, *27* (5), 418–436.
33. Ridgway, K.; Lalljie, S. P. D.; Smith, R. M., Sample preparation techniques for the determination of trace residues and contaminants in foods. *J. Chromatogr. A* 2007, *1153* (1–2), 36–53.
34. Bicchi, C.; Cordero, C.; Liberto, E.; Sgorbini, B.; Rubiolo, P., Headspace sampling of the volatile fraction of vegetable matrices. *J. Chromatogr. A* 2008, *1184* (1–2), 220–233.
35. Hyötyläinen, T.; Riekkola, M. L., Potential of effective extraction techniques and new analytical systems for profiling the marine environment. *Trends Anal. Chem.* 2007, *26* (8), 788–808.
36. Flanagan, R. J.; Morgan, P. E.; Spencer, E. P.; Whelpton, R., Micro-extraction techniques in analytical toxicology: short review. *Biomed. Chromatogr.* 2006, *20* (6–7), 530–538.
37. Wille, S. M. R.; Lambert, W. E. E., Recent developments in extraction procedures relevant to analytical toxicology. *Anal. Bioanal. Chem.* 2007, *388* (7), 1381–1391.
38. Baghdadi, M.; Shemirani, F., Cold-induced aggregation microextraction: a novel sample preparation technique based on ionic liquids. *Anal. Chim. Acta* 2008, *613* (1), 56–63.

39. Berijani, S.; Assadi, Y.; Anbia, M.; Milani Hosseini, M.-R.; Aghaee, E., Dispersive liquid–liquid microextraction combined with gas chromatography–flame photometric detection: very simple, rapid and sensitive method for the determination of organophosphorus pesticides in water. *J. Chromatogr. A* 2006, *1123* (1), 1–9.

40. Bidari, A.; Zeini Jahromi, E.; Assadi, Y.; Milani Hosseini, M. R., Monitoring of selenium in water samples using dispersive liquid–liquid microextraction followed by iridium-modified tube graphite furnace atomic absorption spectrometry. *Microchem. J.* 2007, *87* (1), 6–12.

41. Birjandi, A. P.; Bidari, A.; Rezaei, F.; Hosseini, M. R. M.; Assadi, Y., Speciation of butyl and phenyltin compounds using dispersive liquid–liquid microextraction and gas chromatography–flame photometric detection. *J. Chromatogr. A* 2008, *1193* (1–2), 19–25.

42. Chiang, J. S.; Huang, S. D., Simultaneous derivatization and extraction of anilines in waste water with dispersive liquid–liquid microextraction followed by gas chromatography-mass spectrometric detection. *Talanta* 2008, *75* (1), 70–75.

43. Farahani, H.; Norouzi, P.; Dinarvand, R.; Ganjali, M. R., Development of dispersive liquid–liquid microextraction combined with gas chromatography–mass spectrometry as a simple, rapid and highly sensitive method for the determination of phthalate esters in water samples. *J. Chromatogr. A* 2007, *1172* (2), 105–112.

44. Farajzadeh, M. A.; Bahram, M.; Jönsson, J. Å., Dispersive liquid–liquid microextraction followed by high-performance liquid chromatography–diode-array detection as an efficient and sensitive technique for determination of antioxidants. *Anal. Chim. Acta* 2007, *591* (1), 69–79.

45. Farajzadeh, M. A.; Bahram, M.; Mehr, B. G.; Jönsson, J. Å., Optimization of dispersive liquid–liquid microextraction of copper(II) by atomic absorption spectrometry as its oxinate chelate: application to determination of copper in different water samples. *Talanta* 2008, *75* (3), 832–840.

46. Fariña, L.; Boido, E.; Carrau, F.; Dellacassa, E., Determination of volatile phenols in red wines by dispersive liquid–liquid microextraction and gas chromatography–mass spectrometry detection. *J. Chromatogr. A* 2007, *1157* (1–2), 46–50.

47. Fattahi, N.; Assadi, Y.; Milani Hosseini, M. R.; Zeini Jahromi, E., Determination of chlorophenols in water samples using simultaneous dispersive liquid–liquid microextraction and derivatization followed by gas chromatography–electron-capture detection. *J. Chromatogr. A* 2007, *1157* (1–2), 23–29.

48. Fattahi, N.; Samadi, S.; Assadi, Y.; Milani Hosseini, M. R., Solid-phase extraction combined with dispersive liquid–liquid microextraction–ultra preconcentration of chlorophenols in aqueous samples. *J. Chromatogr. A* 2007, *1169* (1–2), 63–69.

49. García-López, M.; Rodríguez, I.; Cela, R., Development of a dispersive liquid–liquid microextraction method for organophosphorus flame retardants and plasticizers determination in water samples. *J. Chromatogr. A* 2007, *1166* (1–2), 9–15.

50. Han, F.; Yin, R.; Shi, X.; Jia, Q.; Liu, H.; Yao, H.; Xu, L.; Li, S., Cloud point extraction–HPLC method for determination and pharmacokinetic study of flurbiprofen in rat plasma after oral and transdermal administration. *J. Chromatogr. B* 2008, *868* (1–2), 64–69.

51. Jiang, H.; Qin, Y.; Hu, B., Dispersive liquid phase microextraction (DLPME) combined with graphite furnace atomic absorption spectrometry (GFAAS) for determination of trace Co and Ni in environmental water and rice samples. *Talanta* 2008, *74* (5), 1160–1165.

52. Kozani, R. R.; Assadi, Y.; Shemirani, F.; Milani Hosseini, M. R.; Jamali, M. R., Determination of trihalomethanes in drinking water by dispersive liquid–liquid microextraction then gas chromatography with electron-capture detection. *Chromatographia* 2007, *66* (1–2), 81–86.

53. Kozani, R. R.; Assadi, Y.; Shemirani, F.; Milani Hosseini, M.-R.; Jamali, M. R., Part-per-trillion determination of chlorobenzenes in water using dispersive liquid–liquid microextraction combined gas chromatography–electron capture detection. *Talanta* 2007, *72* (2), 387–393.

54. Li, Y.; Wei, G.; Hu, J.; Liu, X.; Zhao, X.; Wang, X., Dispersive liquid–liquid micro-extraction followed by reversed phase-high performance liquid chromatography for the determination of polybrominated diphenyl ethers at trace levels in landfill leachate and environmental water samples. *Anal. Chim. Acta* 2008, *615* (1), 96–103.

55. Liang, P.; Xu, J.; Li, Q., Application of dispersive liquid–liquid microextraction and high-performance liquid chromatography for the determination of three phthalate esters in water samples. *Anal. Chim. Acta* 2008, *609* (1), 53–58.

56. Melwanki, M. B.; Fuh, M. R., Dispersive liquid–liquid microextraction combined with semi-automated in-syringe back extraction as a new approach for the sample preparation of ionizable organic compounds prior to liquid chromatography. *J. Chromatogr. A* 2008, *1198–1199*, 1–6.

57. Nagaraju, D.; Huang, S. D., Determination of triazine herbicides in aqueous samples by dispersive liquid–liquid microextraction with gas chromatography–ion trap mass spectrometry. *J. Chromatogr. A* 2007, *1161* (1–2), 89–97.

58. Naseri, M. T.; Hemmatkhah, P.; Hosseini, M. R. M.; Assadi, Y., Combination of dispersive liquid–liquid microextraction with flame atomic absorption spectrometry using microsample introduction for determination of lead in water samples. *Anal. Chim. Acta* 2008, *610* (1), 135–141.

59. Naseri, M. T.; Hosseini, M. R. M.; Assadi, Y.; Kiani, A., Rapid determination of lead in water samples by dispersive liquid–liquid microextraction coupled with electrothermal atomic absorption spectrometry. *Talanta* 2008, *75* (1), 56–62.

60. Regueiro, J.; Llompart, M.; Garcia-Jares, C.; Garcia-Monteagudo, J. C.; Cela, R., Ultrasound-assisted emulsification-microextraction of emergent contaminants and pesticides in environmental waters. *J. Chromatogr. A* 2008, *1190* (1–2), 27–38.

61. Rezaee, M.; Assadi, Y.; Milani Hosseini, M.-R.; Aghaee, E.; Ahmadi, F.; Berijani, S., Determination of organic compounds in water using dispersive liquid–liquid micro-extraction. *J. Chromatogr. A* 2006, *1116* (1–2), 1–9.

62. Shokoufi, N.; Shemirani, F.; Assadi, Y., Fiber optic–linear array detection spectro-photometry in combination with dispersive liquid–liquid microextraction for simultaneous preconcentration and determination of palladium and cobalt. *Anal. Chim. Acta* 2007, *597* (2), 349–356.

63. Wei, G.; Li, Y.; Wang, X., Application of dispersive liquid–liquid microextraction combined with high-performance liquid chromatography for the determination of methomyl in natural waters. *J. Sep. Sci.* 2007, *30* (18), 3262–3267.

64. Xiong, J.; Hu, B., Comparison of hollow fiber liquid phase microextraction and dispersive liquid–liquid microextraction for the determination of organosulfur pesticides in environmental and beverage samples by gas chromatography with flame photometric detection. *J. Chromatogr. A* 2008, *1193* (1–2), 7–18.

65. Yazdi, A. S.; Razavi, N.; Yazdinejad, S. R., Separation and determination of amitriptyline and nortriptyline by dispersive liquid–liquid microextraction combined with gas chromatography flame ionization detection. *Talanta* 2008, *75* (5), 1293–1299.

66. Zeini Jahromi, E.; Bidari, A.; Assadi, Y.; Milani Hosseini, M. R.; Jamali, M. R., Dispersive liquid–liquid microextraction combined with graphite furnace atomic absorption spectrometry: ultra trace determination of cadmium in water samples. *Anal. Chim. Acta* 2007, *585* (2), 305–311.

67. Zhao, E.; Zhao, W.; Han, L.; Jiang, S.; Zhou, Z., Application of dispersive liquid–liquid microextraction for the analysis of organophosphorus pesticides in watermelon and cucumber. *J. Chromatogr. A* 2007, *1175* (1), 137–140.

68. Zhou, Q.; Bai, H.; Xie, G.; Xiao, J., Temperature-controlled ionic liquid dispersive liquid phase micro-extraction. *J. Chromatogr. A* 2008, *1177* (1), 43–49.

69. Zhou, Q.; Bai, H.; Xie, G.; Xiao, J., Trace determination of organophosphorus pesticides in environmental samples by temperature-controlled ionic liquid dispersive liquid-phase microextraction. *J. Chromatogr. A* 2008, *1188* (2), 148–153.

70. Zhou, Q. X.; Xie, G. H.; Pang, L., Rapid determination of atrazine in environmental water samples by a novel liquid phase microextraction. *Chin. Chem. Lett.* 2008, *19* (1), 89–91.

71. Akiyama, R.; Takagai, Y.; Igarashi, S., Determination of lower sub ppt levels of environmental analytes using high-powered concentration system and high-performance liquid chromatography with fluorescence detection. *Analyst* 2004, *129* (5), 396–397.

72. Ebrahimzadeh, H.; Yamini, Y.; Kamarei, F.; Shariati, S., Homogeneous liquid–liquid extraction of trace amounts of mononitrotoluenes from waste water samples. *Anal. Chim. Acta* 2007, *594* (1), 93–100.

73. Ghiasvand, A. R.; Mohagheghzadeh, E., Homogeneous liquid–liquid extraction of uranium(VI) using tri-*n*-octylphosphine oxide. *Anal. Sci.* 2004, *20* (6), 917–919.

74. Ghiasvand, A. R.; Moradi, F.; Sharghi, H.; Hasaninejad, A. R., Determination of silver(I) by electrothermal-AAS in a microdroplet formed from a homogeneous liquid–liquid extraction system using tetraspirocyclohexylcalix[4]pyrroles. *Anal. Sci.* 2005, *21* (4), 387–390.

75. Ghiasvand, A. R.; Shadabi, S.; Mohagheghzadeh, E.; Hashemi, P., Homogeneous liquid–liquid extraction method for the selective separation and preconcentration of ultra trace molybdenum. *Talanta* 2005, *66* (4), 912–916.

76. Igarashi, S.; Ide, N.; Takagai, Y., High-performance liquid chromatographic–spectrophotometric determination of copper(II) and palladium(II) with 5,10,15,20-tetrakis(4*N*-pyridyl)porphine following homogeneous liquid–liquid extraction in the water–acetic acid–chloroform ternary solvent system. *Anal. Chim. Acta* 2000, *424* (2), 263–269.

77. Igarashi, S.; Ide, N.; Takahata, K.; Takagai, Y., HPLC spectrophotometric determination of metal–porphyrin complexes following a preconcentration method by homogeneous liquid–liquid extraction in a water/pyridine/ethyl chloroacetate ternary component system. *Bunseki Kagaku* 1999, *48* (12), 1115–1122.

78. Igarashi, S.; Takahashi, A.; Ueki, Y.; Yamaguchi, H., Homogeneous liquid–liquid extraction followed by x-ray fluorescence spectrometry of a microdroplet on filter-paper for the simultaneous determination of small amounts of metals. *Analyst* 2000, *125* (5), 797–798.

79. Oshite, S.; Furukawa, M.; Igarashi, S., Homogeneous liquid–liquid extraction method for the selective spectrofluorimetric determination of trace amounts of tryptophan. *Analyst* 2001, *126* (5), 703–706.

80. Sudo, T.; Igarashi, S., Homogeneous liquid–liquid extraction method for spectrofluorimetric determination of chlorophyll *a*. *Talanta* 1996, *43* (2), 233–237.

81. Takagai, Y.; Akiyama, R.; Igarashi, S., Powerful preconcentration method for capillary electrophoresis and its application to analysis of ultratrace amounts of polycyclic aromatic hydrocarbons. *Anal. Bioanal. Chem.* 2006, *385* (5), 888–894.

82. Takagai, Y.; Igarashi, S., UV-detection capillary electrophoresis for benzo[*a*]pyrene and pyrene following a two-step concentration system using homogeneous liquid–liquid extraction and a sweeping method. *Analyst* 2001, *126* (5), 551–552.

83. Takagai, Y.; Igarashi, S., Homogeneous liquid–liquid extraction and micellar electrokinetic chromatography using sweeping effect concentration system for determination of trace amounts of several polycyclic aromatic hydrocarbons. *Anal. Bioanal. Chem.* 2002, *373* (1), 87–92.

84. Takagai, Y.; Igarashi, S., Homogeneous liquid–liquid extraction as a simple and powerful preconcentration method for capillary gas chromatography and capillary electrophoresis. *Am. Lab.* 2002, *34* (15), 29–30.

85. Takagai, Y.; Igarashi, S., Determination of ppb levels of tryptophan derivatives by capillary electrophoresis with homogeneous liquid–liquid extraction and sweeping method. *Chem. Pharm. Bull.* 2003, *51* (4), 373–377.

86. Takagai, Y.; Maekoya, C.; Igarashi, S., Determination of chlorophenol derivatives using the homogeneous liquid–liquid extraction in ternary component system–GC/MS method. *J. Chem. Soc. Japan* 2000, (4), 291–293.

87. Takahashi, A.; Igarashi, S.; Ueki, Y.; Yamaguchi, H., X-ray fluorescence analysis of trace metal ions following a preconcentration of metal–diethyldithiocarbamate complexes by homogeneous liquid–liquid extraction. *Fresenius' J. Anal. Chem.* 2000, *368* (6), 607–610.

88. Takahashi, A.; Ueki, Y.; Igarashi, S., Homogeneous liquid–liquid extraction of uranium (VI) from acetate aqueous solution. *Anal. Chim. Acta* 1999, *387* (1), 71–75.

89. Tavakoli, L.; Yamini, Y.; Ebrahimzadeh, H.; Shariati, S., Homogeneous liquid–liquid extraction for preconcentration of polycyclic aromatic hydrocarbons using a water/methanol/chloroform ternary component system. *J. Chromatogr. A* 2008, *1196–1197*, 133–138.

90. Wang, X.; Zhao, X.; Liu, X.; Li, Y.; Fu, L.; Hu, J.; Huang, C., Homogeneous liquid–liquid extraction combined with gas chromatography–electron capture detector for the determination of three pesticide residues in soils. *Anal. Chim. Acta* 2008, *620* (1–2), 162–169.

91. Zhu, H. Z.; Liu, W.; Mao, J. W.; Yang, M. M., Cloud point extraction and determination of trace trichlorfon by high performance liquid chromatography with ultraviolet-detection based on its catalytic effect on benzidine oxidizing. *Anal. Chim. Acta* 2008, *614* (1), 58–62.

92. Batlle, R.; López, P.; Nerín, C.; Crescenzi, C., Active single-drop microextraction for the determination of gaseous diisocyanates. *J. Chromatogr. A* 2008, *1185* (2), 155–160.

93. Bishop, E. J.; Mitra, S., Measurement of nitrophenols in air samples by impinger sampling and supported liquid membrane micro-extraction. *Anal. Chim. Acta* 2007, *583* (1), 10–14.

94. Esteve-Turrillas, F. A.; Pastor, A.; de la Guardia, M., Assessing air quality inside vehicles and at filling stations by monitoring benzene, toluene, ethylbenzene and xylenes with the use of semipermeable devices. *Anal. Chim. Acta* 2007, *593* (1), 108–116.

95. Xie, J. P.; Sun, S. H.; Wang, H.-Y.; Zong, Y. L.; Nie, C.; Guo, Y. L., Determination of nicotine in mainstream smoke on the single puff level by liquid-phase microextraction coupled to matrix-assisted laser desorption/ionization Fourier transform mass spectrometry. *Rapid Commun. Mass Spectrom.* 2006, *20* (17), 2573–2578.

96. Curyło, J.; Wardencki, W., Application of single drop extraction (SDE) gas chromatography method for the determination of carbonyl compounds in spirits and vodkas. *Anal. Lett.* 2006, *39* (13), 2629–2642.

97. Michulec, M.; Wardencki, W., The application of single drop extraction technique for chromatographic determination of solvent residues in edible oils and pharmaceutical products. *Chromatographia* 2006, *64* (3–4), 191–197.

98. Zalieckaitė, R.; Adomavičiūtė, E.; Vičkačkaitė, V., Single-drop microextraction for the determination of phthalate esters. *Chemija* (*Vilnius*) 2007, *18* (3), 25–29.

99. Kokosa, J. M.; Przyjazny, A., Headspace microdrop analysis: an alternative test method for gasoline diluent and benzene, toluene, ethylbenzene and xylenes in used engine oils. *J. Chromatogr. A* 2003, *983* (1–2), 205–214.

100. Ouyang, G.; Zhao, W.; Pawliszyn, J., Automation and optimization of liquid-phase microextraction by gas chromatography. *J. Chromatogr. A* 2007, *1138* (1–2), 47–54.

101. Chen, C. C.; Melwanki, M. B.; Huang, S. D., Liquid–liquid–liquid microextraction with automated movement of the acceptor and the donor phase for the extraction of phenoxyacetic acids prior to liquid chromatography detection. *J. Chromatogr. A* 2006, *1104* (1–2), 33–39.

102. Chen, P. S.; Huang, S. D., Coupled two-step microextraction devices with derivatizations to identify hydroxycarbonyls in rain samples by gas chromatography–mass spectrometry. *J. Chromatogr. A* 2006, *1118* (2), 161–167.

103. Chen, P. S.; Huang, S. D., Determination of ethoprop, diazinon, disulfoton and fenthion using dynamic hollow fiber–protected liquid-phase microextraction coupled with gas chromatography–mass spectrometry. *Talanta* 2006, *69* (3), 669–675.

104. Chiang, J. S.; Huang, S. D., Determination of haloethers in water with dynamic hollow fiber liquid-phase microextraction using GC-FID and GC-ECD *Talanta* 2007, *71* (2), 882–886.

105. Chiang, J.-S.; Huang, S.-D., Simultaneous derivatization and extraction of amphetamine and methylenedioxyamphetamine in urine with headspace liquid-phase microextraction followed by gas chromatography–mass spectrometry. *J. Chromatogr. A* 2008, *1185* (1), 19–22.

106. Hou, L.; Lee, H. K., Dynamic Three-phase microextraction as a sample preparation technique prior to capillary electrophoresis. *Anal. Chem.* 2003, *75* (11), 2784–2789.

107. Hou, L.; Lee, H. K., Determination of pesticides in soil by liquid-phase microextraction and gas chromatography–mass spectrometry. *J. Chromatogr. A* 2004, *1038* (1–2), 37–42.

108. Hou, L.; Shen, G.; Lee, H. K., Automated hollow fiber–protected dynamic liquid-phase microextraction of pesticides for gas chromatography–mass spectrometric analysis. *J. Chromatogr. A* 2003, *985* (1–2), 107–116.

109. Huang, S. P.; Huang, S. D., Dynamic hollow fiber protected liquid phase microextraction and quantification using gas chromatography combined with electron capture detection of organochlorine pesticides in green tea leaves and ready-to-drink tea. *J. Chromatogr. A* 2006, *1135* (1), 6–11.

110. Huang, S. P.; Huang, S. D., Determination of organochlorine pesticides in water using solvent cooling assisted dynamic hollow-fiber-supported headspace liquid-phase microextraction. *J. Chromatogr. A* 2007, *1176* (1–2), 19–25.

111. Jiang, X.; Basheer, C.; Zhang, J.; Lee, H. K., Dynamic hollow fiber–supported headspace liquid-phase microextraction. *J. Chromatogr. A* 2005, *1087* (1–2), 289–294.

112. Jiang, X.; Oh, S. Y.; Lee, H. K., Dynamic liquid–liquid–liquid microextraction with automated movement of the acceptor phase. *Anal. Chem.* 2005, *77* (6), 1689–1695.

113. Lin, C. Y.; Huang, S. D., Application of liquid–liquid–liquid microextraction and high-performance liquid–chromatography for the determination of sulfonamides in water. *Anal. Chim. Acta* 2008, *612* (1), 37–43.

114. Pezo, D.; Salafranca, J.; Nerín, C., Development of an automatic multiple dynamic hollow fibre liquid-phase microextraction procedure for specific migration analysis of new active food packagings containing essential oils. *J. Chromatogr. A* 2007, *1174* (1–2), 85–94.

115. Wu, J.; Ee, K. H.; Lee, H. K., Automated dynamic liquid–liquid–liquid microextraction followed by high-performance liquid chromatography–ultraviolet detection for the

determination of phenoxy acid herbicides in environmental waters. *J. Chromatogr. A* 2005, *1082* (2), 121–127.

116. Wu, J.; Lee, H. K., Ion-pair dynamic liquid-phase microextraction combined with injection-port derivatization for the determination of long-chain fatty acids in water samples. *J. Chromatogr. A* 2006, *1133* (1–2), 13–20.

117. Zhao, L.; Lee, H. K., Liquid-phase microextraction combined with hollow fiber as a sample preparation technique prior to gas chromatography/mass spectrometry. *Anal. Chem.* 2002, *74* (11), 2486–2492.

118. Mohammadi, A.; Alizadeh, N., Automated dynamic headspace organic solvent film microextraction for benzene, toluene, ethylbenzene and xylene: renewable liquid film as a sampler by a programmable motor. *J. Chromatogr. A* 2006, *1107* (1–2), 19–28.

119. Myung, S. W.; Yoon, S. H.; Kim, M., Analysis of benzene ethylamine derivatives in urine using the programmable dynamic liquid-phase microextraction (LPME) device. *Analyst* 2003, *128* (12), 1443–1446.

120. Saraji, M., Dynamic headspace liquid-phase microextraction of alcohols. *J. Chromatogr. A* 2005, *1062* (1), 15–21.

121. Liang, P.; Xu, J.; Guo, L.; Song, F., Dynamic liquid-phase microextraction with HPLC for the determination of phoxim in water samples. *J. Sep. Sci.* 2006, *29* (3), 366–370.

122. Shen, G.; Lee, H. K., Headspace liquid-phase microextraction of chlorobenzenes in soil with gas chromatography–electron capture detection. *Anal. Chem.* 2003, *75* (1), 98–103.

123. Shrivas, K.; Wu, H. F., Quantitative bioanalysis of quinine by atmospheric pressure–matrix assisted laser desorption/ionization mass spectrometry combined with dynamic drop-to-drop solvent microextraction. *Anal. Chim. Acta* 2007, *605* (2), 153–158.

124. Wang, Y.; Kwok, Y. C.; He, Y.; Lee, H. K., Application of dynamic liquid-phase microextraction to the analysis of chlorobenzenes in water by using a conventional microsyringe. *Anal. Chem.* 1998, *70* (21), 4610–4614.

125. Xu, J.; Liang, P.; Zhang, T., Dynamic liquid-phase microextraction of three phthalate esters from water samples and determination by gas chromatography. *Anal. Chim. Acta* 2007, *597* (1), 1–5.

126. Grefslie Ugland, H.; Krogh, M.; Reubsaet, L., Three-phase liquid-phase microextraction of weakly basic drugs from whole blood. *J. Chromatogr. B* 2003, *798* (1), 127–135.

127. Halvorsen, T. G.; Pedersen-Bjergaard, S.; Rasmussen, K. E., Reduction of extraction times in liquid-phase microextraction. *J. Chromatogr. B* 2001, *760* (2), 219–226.

128. Halvorsen, T. G.; Pedersen-Bjergaard, S.; Reubsaet, J. L. E.; Rasmussen, K. E., Liquid-phase microextraction combined with flow-injection tandem mass spectrometry: rapid screening of amphetamines from biological matrices. *J. Sep. Sci.* 2001, *24* (7), 615–622.

129. Halvorsen, T. G.; Pedersen-Bjergaard, S.; Reubsaet, J. L. E.; Rasmussen, K. E., Liquid-phase microextraction combined with liquid chromatography–mass spectrometry: extraction from small volumes of biological samples. *J. Sep. Sci.* 2003, *26* (17), 1520–1526.

130. Agrawal, K.; Wu, H. F., Drop-to-drop solvent microextraction coupled with gas chromatography/mass spectrometry for rapid determination of trimeprazine in urine and blood of rats: application to pharmacokinetic studies. *Rapid Commun. Mass Spectrom.* 2007, *21* (20), 3352–3356.

131. Andersen, S.; Halvorsen, T. G.; Pedersen-Bjergaard, S.; Rasmussen, K. E., Liquid-phase microextraction combined with capillary electrophoresis, a promising tool for the determination of chiral drugs in biological matrices. *J. Chromatogr. A* 2002, *963* (1–2), 303–312.

132. Andersen, S.; Halvorsen, T. G.; Pedersen-Bjergaard, S.; Rasmussen, K. E.; Tanum, L.; Refsum, H., Stereospecific determination of citalopram and desmethylcitalopram by capillary electrophoresis and liquid-phase microextraction. *J. Pharm. Biomed. Anal.* 2003, *33* (2), 263–273.

133. Bagheri, H.; Khalilian, F.; Babanezhad, E.; Es-haghi, A.; Rouini, M. R., Modified solvent microextraction with back extraction combined with liquid chromatography–fluorescence detection for the determination of citalopram in human plasma. *Anal. Chim. Acta* 2008, *610* (2), 211–216.

134. Basheer, C.; Lee, H. K.; Obbard, J. P., Application of liquid-phase microextraction and gas chromatography–mass spectrometry for the determination of polychlorinated biphenyls in blood plasma. *J. Chromatogr. A* 2004, *1022* (1–2), 161–169.

135. de Santana, F. J. M.; Bonato, P. S., Enantioselective analysis of mirtazapine and its two major metabolites in human plasma by liquid chromatography–mass spectrometry after three-phase liquid-phase microextraction. *Anal. Chim. Acta* 2007, *606* (1), 80–91.

136. de Santana, F. J. M.; de Oliveira, A. R. M.; Bonato, P. S., Chiral liquid chromatographic determination of mirtazapine in human plasma using two-phase liquid-phase micro-extraction for sample preparation. *Anal. Chim. Acta* 2005, *549* (1–2), 96–103.

137. Dong, L.; Deng, C.; Wang, B.; Shen, X., Fast determination of Z-ligustilide in plasma by gas chromatography/mass spectrometry following headspace single-drop micro-extraction. *J. Sep. Sci.* 2007, *30* (9), 1318–1325.

138. Dziarkowska, K.; Jönsson, J. Å.; Wieczorek, P. P., Single hollow fiber SLM extraction of polyamines followed by tosyl chloride derivatization and HPLC determination. *Anal. Chim. Acta* 2008, *606* (2), 184–193.

139. Ebrahimzadeh, H.; Yamini, Y.; Sedighi, A.; Rouini, M. R., Determination of tramadol in human plasma and urine samples using liquid phase microextraction with back extraction combined with high performance liquid chromatography. *J. Chromatogr. B* 2008, *863* (2), 229–234.

140. Esrafili, A.; Yamini, Y.; Shariati, S., Hollow fiber-based liquid phase microextraction combined with high-performance liquid chromatography for extraction and determination of some antidepressant drugs in biological fluids. *Anal. Chim. Acta* 2007, *604* (2), 127–133.

141. Fiamegos, Y. C.; Stalikas, C. D., Gas chromatographic determination of carbonyl compounds in biological and oil samples by headspace single-drop microextraction with in-drop derivatisation. *Anal. Chim. Acta* 2008, *609* (2), 175–183.

142. Grefslie Ugland, H.; Krogh, M.; Rasmussen, K. E., Liquid-phase microextraction as a sample preparation technique prior to capillary gas chromatographic-determination of benzodiazepines in biological matrices. *J. Chromatogr. B* 2000, *749* (1), 85–92.

143. Halvorsen, T. G.; Pedersen-Bjergaard, S.; Rasmussen, K. E., Liquid-phase micro-extraction and capillary electrophoresis of citalopram, an antidepressant drug. *J. Chromatogr. A* 2001, *909* (1), 87–93.

144. Ho, T. S.; Halvorsen, T. G.; Pedersen-Bjergaard, S.; Rasmussen, K. E., Liquid-phase microextraction of hydrophilic drugs by carrier-mediated transport. *J. Chromatogr. A* 2003, *998* (1–2), 61–72.

145. Ho, T. S.; Pedersen-Bjergaard, S.; Rasmussen, K. E., Recovery, enrichment and selectivity in liquid-phase microextraction: comparison with conventional liquid–liquid extraction *J. Chromatogr. A* 2002, *963* (1–2), 3–17.

146. Ho, T. S.; Pedersen-Bjergaard, S.; Rasmussen, K. E., Liquid-phase microextraction of protein-bound drugs under non-equilibrium conditions. *Analyst* 2002, *127* (5), 608–613.

147. Ho, T. S.; Pedersen-Bjergaard, S.; Rasmussen, K. E., Experiences with carrier-mediated transport in liquid-phase microextraction. *J. Chromatogr. Sci.* 2006, *44* (6), 308–316.

148. Ho, T. S.; Reubsaet, J. L. E.; Anthonsen, H. S.; Pedersen-Bjergaard, S.; Rasmussen, K. E., Liquid-phase microextraction based on carrier mediated transport combined with liquid chromatography–mass spectrometry: new concept for the determination of polar drugs in a single drop of human plasma. *J. Chromatogr. A* 2005, *1072* (1), 29–36.

149. Kjelsen, I. J. O.; Gjelstad, A.; Rasmussen, K. E.; Pedersen-Bjergaard, S., Low-voltage electromembrane extraction of basic drugs from biological samples. *J. Chromatogr. A* 2008, *1180* (1–2), 1–9.

150. Li, L.; Hu, B., Hollow-fibre liquid phase microextraction for separation and preconcentration of vanadium species in natural waters and their determination by electrothermal vaporization-ICP-OES. *Talanta* 2007, *72* (2), 472–479.

151. Li, L.; Hu, B.; Xia, L.; Jiang, Z., Determination of trace Cd and Pb in environmental and biological samples by ETV-ICP-MS after single-drop microextraction. *Talanta* 2006, *70* (2), 468–473.

152. Magalhães, I. R. S.; Bonato, P. S., Liquid-phase microextraction combined with high-performance liquid chromatography for the enantioselective analysis of mefloquine in plasma samples. *J. Pharm. Biomed. Anal.* 2008, *46* (5), 929–936.

153. Pedersen-Bjergaard, S.; Ho, T. S.; Rasmussen, K. E., Fundamental studies on selectivity in 3-phase liquid-phase microextraction (LPME) of basic drugs. *J. Sep. Sci.* 2002, *25* (3), 141–146.

154. Pedersen-Bjergaard, S.; Rasmussen, K. E., Liquid–liquid–liquid microextraction for sample preparation of biological fluids prior to capillary electrophoresis. *Anal. Chem.* 1999, *71* (14), 2650–2656.

155. Pedersen-Bjergaard, S.; Rasmussen, K. E., Liquid-phase microextraction utilising plant oils as intermediate extraction medium: towards elimination of synthetic organic solvents in sample preparation. *J. Sep. Sci.* 2004, *27* (17–18), 1511–1516.

156. Pedersen-Bjergaard, S.; Rasmussen, K. E., Electrokinetic migration across artificial liquid membranes: new concept for rapid sample preparation of biological fluids. *J. Chromatogr. A* 2006, *1109* (2), 183–190.

157. Pedersen-Bjergaard, S.; Rasmussen, K. E.; Brekke, A.; Ho, T. S.; Halvorsen, T. G., Liquid-phase microextraction of basic drugs: selection of extraction mode based on computer calculated solubility data. *J. Sep. Sci.* 2005, *28* (11), 1195–1203.

158. Rasmussen, K. E.; Pedersen-Bjergaard, S.; Krogh, M.; Grefslie Ugland, H.; Grønhaug, T., Development of a simple in-vial liquid-phase microextraction device for drug analysis compatible with capillary gas chromatography, capillary electrophoresis and high-performance liquid chromatography. *J. Chromatogr. A* 2000, *873* (1), 3–11.

159. Shariati, S.; Yamini, Y.; Darabi, M.; Amini, M., Three phase liquid phase microextraction of phenylacetic acid and phenylpropionic acid from biological fluids. *J. Chromatogr. B* 2007, *855* (2), 228–235.

160. Tan, L.; Zhao, X. P.; Liu, X. Q.; Ju, H. X.; Li, J. S., Headspace liquid-phase microextraction of short-chain fatty acids in plasma, and gas chromatography with flame ionization detection. *Chromatographia* 2005, *62* (5–6), 305–309.

161. Trtić-Petrović, T.; Liu, J. F.; Jönsson, J. Å., Equilibrium sampling through membrane based on a single hollow fibre for determination of drug–protein binding and free drug concentration in plasma. *J. Chromatogr. B* 2005, *826* (1–2), 169–176.

162. Xia, L.; Hu, B.; Jiang, Z.; Wu, Y.; Chen, R.; Li, L., Hollow fiber liquid phase microextraction combined with electrothermal vaporization ICP-MS for the speciation of inorganic selenium in natural waters. *J. Anal. At. Spectrom.* 2006, *21* (3), 362–365.

163. Yang, C.; Guo, L.; Liu, X.; Zhang, H.; Liu, M., Determination of tetrandrine and fangchinoline in plasma samples using hollow fiber liquid-phase microextraction combined with high-performance liquid chromatography. *J. Chromatogr. A* 2007, *1164* (1–2), 56–64.

164. Yang, X.; Luo, M.; Tang, Y., Novel approach to enrich nicotine in plasma for rapid high performance liquid chromatographic analysis using three-phase hollow fiber based liquid phase microextraction. *Chin. J. Chromatogr.* 2006, *24* (6), 555–559.

165. Zhang, Z.; Zhao, Q.; Kang, S.; Chen, B.; Ma, M.; Yao, S., Determination of local anesthetics in human plasma by liquid–liquid–liquid microextraction coupled with high performance liquid chromatography. *Chin. J. Anal. Chem.* 2006, *34* (2), 165–169.

166. Fu, H.; Guan, J.; Bao, J. J., A hollow fiber solvent microextraction approach to measure drug–protein binding. *Anal. Sci.* 2006, *22* (12), 1565–1569.

167. Liu, B. M.; Malik, P. K.; Wu, H.-F., Single-drop microextraction and gas chromatography/mass spectrometric determination of anisaldehyde isomers in human urine and blood serum. *Rapid Commun. Mass Spectrom.* 2004, *18* (18), 2059–2064.

168. Shahdousti, P.; Mohammadi, A.; Alizadeh, N., Determination of valproic acid in human serum and pharmaceutical preparations by headspace liquid-phase microextraction gas chromatography–flame ionization detection without prior derivatization. *J. Chromatogr. B* 2007, *850* (1–2), 128–133.

169. Sobhi, H. R.; Yamini, Y.; Abadi, R. H. H. B., Extraction and determination of trace amounts of chlorpromazine in biological fluids using hollow fiber liquid phase microextraction followed by high-performance liquid chromatography. *J. Pharm. Biomed. Anal.* 2007, *45* (5), 769–774.

170. Xia, L.; Wu, Y.; Hu, B., Hollow-fiber liquid-phase microextraction prior to low-temperature electrothermal vaporization ICP-MS for trace element analysis in environmental and biological samples. *J. Mass Spectrom.* 2007, *42* (6), 803–810.

171. Xiao, Q.; Hu, B.; Duan, J.; He, M.; Zu, W., Analysis of PBDEs in soil, dust, spiked lake water, and human serum samples by hollow fiber–liquid phase microextraction combined with GC-ICP-MS. *J. Am. Soc. Mass Spectrom.* 2007, *18* (10), 1740–1748.

172. Zhao, G.; Liu, J. F.; Nyman, M.; Jönsson, J. Å., Determination of short-chain fatty acids in serum by hollow fiber supported liquid membrane extraction coupled with gas chromatography. *J. Chromatogr. B* 2007, *846* (1–2), 202–208.

173. Abdel-Rehim, M.; Dahlgren, M.; Blomberg, L., Quantification of ropivacaine and its major metabolites in human urine samples utilizing microextraction in a packed syringe automated with liquid chromatography–tandem mass spectrometry (MEPS-LC-MS/MS). *J. Sep. Sci.* 2006, *29* (11), 1658–1661.

174. Casari, C.; Andrews, A. R. J., Application of solvent microextraction to the analysis of amphetamines and phencyclidine in urine. *Forensic Sci. Intern.* 2001, *120* (3), 165–171.

175. de Jager, L.; Andrews, A. R. J., Development of a screening method for cocaine and cocaine metabolites in urine using solvent microextraction in conjunction with gas chromatography. *J. Chromatogr. A* 2001, *911* (1), 97–105.

176. de Jager, L.; Andrews, A. R. J., Preliminary studies of a fast screening method for cocaine and cocaine metabolites in urine using hollow fibre membrane solvent microextraction (HFMSME). *Analyst* 2001, *126* (8), 1298–1303.

177. de Oliveira, A. R. M.; Cardoso, C. D.; Bonato, P. S., Stereoselective determination of hydroxychloroquine and its metabolites in human urine by liquid-phase microextraction and CE. *Electrophoresis* 2007, *28* (7), 1081–1091.

178. El-Beqqali, A.; Abdel-Rehim, M., Quantitative analysis of methadone in human urine samples by microextraction in packed syringe–gas chromatography–mass spectrometry (MEPS-GC-MS). *J. Sep. Sci.* 2007, *30* (15), 2501–2505.

179. El-Beqqali, A.; Kussak, A.; Abdel-Rehim, M., Determination of dopamine and serotonin in human urine samples utilizing microextraction online with liquid chromatography/electrospray tandem mass spectrometry. *J. Sep. Sci.* 2007, *30* (3), 421–424.

180. Fiamegos, Y. C.; Stalikas, C. D., In-drop derivatisation liquid-phase microextraction assisted by ion-pairing transfer for the gas chromatographic determination of phenolic endocrine disruptors. *Anal. Chim. Acta* 2007, *597* (1), 32–40.

181. He, Y.; Kang, Y. J., Single drop liquid–liquid–liquid microextraction of methamphetamine and amphetamine in urine. *J. Chromatogr. A* 2006, *1133* (1–2), 35–40.

182. He, Y.; Vargas, A.; Kang, Y.-J., Headspace liquid-phase microextraction of metamphetamine and amphetamine in urine by an aqueous drop. *Anal. Chim. Acta* 2007, *589* (2), 225–230.

183. Hou, L.; Wen, X.; Tu, C.; Lee, H. K., Combination of liquid-phase microextraction and on-column stacking for trace analysis of amino alcohols by capillary electrophoresis. *J. Chromatogr. A* 2002, *979* (1–2), 163–169.

184. Jermak, S.; Pranaitytė, B.; Padarauskas, A., Headspace single-drop microextraction with in-drop derivatization and capillary electrophoretic determination for free cyanide analysis. *Electrophoresis* 2006, *27* (22), 4538–4544.

185. Kawaguchi, M.; Ito, R.; Okanouchi, N.; Saito, K.; Nakazawa, H., Miniaturized hollow fiber assisted liquid-phase microextraction with in situ derivatization and gas chromatography–mass spectrometry for analysis of bisphenol A in human urine sample. *J. Chromatogr. B* 2008, *870* (1), 98–102.

186. Kramer, K. E.; Andrews, A. R. J., Screening method for 11-nor-delta9-tetrahydrocannabinol-9-carboxylic acid in urine using hollow fiber membrane solvent microextraction with in-tube derivatization. *J. Chromatogr. B* 2001, *760* (1), 27–36.

187. Kuuranne, T.; Kotiaho, T.; Pedersen-Bjergaard, S.; Rasmussen, K. E.; Leinonen, A.; Westwood, S.; Kostiainen, R., Feasibility of a liquid-phase microextraction sample cleanup and liquid chromatographic/mass spectrometric screening method for selected anabolic steroid glucuronides in biological samples. *J. Mass Spectrom.* 2003, *38* (1), 16–26.

188. Lai, B. W.; Liu, B. M.; Malik, P. K.; Wu, H. F., Combination of liquid-phase hollow fiber membrane microextraction with gas chromatography–negative chemical ionization mass spectrometry for the determination of dichlorophenol isomers in water and urine. *Anal. Chim. Acta* 2006, *576* (1), 61–66.

189. Leinonen, A.; Vuorensola, K.; Lepola, L. M.; Kuuranne, T.; Kotiaho, T.; Ketola, R. A.; Kostiainen, R., Liquid-phase microextraction for sample preparation in analysis of unconjugated anabolic steroids in urine. *Anal. Chim. Acta* 2006, *559* (2), 166–172.

190. Melwanki, M. B.; Hsu, W. H.; Huang, S. D., Determination of clenbuterol in urine using headspace solid phase microextraction or liquid–liquid–liquid microextraction. *Anal. Chim. Acta* 2005, *552* (1–2), 67–75.

191. Melwanki, M. B.; Huang, S. D.; Fuh, M. R., Three-phase solvent bar microextraction and determination of trace amounts of clenbuterol in human urine by liquid chromatography and electrospray tandem mass spectrometry. *Talanta* 2007, *72* (2), 373–377.

192. Pedersen-Bjergaard, S.; Rasmussen, K. E., Liquid-phase microextraction and capillary electrophoresis of acidic drugs. *Electrophoresis* 2000, *21* (3), 579–585.

193. Sarafraz-Yazdi, A.; Raouf-Yazdinejad, S.; Es'haghi, Z., Directly suspended droplet microextraction and analysis of amitriptyline and nortriptyline by GC. *Chromatographia* 2007, *66* (7–8), 613–617.

194. Sekar, R.; Wu, H. F., Quantitative method for analysis of monensin in soil, water, and urine by direct combination of single-drop microextraction with atmospheric pressure matrix-assisted laser desorption/ionization mass spectrometry. *Anal. Chem.* 2006, *78* (18), 6306–6313.

195. Tsai, T. F.; Lee, M. R., Liquid-phase microextration combined with liquid chromatography–electrospray tandem mass spectrometry for detecting diuretics in urine. *Talanta* 2008, *75* (3), 658–665.

196. Vidal, L.; Canals, A.; Salvador, A., Sensitive determination of free benzophenone-3 in human urine samples based on an ionic liquid as extractant phase in single-drop microextraction prior to liquid chromatography analysis. *J. Chromatogr. A* 2007, *1174* (1–2), 95–103.

197. Wang, C.; Li, C.; Zang, X.; Han, D.; Liu, Z.; Wang, Z., Hollow fiber–based liquid-phase microextraction combined with on-line sweeping for trace analysis of *Strychnos* alkaloids in urine by micellar electrokinetic chromatography. *J. Chromatogr. A* 2007, *1143* (1–2), 270–275.

198. Wu, H. F.; Lin, C. H., Direct combination of immersed single-drop microextraction with atmospheric pressure matrix-assisted laser desorption/ionization tandem mass spectrometry for rapid analysis of a hydrophilic drug via hydrogen-bonding interaction and comparison with liquid–liquid extraction and liquid-phase microextraction using a dual gauge microsyringe with a hollow fiber. *Rapid Commun. Mass Spectrom.* 2006, *20* (16), 2511–2515.

199. Yamini, Y.; Reimann, C. T.; Vatanara, A.; Jönsson, J. Å., Extraction and pre-concentration of salbutamol and terbutaline from aqueous samples using hollow fiber supported liquid membrane containing anionic carrier. *J. Chromatogr. A* 2006, *1124* (1–2), 57–67.

200. Zang, X. H.; Li, C. R.; Wu, Q. H.; Han, D. D.; Wang, Z., Combination of hollow fiber–based liquid-phase microextraction with sweeping techniques in micellar electrokinetic chromatography for the determination of *Strychnos* alkaloids in human urine. *Chin. Chem. Lett.* 2007, *18* (3), 316–318.

201. Zhang, Z.; Zhang, C.; Su, X.; Ma, M.; Chen, B.; Yao, S., Carrier-mediated liquid phase microextraction coupled with high performance liquid chromatography for determination of illicit drugs in human urine. *Anal. Chim. Acta* 2008, *621* (2), 185–192.

202. Colombini, V.; Bancon-Montigny, C.; Yang, L.; Maxwell, P.; Sturgeon, R. E.; Mester, Z., Headspace single drop microextraction for the detection of organotin compounds. *Talanta* 2004, *63* (3), 555–560.

203. Gil, S.; Fragueiro, S.; Lavilla, I.; Bendicho, C., Determination of methylmercury by electrothermal atomic absorption spectrometry using headspace single-drop microextraction with in situ hydride generation. *Spectrochim. Acta B* 2005, *60* (1), 145–150.

204. Maltez, H. F.; Borges, D. L. G.; Carasek, E.; Welz, B.; Curtius, A. J., Single drop microextraction with *O,O*-diethyl dithiophosphate for the determination of lead by electrothermal atomic absorption spectrometry. *Talanta* 2008, *74* (4), 800–805.

205. Xia, L.; Hu, B.; Wu, Y., Hollow fiber–based liquid–liquid–liquid microextraction combined with high-performance liquid chromatography for the speciation of organomercury. *J. Chromatogr. A* 2007, *1173* (1–2), 44–51.

206. Ouyang, G.; Pawliszyn, J., Kinetic calibration for automated hollow fiber–protected liquid-phase microextraction. *Anal. Chem.* 2006, *78* (16), 5783–5788.

207. Ouyang, G.; Zhao, W.; Pawliszyn, J., Kinetic calibration for automated headspace liquid-phase microextraction. *Anal. Chem.* 2005, *77* (24), 8122–8128.

208. Chung, L. W.; Lee, M. R., Evaluation of liquid-phase microextraction conditions for determination of chlorophenols in environmental samples using gas chromatography–mass spectrometry without derivatization. *Talanta* 2008, *76* (1), 154–160.

209. Jiang, X.; Lee, H. K., Solvent bar microextraction. *Anal. Chem.* 2004, *76* (18), 5591–5596.

210. King, S.; Meyer, J. S.; Andrews, A. R. J., Screening method for polycyclic aromatic hydrocarbons in soil using hollow fiber membrane solvent microextraction. *J. Chromatogr. A* 2002, *982* (2), 201–208.

211. Kuosmanen, K.; Hyötyläinen, T.; Hartonen, K.; Jönsson, J. Å.; Riekkola, M. L., Analysis of PAH compounds in soil with on-line coupled pressurised hot water extraction–microporous membrane liquid–liquid extraction–gas chromatography. *Anal. Bioanal. Chem.* 2003, *375* (3), 389–399.

212. Kuosmanen, K.; Hyötyläinen, T.; Hartonen, K.; Riekkola, M. L., Analysis of polycyclic aromatic hydrocarbons in soil and sediment with on-line coupled pressurised hot water extraction, hollow fibre microporous membrane liquid–liquid extraction and gas chromatography. *Analyst* 2003, *128* (5), 434–439.

213. Miró, M.; Frenzel, W., The potential of microdialysis as an automatic sample-processing technique for environmental research. *Trends Anal. Chem.* 2005, *24* (4), 324–333.

214. Pardasani, D.; Kanaujia, P. K.; Gupta, A. K.; Tak, V.; Shrivastava, R. K.; Dubey, D. K., In situ derivatization hollow fiber mediated liquid phase microextraction of alkylphosphonic acids from water. *J. Chromatogr. A* 2007, *1141* (2), 151–157.

215. Shen, G.; Lee, H. K., Hollow fiber–protected liquid-phase microextraction of triazine herbicides. *Anal. Chem.* 2002, *74* (3), 648–654.

216. Xu, H.; Pan, W.; Song, D.; Yang, G., Development of an improved liquid phase microextraction technique and its application in the analysis of flumetsulam and its two analogous herbicides in soil. *J. Agric. Food. Chem.* 2007, *55* (23), 9351–9356.

217. Zhang, H.; Andrews, A. R. J., Preliminary studies of a fast screening method for polycyclic aromatic hydrocarbons in soil by using solvent microextraction–gas chromatography *J. Environ. Monit.* 2000, *2* (6), 656–661.

218. Basheer, C.; Obbard, J. P.; Lee, H. K., Analysis of persistent organic pollutants in marine sediments using a novel microwave assisted solvent extraction and liquid-phase microextraction technique. *J. Chromatogr. A* 2005, *1068* (2), 221–228.

219. Lambropoulou, D. A.; Albanis, T. A., Sensitive trace enrichment of environmental andiandrogen vinclozolin from natural waters and sediment samples using hollow-fiber liquid-phase microextraction. *J. Chromatogr. A* 2004, *1061* (1), 11–18.

220. Jermak, S.; Pranaityté, B.; Padarauskas, A., Ligand displacement, headspace single-drop microextraction, and capillary electrophoresis for the determination of weak acid dissociable cyanide. *J. Chromatogr. A* 2007, *1148* (1), 123–127.

221. Lin, M. Y.; Whang, C. W., Microwave-assisted derivatization and single-drop microextraction for gas chromatographic determination of chromium(III) in water. *J. Chromatogr. A* 2007, *1160* (1–2), 336–339.

222. Liu, X.; Chen, X.; Yang, S.; Wang, X., Comparison of continuous-flow microextraction and static liquid-phase microextraction for the determination of *p*-toluidine in *Chlamydomonas reinhardtii. J. Sep. Sci.* 2007, *30* (15), 2506–2512.

223. Liu, X.; Wang, X.; Chen, X.; Yang, S., LC determination of 4-bromoaniline in green algae *Chlamydomonas reinhardtii* after continuous-flow microextraction. *Chromatographia* 2007, *65* (7–8), 447–451.

224. Liu, X. J.; Chen, X. W.; Yang, S.; Wang, X. D., Continuous-flow microextraction coupled with HPLC for the determination of 4-chloroaniline in *Chlamydomonas reinhardtii*. *Bull. Environ. Contam. Toxicol.* 2007, *78* (5), 368–372.

225. Sarafraz-Yazdi, A.; Es'haghi, Z., Surfactant enhanced liquid-phase microextraction of basic drugs of abuse in hair combined with high performance liquid chromatography. *J. Chromatogr. A* 2005, *1094* (1–2), 1–8.

226. Besharati-Seidani, A.; Jabbari, A.; Yamini, Y., Headspace solvent microextraction: a very rapid method for identification of volatile components of Iranian *Pimpinella anisum* seed. *Anal. Chim. Acta* 2005, *530* (1), 155–161.

227. Besharati-Seidani, A.; Jabbari, A.; Yamini, Y.; Saharkhiz, M. J., Rapid extraction and analysis of volatile organic compounds of Iranian feverfew (*Tanacetum parthenium*) using headspace solvent microextraction (HSME), and gas chromatography/mass spectrometry. *Flavour Fragr. J.* 2006, *21* (3), 502–509.

228. Cao, J.; Qi, M.; Zhang, Y.; Zhou, S.; Shao, Q.; Fu, R., Analysis of volatile compounds in *Curcuma wenyujin* Y.H. Chen et C. Ling by headspace solvent microextraction–gas chromatography–mass spectrometry. *Anal. Chim. Acta* 2006, *561* (1–2), 88–95.

229. Deng, C.; Mao, Y.; Hu, F.; Zhang, X., Development of gas chromatography–mass spectrometry following microwave distillation and simultaneous headspace single-drop microextraction for fast determination of volatile fraction in Chinese herb. *J. Chromatogr. A* 2007, *1152* (1–2), 193–198.

230. Deng, C.; Yang, X.; Zhang, X., Rapid determination of panaxynol in a traditional Chinese medicine of *Saposhnikovia divaricata* by pressurized hot water extraction followed by liquid-phase microextraction and gas chromatography–mass spectrometry. *Talanta* 2005, *68* (1), 6–11.

231. Deng, C.; Yao, N.; Wang, A.; Zhang, X., Determination of essential oil in a traditional Chinese medicine, *Fructus amomi*, by pressurized hot water extraction followed by liquid-phase microextraction and gas chromatography–mass spectrometry. *Anal. Chim. Acta* 2005, *536* (1–2), 237–244.

232. Deng, C.; Yao, N.; Wang, B.; Zhang, X., Development of microwave-assisted extraction followed by headspace single-drop microextraction for fast determination of paeonol in traditional Chinese medicines. *J. Chromatogr. A* 2006, *1103* (1), 15–21.

233. Fakhari, A. R.; Salehi, P.; Heydari, R.; Ebrahimi, S. N.; Haddad, P. R., Hydro-distillation–headspace solvent microextraction, a new method for analysis of the essential oil components of *Lavandula angustifolia* Mill. *J. Chromatogr. A* 2005, *1098* (1–2), 14–18.

234. Fang, L.; Qi, M.; Li, T.; Shao, Q.; Fu, R., Headspace solvent microextraction–gas chromatography–mass spectrometry for the analysis of volatile compounds from *Foeniculum vulgare* Mill. *J. Pharm. Biomed. Anal.* 2006, *41* (3), 791–797.

235. Hashemi, P.; Abolghasemi, M. M.; Fakhari, A. R.; Ebrahimi, S. N.; Ahmadi, S., Hydrodistillation–solvent microextraction and GC–MS identification of volatile components of *Artemisia aucheri*. *Chromatographia* 2007, *66* (3–4), 283–286.

236. Jalali Heravi, M.; Sereshti, H., Determination of essential oil components of *Artemisia haussknechtii* Boiss. using simultaneous hydrodistillation–static headspace liquid phase microextraction–gas chromatography mass spectrometry. *J. Chromatogr. A* 2007, *1160* (1–2), 81–89.

237. Jung, M. J.; Shin, Y. J.; Oh, S. Y.; Kim, N. S.; Kim, K.; Lee, D. S., Headspace hanging drop liquid phase microextraction and gas chromatography–mass spectrometry for the analysis of flavors from clove buds. *Bull. Korean Chem. Soc.* 2006, *27* (2), 231–236.

238. Kim, N. S.; Jung, M. J.; Yoo, Z. W.; Lee, S. N.; Lee, D. S., Headspace hanging drop liquid phase microextraction and GC-MS for the determination of linalool from evening primrose flowers. *Bull. Korean Chem. Soc.* 2005, *26* (12), 1996–2000.

239. López, P.; Sánchez, C.; Batlle, R.; Nerín, C., Vapor-phase activities of cinnamon, thyme, and oregano essential oils and key constituents against foodborne microorganisms. *J. Agric. Food. Chem.* 2007, *55* (11), 4348–4356.

240. Ratola, N.; Alves, A.; Kalogerakis, N.; Psillakis, E., Hollow-fibre liquid-phase microextraction: a simple and fast cleanup step used for PAHs determination in pine needles. *Anal. Chim. Acta* 2008, *618* (1), 70–78.

241. Salehi, P.; Asghari, B.; Mohammadi, F., Hydrodistillation–headspace solvent microextraction: an efficient method for analysis of the essential oil from the seeds of *Foeniculum vulgare* Mill. *Chromatographia* 2007, *65* (1–2), 119–122.

242. Salehi, P.; Fakhari, A. R.; Ebrahimi, S. N.; Heydari, R., Rapid essential oil screening of *Rosmarinus officinalis* L. by hydrodistillation–headspace solvent microextraction. *Flavour Fragr. J.* 2007, *22* (4), 280–285.

243. Sun, S.; Cheng, Z.; Xie, J.; Zhang, J.; Liao, Y.; Wang, H.; Guo, Y., Identification of volatile basic components in tobacco by headspace liquid-phase microextraction coupled to matrix-assisted laser desorption/ionization with Fourier transform mass spectrometry. *Rapid Commun. Mass Spectrom.* 2005, *19* (8), 1025–1030.

244. Wang, G.; Dong, C.; Sun, Y. A.; Xie, K.; Zheng, H., Characterization of volatile components in dry chrysanthemum flowers using headspace liquid-phase microextraction–gas chromatography. *J. Chromatogr. Sci.* 2008, *46* (2), 127–132.

245. Aubert, C.; Baumann, S.; Arguel, H., Optimization of the analysis of flavor volatile compounds by liquid–liquid microextraction (LLME): application to the aroma analysis of melons, peaches, grapes, strawberries, and tomatoes. *J. Agric. Food. Chem.* 2005, *53* (23), 8881–8895.

246. Aubert, C.; Chanforan, C., Postharvest changes in physicochemical properties and volatile constituents of apricot (*Prunus armeniaca* L.): characterization of 28 cultivars. *J. Agric. Food Chem.* 2007, *55* (8), 3074–3082.

247. Aubert, C.; Pitrat, M., Volatile compounds in the skin and pulp of Queen Anne's pocket melon. *J. Agric. Food Chem.* 2006, *54* (21), 8177–8182.

248. Das, P.; Gupta, M.; Jain, A.; Verma, K. K., Single drop microextraction or solid phase microextraction–gas chromatography–mass spectrometry for the determination of iodine in pharmaceuticals, iodized salt, milk powder and vegetables involving conversion into 4-iodo-*N,N*-dimethylaniline. *J. Chromatogr. A* 2004, *1023* (1), 33–39.

249. Romero-González, R.; Pastor-Montoro, E.; Martínez-Vidal, J. L.; Garrido-Frenich, A., Application of hollow fiber supported liquid membrane extraction to the simultaneous determination of pesticide residues in vegetables by liquid chromatography/mass spectrometry. *Rapid Commun. Mass Spectrom.* 2006, *20* (18), 2701–2708.

250. Saraji, M.; Mousavinia, F., Single-drop microextraction followed by in-syringe derivatization and gas chromatography–mass spectrometric detection for determination of organic acids in fruits and fruit juices. *J. Sep. Sci.* 2006, *29* (9), 1223–1229.

251. Zhang, M.; Huang, J.; Wei, C.; Yu, B.; Yang, X.; Chen, X., Mixed liquids for single-drop microextraction of organochlorine pesticides in vegetables. *Talanta* 2008, *74* (4), 599–604.

252. Hansson, E.; Hakkarainen, M., Multiple headspace single-drop microextraction: a new technique for quantitative determination of styrene in polystyrene. *J. Chromatogr. A* 2006, *1102* (1–2), 91–95.

253. Arab, J.; Yamini, Y.; Hosseini, M. H.; Shamsipur, M., Headspace solvent microextraction of trace amounts of BTX from water samples. *Chem. Anal. (Warsaw)* 2004, *49* (1), 129.

254. Bahramifar, N.; Yamini, Y.; Shariati-Feizabadi, S.; Shamsipur, M., Trace analysis of methyl *tert*-butyl ether in water samples using headspace solvent microextraction and gas chromatography–flame ionization detection. *J. Chromatogr. A* 2004, *1042* (1–2), 211–217.

255. Khajeh, M.; Yamini, Y.; Hassan, J., Trace analysis of chlorobenzenes in water samples using headspace solvent microextraction and gas chromatography/electron capture detection. *Talanta* 2006, *69* (5), 1088–1094.

256. Shamsipur, M.; Hassan, J.; Salar-Amoli, J.; Yamini, Y., Headspace solvent microextraction–gas chromatographic thermionic specific detector determination of amitraz in honey after hydrolysis to 2,4-dimethylaniline. *J. Food Compos. Anal.* 2008, *21* (3), 264–270.

257. Shariati-Feizabadi, S.; Yamini, Y.; Bahramifar, N., Headspace solvent microextraction and gas chromatographic determination of some polycyclic aromatic hydrocarbons in water samples. *Anal. Chim. Acta* 2003, *489* (1), 21–31.

258. Yamini, Y.; Hojjati, M.; Haji-Hosseini, M.; Shamsipur, M., Headspace solvent microextraction: a new method applied to the preconcentration of 2-butoxyethanol from aqueous solutions into a single microdrop. *Talanta* 2004, *62* (2), 265–270.

259. Yamini, Y.; Hosseini, M. H.; Hojaty, M.; Arab, J., Headspace solvent microextraction of trihalomethane compounds into a single drop. *J. Chromatogr. Sci.* 2004, *42* (1), 32–36.

260. Zanjani, M. K.; Yamini, Y.; Shariati, S., Analysis of *n*-alkanes in water samples by means of headspace solvent microextraction and gas chromatography. *J. Hazard. Mater.* 2006, *136* (3), 714–720.

261. Pawliszyn, J., *Solid Phase Microextraction: Theory and Practice*, New York, Wiley-VCH, 1997.

262. Aguilera-Herrador, E.; Lucena, R.; Cárdenas, S.; Valcárcel, M., Direct coupling of ionic liquid based single-drop microextraction and GC/MS. *Anal. Chem.* 2008, *80* (3), 793–800.

263. Aguilera-Herrador, E.; Lucena, R.; Cárdenas, S.; Valcárcel, M., Ionic liquid–based single-drop microextraction/gas chromatographic/mass spectrometric determination of benzene, toluene, ethylbenzene and xylene isomers in waters. *J. Chromatogr. A* 2008, *1201* (1), 106–111.

264. Kaykhaii, M.; Moradi, M., Direct screening of water samples for benzene hydrocarbon compounds by headspace liquid-phase microextraction-gas chromatography. *J. Chromatogr. Sci.* 2008, *46* (5), 413–418.

265. Liu, J. F.; Chi, Y. G.; Jiang, G. B., Screening the extractability of some typical environmental pollutants by ionic liquids in liquid-phase microextraction. *J. Sep. Sci.* 2005, *28* (1), 87–91.

266. Przyjazny, A.; Kokosa, J. M., Analytical characteristics of the determination of benzene, toluene, ethylbenzene and xylenes in water by headspace solvent microextraction. *J. Chromatogr. A* 2002, *977* (2), 143–153.

267. Theis, A. L.; Waldack, A. J.; Hansen, S. M.; Jeannot, M. A., Headspace solvent microextraction. *Anal. Chem.* 2001, *73* (23), 5651–5654.

268. Sarafraz-Yazdi, A.; Amiri, A. H.; Es'haghi, Z., BTEX determination in water matrices using HF-LPME with gas chromatography–flame ionization detector. *Chemosphere* 2008, *71* (4), 671–676.

269. Li, X.; Xu, X.; Wang, X.; Ma, L., Headspace single-drop microextraction with gas chromatography for determination of volatile halocarbons in water samples. *Int. J. Environ. Anal. Chem.* 2004, *84* (9), 633–645.

270. Zhang, T.; Chen, X.; Li, Y.; Liang, P., Application of headspace liquid-phase microextraction to the analysis of volatile halocarbons in water. *Chromatographia* 2006, *63* (11–12), 633–637.

271. Buszewski, B.; Ligor, T., Single-drop extraction versus solid-phase microextraction for the analysis of VOCs in water. *LC•GC Eur.* 2002, *15* (2), 92–97.

272. Li, Y.; Zhang, T.; Liang, P., Application of continuous-flow liquid-phase microextraction to the analysis of volatile halohydrocarbons in water. *Anal. Chim. Acta* 2005, *536* (1–2), 245–249.

273. Schellin, M.; Popp, P., Miniaturized membrane-assisted solvent extraction combined with gas chromatography/electron-capture detection applied to the analysis of volatile organic compounds. *J. Chromatogr. A* 2006, *1103* (2), 211–218.

274. Quintana, J. B.; Rodriguez, I., Strategies for the microextraction of polar organic contaminants in water samples. *Anal. Bioanal. Chem.* 2006, *384* (7–8), 1447–1461.

275. Zhao, R. S.; Lao, W. J.; Xu, X. B., Headspace liquid-phase microextraction of trihalomethanes in drinking water and their gas chromatographic determination. *Talanta* 2004, *62* (4), 751–756.

276. Ligor, T.; Buszewski, B., Extraction of trace organic pollutants from aqueous samples by a single drop method. *Chromatographia* 2000, *51* (1), S279–S282.

277. Tor, A.; Aydin, M. E., Application of liquid-phase microextraction to the analysis of trihalomethanes in water. *Anal. Chim. Acta* 2006, *575* (1), 138–143.

278. Vora-adisak, N.; Varanusupakul, P., A simple supported liquid hollow fiber membrane microextraction for sample preparation of trihalomethanes in water samples. *J. Chromatogr. A* 2006, *1121* (2), 236–241.

279. Varanusupakul, P.; Vora-adisak, N.; Pulpoka, B., In situ derivatization and hollow fiber membrane microextraction for gas chromatographic determination of haloacetic acids in water. *Anal. Chim. Acta* 2007, *598* (1), 82–86.

280. Stalikas, C. D.; Fiamegos, Y. C., Microextraction combined with derivatization. *Trends Anal. Chem.* 2008, *27* (6), 533–542.

281. Deng, C.; Yao, N.; Li, N.; Zhang, X., Headspace single-drop microextraction with in-drop derivatization for aldehyde analysis. *J. Sep. Sci.* 2005, *28* (17), 2301–2305.

282. Fiamegos, Y. C.; Stalikas, C. D., Theoretical analysis and experimental evaluation of headspace in-drop derivatisation single-drop microextraction using aldehydes as model analytes. *Anal. Chim. Acta* 2007, *599* (1), 76–83.

283. Tankeviciute, A.; Bobnis, R.; Kazlauskas, R.; Vickackaite, V., Single drop microextraction of esters: comparison of headspace and direct extractions. *Chem. Anal. (Warsaw)* 2005, *50* (3), 539–549.

284. Nazarenko, A. Y., Liquid-phase headspace microextraction into a single drop. *Am. Lab.* 2004, *36* (16), 30–33.

285. Zhang, J.; Su, T.; Lee, H. K., Headspace water-based liquid-phase microextraction. *Anal. Chem.* 2005, *77* (7), 1988–1992.

286. Liu, J.-F.; Jönsson, J. Å.; Jiang, G.-B., Application of ionic liquids in analytical chemistry. *Trends Anal. Chem.* 2005, *24* (1), 20–27.

287. Liu, J. F.; Jiang, G. B.; Chi, Y. G.; Cai, Y.-Q.; Zhou, Q. X.; Hu, J. T., Use of ionic liquids for liquid-phase microextraction of polycyclic aromatic hydrocarbons. *Anal. Chem.* 2003, *75* (21), 5870–5876.

288. Peng, J.-F.; Liu, J.-F.; Jiang, G.-B.; Tai, C.; Huang, M.-J., Ionic liquid for high temperature headspace liquid-phase microextraction of chlorinated anilines in environmental water samples. *J. Chromatogr. A* 2005, *1072* (1), 3–6.

289. Vidal, L.; Domini, C. E.; Grané, N.; Psillakis, E.; Canals, A., Microwave-assisted headspace single-drop microextraction of chlorobenzenes from water samples. *Anal. Chim. Acta* 2007, *592* (1), 9–15.

290. Vidal, L.; Psillakis, E.; Domini, C. E.; Grané, N.; Marken, F.; Canals, A., An ionic liquid as a solvent for headspace single drop microextraction of chlorobenzenes from water samples. *Anal. Chim. Acta* 2007, *584* (1), 189–195.

291. Ye, C.; Zhou, Q.; Wang, X.; Xiao, J., Determination of phenols in environmental water samples by ionic liquid–based headspace liquid-phase microextraction coupled with high-performance liquid chromatography. *J. Sep. Sci.* 2007, *30* (1), 42–47.

292. Ye, C.-L.; Zhou, Q.-X.; Wang, X.-M., Headspace liquid-phase microextraction using ionic liquid as extractant for the preconcentration of dichlorodiphenyltrichloroethane and its metabolites at trace levels in water samples. *Anal. Chim. Acta* 2006, *572* (2), 165–171.

293. Fu, X.; Dai, S.; Zhang, Y., 1-Butyl-3-methylimidazolium hexafluorophosphate ionic liquid–based liquid–liquid microextraction for the determination of 4-nonylphenol and 4-*tert*-octylphenol in environmental waters. *Int. J. Environ. Anal. Chem.* 2006, *86* (13), 985–993.

294. Liu, J.; Peng, J. F.; Chi, Y. G.; Jiang, G. B., Determination of formaldehyde in shiitake mushroom by ionic liquid–based liquid-phase microextraction coupled with liquid chromatography. *Talanta* 2005, *65* (3), 705–709.

295. Liu, J. F.; Chi, Y. G.; Jiang, G. B.; Tai, C.; Peng, J. F.; Hu, J. T., Ionic liquid–based liquid-phase microextraction, a new sample enrichment procedure for liquid chromatography. *J. Chromatogr. A* 2004, *1026* (1–2), 143–147.

296. Cruz-Vera, M.; Lucena, R.; Cárdenas, S.; Valcárcel, M., Ionic liquid–based dynamic liquid-phase microextraction: application to the determination of anti-inflammatory drugs in urine samples. *J. Chromatogr. A* 2008, *1202* (1), 1–7.

297. Peng, J. F.; Liu, J. F.; Hu, X. L.; Jiang, G. B., Direct determination of chlorophenols in environmental water samples by hollow fiber supported ionic liquid membrane extraction coupled with high-performance liquid chromatography. *J. Chromatogr. A* 2007, *1139* (2), 165–170.

298. Sarafraz Yazdi, A.; Assadi, Y., Determination of trace of methyl *tert*-butyl ether in water using liquid drop headspace sampling and GC. *Chromatographia* 2004, *60* (11–12), 699–702.

299. Rawa-Adkonis, M.; Wolska, L.; Przyjazny, A.; Namieśnik, J., Sources of errors associated with the determination of PAH and PCB analytes in water samples. *Anal. Lett.* 2006, *39* (11), 2317–2331.

300. Ackerman, A. H.; Hurtubise, R. J., The effects of adsorption of solutes on glassware and teflon in the calculation of partition coefficients for solid-phase microextraction with 1PS paper. *Talanta* 2000, *52* (5), 853–861.

301. Hou, L.; Lee, H. K., Application of static and dynamic liquid-phase microextraction in the determination of polycyclic aromatic hydrocarbons. *J. Chromatogr. A* 2002, *976* (1–2), 377–385.

302. Basheer, C.; Balasubramanian, R.; Lee, H. K., Determination of organic micropollutants in rainwater using hollow fiber membrane/liquid-phase microextraction combined with gas chromatography–mass spectrometry. *J. Chromatogr. A* 2003, *1016* (1), 11–20.

303. Charalabaki, M.; Psillakis, E.; Mantzavinos, D.; Kalogerakis, N., Analysis of polycyclic aromatic hydrocarbons in wastewater treatment plant effluents using hollow fibre liquid-phase microextraction. *Chemosphere* 2005, *60* (5), 690–698.

304. Liu, Y.; Hashi, Y.; Lin, J.-M., Continuous-flow microextraction and gas chromatographic–mass spectrometric determination of polycyclic aromatic hydrocarbon compounds in water. *Anal. Chim. Acta* 2007, *585* (2), 294–299.

305. Marlow, M.; Hurtubise, R. J., Liquid–liquid–liquid microextraction for the enrichment of polycyclic aromatic hydrocarbon metabolites investigated with fluorescence spectroscopy and capillary electrophoresis. *Anal. Chim. Acta* 2004, *526* (1), 41–49.

306. Wu, Y.; Xia, L.; Chen, R.; Hu, B., Headspace single drop microextraction combined with HPLC for the determination of trace polycyclic aromatic hydrocarbons in environmental samples. *Talanta* 2008, *74* (4), 470–477.

307. Zanjani, M. R. K.; Yamini, Y.; Shariati, S.; Jönsson, J. Å., A new liquid-phase microextraction method based on solidification of floating organic drop. *Anal. Chim. Acta* 2007, *585* (2), 286–293.

308. Basheer, C.; Vetrichelvan, M.; Valiyaveettil, S.; Lee, H. K., On-site polymer-coated hollow fiber membrane microextraction and gas chromatography–mass spectrometry of polychlorinated biphenyls and polybrominated diphenyl ethers. *J. Chromatogr. A* 2007, *1139* (2), 157–164.

309. Schellin, M.; Popp, P., Membrane-assisted solvent extraction of polychlorinated biphenyls in river water and other matrices combined with large volume injection–gas chromatography–mass spectrometric detection. *J. Chromatogr. A* 2003, *1020* (2), 153–160.

310. Fontanals, N.; Barri, T.; Bergström, S.; Jönsson, J. Å., Determination of polybrominated diphenyl ethers at trace levels in environmental waters using hollow-fiber microporous membrane liquid–liquid extraction and gas chromatography–mass spectrometry. *J. Chromatogr. A* 2006, *1133* (1–2), 41–48.

311. Psillakis, E.; Kalogerakis, N., Hollow-fibre liquid-phase microextraction of phthalate esters from water. *J. Chromatogr. A* 2003, *999* (1–2), 145–153.

312. Yao, J.; Xu, H.; Lv, L.; Song, D.; Cui, Y.; Zhang, T.; Feng, Y. Q., A novel liquid-phase microextraction method combined with high performance liquid chromatography for analysis of phthalate esters in landfill leachates. *Anal. Chim. Acta* 2008, *616* (1), 42–48.

313. Farahani, H.; Ganjali, M. R.; Dinarvand, R.; Norouzi, P., Screening method for phthalate esters in water using liquid-phase microextraction based on the solidification of a floating organic microdrop combined with gas chromatography–mass spectrometry. *Talanta* 2008, *76* (4), 718–723.

314. Xu, H.; Liao, Y.; Yao, J., Development of a novel ultrasound-assisted headspace liquid-phase microextraction and its application to the analysis of chlorophenols in real aqueous samples. *J. Chromatogr. A* 2007, *1167* (1), 1–8.

315. Zhao, L.; Zhu, L.; Lee, H. K., Analysis of aromatic amines in water samples by liquid–liquid–liquid microextraction with hollow fibers and high-performance liquid chromatography. *J. Chromatogr. A* 2002, *963* (1–2), 239–248.

316. Sarafraz-Yazdi, A.; Beiknejad, D.; Es'haghi, Z., LC Determination of mono-substituted phenols in water using liquid–liquid–liquid phase microextraction. *Chromatographia* 2005, *62* (1–2), 49–54.

317. Yazdi, A. S.; Es'haghi, Z., Liquid–liquid–liquid phase microextraction of aromatic amines in water using crown ethers by high-performance liquid chromatography with monolithic column. *Talanta* 2005, *66* (3), 664–669.

318. Fan, Z.; Liu, X., Determination of methylmercury and phenylmercury in water samples by liquid–liquid–liquid microextraction coupled with capillary electrophoresis. *J. Chromatogr. A* 2008, *1180* (1–2), 187–192.

319. Zhu, L.; Tay, C. B.; Lee, H. K., Liquid–liquid–liquid microextraction of aromatic amines from water samples combined with high-performance liquid chromatography. *J. Chromatogr. A* 2002, *963* (1–2), 231–237.

320. Almeda, S.; Nozal, L.; Arce, L.; Valcárcel, M., Direct determination of chlorophenols present in liquid samples by using a supported liquid membrane coupled

in-line with capillary electrophoresis equipment. *Anal. Chim. Acta* 2007, *587* (1), 97–103.

321. Berhanu, T.; Liu, J. F.; Romero, R.; Megersa, N.; Jönsson, J. Å., Determination of trace levels of dinitrophenolic compounds in environmental water samples using hollow fiber supported liquid membrane extraction and high performance liquid chromatography. *J. Chromatogr. A* 2006, *1103* (1), 1–8.

322. Lezamiz, J.; Jönsson, J. Å., Development of a simple hollow fibre supported liquid membrane extraction method to extract and preconcentrate dinitrophenols in environmental samples at ng L^{-1} level by liquid chromatography. *J. Chromatogr. A* 2007, *1152* (1–2), 226–233.

323. Zhao, L.; Lee, H. K., Determination of phenols in water using liquid phase microextraction with back extraction combined with high-performance liquid chromatography. *J. Chromatogr. A* 2001, *931* (1–2), 95–105.

324. Zhu, L.; Zhu, L.; Lee, H. K., Liquid–liquid–liquid microextraction of nitrophenols with a hollow fiber membrane prior to capillary liquid chromatography. *J. Chromatogr. A* 2001, *924* (1–2), 407–414.

325. Lin, C. Y.; Huang, S. D., Application of liquid–liquid–liquid microextraction and ion-pair liquid chromatography coupled with photodiode array detection for the determination of chlorophenols in water. *J. Chromatogr. A* 2008, *1193* (1–2), 79–84.

326. Farajzadeh, M. A.; Vardast, M. R.; Jönsson, J. Å., Liquid–gas–liquid microextraction as a simple technique for the extraction of 2,4-di-*tert*-butyl phenol from aqueous samples. *Chromatographia* 2007, *66* (5–6), 415–419.

327. Yazdi, A. S.; Es'haghi, Z., Two-step hollow fiber–based, liquid-phase microextraction combined with high-performance liquid chromatography: a new approach to determination of aromatic amines in water. *J. Chromatogr. A* 2005, *1082* (2), 136–142.

328. Ma, M.; Cantwell, F. F., Solvent microextraction with simultaneous back-extraction for sample cleanup and preconcentration: quantitative extraction *Anal. Chem.* 1998, *70* (18), 3912–3919.

329. Sarafra-Yazdi, A.; Es'haghi, Z., Antibiotic-assisted three-phase liquid-phase microextraction of aromatic amines from aqueous solutions combined with high-performance liquid chromatography. *J. Anal. Chem.* 2006, *61* (8), 787–793.

330. Sarafraz-Yazdi, A.; Es'haghi, Z., Comparison of hollow fiber and single-drop liquid-phase microextraction techniques for HPLC determination of aniline derivatives in water. *Chromatographia* 2006, *63* (11–12), 563–569.

331. Bagheri, H.; Saber, A.; Mousavi, S. R., Immersed solvent microextraction of phenol and chlorophenols from water samples followed by gas chromatography–mass spectrometry. *J. Chromatogr. A* 2004, *1046* (1–2), 27–33.

332. Kawaguchi, M.; Ito, R.; Endo, N.; Okanouchi, N.; Sakui, N.; Saito, K.; Nakazawa, H., Liquid phase microextraction with in situ derivatization for measurement of bisphenol A in river water sample by gas chromatography–mass spectrometry. *J. Chromatogr. A* 2006, *1110* (1–2), 1–5.

333. Chia, K. J.; Huang, S. D., Simultaneous derivatization and extraction of primary amines in river water with dynamic hollow fiber liquid-phase microextraction followed by gas chromatography–mass spectrometric detection. *J. Chromatogr. A* 2006, *1103* (1), 158–161.

334. Reddy-Noone, K.; Jain, A.; Verma, K. K., Liquid-phase microextraction and GC for the determination of primary, secondary and tertiary aromatic amines as their iodo-derivatives. *Talanta* 2007, *73* (4), 684–691.

335. Deng, C.; Li, N.; Wang, L.; Zhang, X., Development of gas chromatography–mass spectrometry following headspace single-drop microextraction and simultaneous derivatization for fast determination of short-chain aliphatic amines in water samples. *J. Chromatogr. A* 2006, *1131* (1–2), 45–50.

336. Fiamegos, Y. C.; Kefala, A. P.; Stalikas, C. D., Ion-pair single-drop microextraction versus phase-transfer catalytic extraction for the gas chromatographic determination of phenols as tosylated derivatives. *J. Chromatogr. A* 2008, *1190* (1–2), 44–51.

337. Saraji, M.; Bakhshi, M., Determination of phenols in water samples by single-drop microextraction followed by in-syringe derivatization and gas chromatography–mass spectrometric detection. *J. Chromatogr. A* 2005, *1098* (1–2), 30–36.

338. Basheer, C.; Lee, H. K., Analysis of endocrine disrupting alkylphenols, chlorophenols and bisphenol-A using hollow fiber–protected liquid-phase microextraction coupled with injection port–derivatization gas chromatography–mass spectrometry. *J. Chromatogr. A* 2004, *1057* (1–2), 163–169.

339. Basheer, C.; Parthiban, A.; Jayaraman, A.; Lee, H. K.; Valiyaveettil, S., Determination of alkylphenols and bisphenol-A: a comparative investigation of functional polymer-coated membrane microextraction and solid-phase microextraction techniques. *J. Chromatogr. A* 2005, *1087* (1–2), 274–282.

340. Chen, X.; Zhang, T.; Liang, P.; Li, Y., Application of continuous-flow liquid-phase microextraction to the analysis of phenolic compounds in wastewater samples. *Microchim. Acta* 2006, *155* (3–4), 415–420.

341. Schellin, M.; Popp, P., Membrane-assisted solvent extraction of seven phenols combined with large volume injection–gas chromatography–mass spectrometric detection. *J. Chromatogr. A* 2005, *1072* (1), 37–43.

342. Carasek, E.; Tonjes, J. W.; Scharf, M., A liquid–liquid microextraction system for Pb and Cd enrichment and determination by flame atomic absorption spectrometry. *Quim. Nova* 2002, *25* (5), 748–752.

343. Dadfarnia, S.; Salmanzadeh, A. M.; Shabani, A. M. H., A novel separation/preconcentration system based on solidification of floating organic drop microextraction for determination of lead by graphite furnace atomic absorption spectrometry. *Anal. Chim. Acta* 2008, *623* (2), 163–167.

344. Parthasarathy, N.; Pelletier, M.; Buffle, J., Hollow fiber based supported liquid membrane: a novel analytical system for trace metal analysis. *Anal. Chim. Acta* 1997, *350* (1–2), 183–195.

345. Sigg, L.; Black, F.; Buffle, J.; Cao, J.; Cleven, R.; Davison, W.; Galceran, J.; Gunkel, P.; Kalis, E.; Kistler, D.; Martin, M.; Noël, S.; Nur, Y.; Odzak, N.; Puy, J.; van Riemsdijk, W.; Temminghoff, E.; Tercier-Waeber, M. L.; Toepperwien, S.; Town, R. M.; Unsworth, E.; Warnken, K. W.; Weng, L.; Xue, H.; Zhang, H., Comparison of analytical techniques for dynamic trace metal speciation in natural freshwaters. *Environ. Sci. Technol.* 2006, *40* (6), 1934–1941.

346. Fan, Z.; Zhou, W., Dithizone–chloroform single drop microextraction system combined with electrothermal atomic absorption spectrometry using Ir as permanent modifier for the determination of Cd in water and biological samples. *Spectrochim. Acta B* 2006, *61* (7), 870–874.

347. Xia, L.; Hu, B.; Jiang, Z.; Wu, Y.; Liang, Y., Single-drop microextraction combined with low-temperature electrothermal vaporization ICPMS for the detcrmination of trace Be, Co, Pd, and Cd in biological samples. *Anal. Chem.* 2004, *76* (10), 2910–2915.

348. Romero, M.; Liu, J. F.; Mayer, P.; Jönsson, J. Å., Equilibrium sampling through membranes of freely dissolved copper concentrations with selective hollow fiber

membranes and the spectrophotometric detection of a metal stripping agent. *Anal. Chem.* 2005, *77* (23), 7605–7611.

349. Chamsaz, M.; Arbab-Zawar, M. H.; Nazari, S., Determination of arsenic by electro-thermal atomic absorption spectrometry using headspace liquid phase microextraction after in situ hydride generation. *J. Anal. At. Spectrom.* 2003, *18* (10), 1279–1282.

350. Fragueiro, S.; Lavilla, I.; Bendicho, C., Headspace sequestration of arsine onto a Pd(II)-containing aqueous drop as a preconcentration method for electrothermal atomic absorption spectrometry. *Spectrochim. Acta B* 2004, *59* (6), 851–855.

351. Fan, Z., Determination of antimony(III) and total antimony by single-drop micro-extraction combined with electrothermal atomic absorption spectroscopy. *Anal. Chim. Acta* 2007, *585* (2), 300–304.

352. Figueroa, R.; García, M.; Lavilla, I.; Bendicho, C., Photoassisted vapor generation in the presence of organic acids for ultrasensitive determination of Se by electrothermal–atomic absorption spectrometry following headspace single-drop microextraction. *Spectrochim. Acta B* 2005, *60* (12), 1556–1563.

353. Fragueiro, S.; Lavilla, I.; Bendicho, C., Hydride generation–headspace single-drop microextraction–electrothermal atomic spectrometry method for determination of selenium in water after photoassisted prereduction. *Talanta* 2006, *68* (4), 1096–1101.

354. Chikama, K.; Negishi, T.; Nakatani, K., Extraction of tributyltin and triphenyltin across a single oil droplet/water interface. *Anal. Chim. Acta* 2004, *514* (2), 145–150.

355. Shioji, H.; Tsunoi, S.; Harino, H.; Tanaka, M., Liquid-phase microextraction of tribu-tyltin and triphenyltin coupled with gas chromatography–tandem mass spectrometry. Comparison between 4-fluorophenyl and ethyl derivatizations. *J. Chromatogr. A* 2004, *1048* (1), 81–88.

356. Pereira, F. J. P.; Bendicho, C.; Kalogerakis, N.; Psillakis, E., Headspace single drop microextraction of methylcyclopentadienyl–manganese tricarbonyl from water samples followed by gas chromatography–mass spectrometry *Talanta* 2007, *74* (1), 47–51.

357. Hashemi, P.; Rahimi, A.; Ghiasvand, A. R.; Abolghasemi, M. M., Headspace micro-extraction of tin into an aqueous microdrop containing Pd(II) and tributyl phosphate for its determination by ETAAS *J. Braz. Chem. Soc.* 2007, *18* (6), 1145–1149.

358. Reddy-Noone, K.; Jain, A.; Verma, K. K., Liquid-phase microextraction–gas chroma-tography–mass spectrometry for the determination of bromate, iodate, bromide and iodide in high-chloride matrix. *J. Chromatogr. A* 2007, *1148* (2), 145–151.

359. Ahmadi, F.; Assadi, Y.; Millani Hosseini, S. M. R.; Rezaee, M., Determination of orga-nophosphorus pesticides in water samples by single drop microextraction and gas chro-matography–flame photometric detector. *J. Chromatogr. A* 2006, *1101* (1–2), 307–312.

360. Bagheri, H.; Khalilian, F., Immersed solvent microextraction and gas chromatography–mass spectrometric detection of *s*-triazine herbicides in aquatic media. *Anal. Chim. Acta* 2005, *537* (1–2), 81–87.

361. Basheer, C.; Alnedhary, A. A.; Rao, B. S. M.; Lee, H. K., Determination of organo-phosphorus pesticides in wastewater samples using binary-solvent liquid-phase micro-extraction and solid-phase microextraction: a comparative study. *Anal. Chim. Acta* 2007, *605* (2), 147–152.

362. Basheer, C.; Lee, H. K.; Obbard, J. P., Determination of organochlorine pesticides in seawater using liquid-phase hollow fibre membrane microextraction and gas chroma-tography–mass spectrometry. *J. Chromatogr. A* 2002, *968* (1–2), 191–199.

363. Basheer, C.; Suresh, V.; Renu, R.; Lee, H. K., Development and application of polymer-coated hollow fiber membrane microextraction to the determination of organochlorine pesticides in water. *J. Chromatogr. A* 2004, *1033* (2), 213–220.

364. de Jager, L.; Andrews, A. R. J., Development of a rapid screening technique for organochlorine pesticides using solvent microextraction (SME) and fast gas chromatography (GC). *Analyst* 2000, *125* (11), 1943–1948.

365. de Jager, L. S.; Andrews, A. R. J., Solvent microextraction of chlorinated pesticides. *Chromatographia* 1999, *50* (11–12), 733–738.

366. Guo, L.; Liang, P.; Zhang, T.; Liu, Y.; Liu, S., Use of continuous-flow microextraction and liquid chromatography for determination of phoxim in water samples. *Chromatographia* 2005, *61* (9–10), 523–526.

367. Hauser, B.; Popp, P.; Kleine-Benne, E., Membrane-assisted solvent extraction of triazines and other semi-volatile contaminants directly coupled to large-volume injection-gas chromatography–mass spectrometric detection. *J. Chromatogr. A* 2002, *963* (1–2), 27–36.

368. Hauser, B.; Schellin, M.; Popp, P., Membrane-assisted solvent extraction of triazines, organochlorine, and organophosphorus compounds in complex samples combined with large-volume injection–gas chromatography/mass spectrometric detection. *Anal. Chem.* 2004, *76* (20), 6029–6038.

369. He, Y.; Lee, H. K., Continuous flow microextraction combined with high-performance liquid chromatography for the analysis of pesticides in natural waters. *J. Chromatogr. A* 2006, *1122* (1–2), 7–12.

370. Khalili-Zanjani, M. R.; Yamini, Y.; Yazdanfar, N.; Shariati, S., Extraction and determination of organophosphorus pesticides in water samples by a new liquid phase microextraction–gas chromatography–flame photometric detection. *Anal. Chim. Acta* 2008, *606* (2), 202–208.

371. Lambropoulou, D. A.; Albanis, T. A., Application of solvent microextraction in a single drop for the determination of new antifouling agents in waters. *J. Chromatogr. A* 2004, *1049* (1–2), 17–23.

372. Lambropoulou, D. A.; Albanis, T. A., Application of hollow fiber liquid phase microextraction for the determination of insecticides in water. *J. Chromatogr. A* 2005, *1072* (1), 55–61.

373. Lambropoulou, D. A.; Psillakis, E.; Albanis, T. A.; Kalogerakis, N., Single-drop microextraction for the analysis of organophosphorus insecticides in water. *Anal. Chim. Acta* 2004, *516* (1–2), 205–211.

374. Liang, P.; Guo, L.; Liu, Y.; Liu, S.; Zhang, T. Z., Application of liquid-phase microextraction for the determination of phoxim in water samples by high performance liquid chromatography with diode array detector. *Microchem. J.* 2005, *80* (1), 19–23.

375. Liu, J. F.; Toräng, L.; Mayer, P.; Jönsson, J. Å., Passive extraction and clean-up of phenoxy acid herbicides in samples from a groundwater plume using hollow fiber supported liquid membranes. *J. Chromatogr. A* 2007, *1160* (1–2), 56–63.

376. López-Blanco, C.; Gómez-Álvarez, S.; Rey-Garrote, M.; Cancho-Grande, B.; Simal-Gándara, J., Determination of pesticides by solid phase extraction followed by gas chromatography with nitrogen–phosphorus detection in natural water and comparison with solvent drop microextraction. *Anal. Bioanal. Chem.* 2006, *384* (4), 1002–1006.

377. López-Blanco, M. C.; Blanco-Cid, S.; Cancho-Grande, B.; Simal-Gándara, J., Application of single-drop microextraction and comparison with solid-phase microextraction and solid-phase extraction for the determination of α- and β-endosulfan in water samples by gas chromatography–electron-capture detection. *J. Chromatogr. A* 2003, *984* (2), 245–252.

378. López-Blanco, M. C.; Gómez-Álvarez, S.; Rey-Garrote, M.; Cancho-Grande, B.; Simal-Gándara, J., Determination of carbamates and organophosphorus pesticides by SDME–GC in natural water. *Anal. Bioanal. Chem.* 2005, *383* (4), 557–561.

379. Lüthje, K.; Hyötyläinen, T.; Riekkola, M. L., On-line coupling of microporous membrane liquid–liquid extraction and gas chromatography in the analysis of organic pollutants in water. *Anal. Bioanal. Chem.* 2004, *378* (8), 1991–1998.

380. Melwanki, M. B.; Huang, S. D., Three-phase system in solvent bar microextraction: an approach for the sample preparation of ionizable organic compounds prior to liquid chromatography. *Anal. Chim. Acta* 2006, *555* (1), 139–145.

381. Peng, J.; Lü, J.; Hu, X.; Liu, J.; Jiang, G., Determination of atrazine, desethyl atrazine and desisopropyl atrazine in environmental water samples using hollow fiber–protected liquid-phase microextraction and high performance liquid chromatography. *Microchim. Acta* 2007, *158* (1–2), 181–186.

382. Qian, L.-L.; He, Y. Z., Funnelform single-drop microextraction for gas chromatography–electron-capture detection. *J. Chromatogr. A* 2006, *1134* (1–2), 32–37.

383. Wu, J.; Lee, H. K., Injection port derivatization following ion-pair hollow fiber–protected liquid-phase microextraction for determining acidic herbicides by gas chromatography/mass spectrometry. *Anal. Chem.* 2006, *78* (20), 7292–7301.

384. Xiao, Q.; Hu, B.; Yu, C.; Xia, L.; Jiang, Z., Optimization of a single-drop microextraction procedure for the determination of organophosphorus pesticides in water and fruit juice with gas chromatography–flame photometric detection. *Talanta* 2006, *69* (4), 848–855.

385. Yan, C. H.; Wu, H. F., A liquid-phase microextraction method, combining a dual gauge microsyringe with a hollow fiber membrane, for the determination of organochlorine pesticides in aqueous solution by gas chromatography/ion trap mass spectrometry. *Rapid Commun. Mass Spectrom.* 2004, *18* (24), 3015–3018.

386. Ye, C.; Zhou, Q.; Wang, X., Improved single-drop microextraction for high sensitive analysis. *J. Chromatogr. A* 2007, *1139* (1), 7–13.

387. Zhang, J.; Lee, H. K., Application of liquid-phase microextraction and on-column derivatization combined with gas chromatography–mass spectrometry to the determination of carbamate pesticides. *J. Chromatogr. A* 2006, *1117* (1), 31–37.

388. Zhao, E.; Shan, W.; Jiang, S.; Liu, Y.; Zhou, Z., Determination of the chloroacetanilide herbicides in waters using single-drop microextraction and gas chromatography. *Microchem. J.* 2006, *83* (2), 105–110.

389. Zhao, L.; Lee, H. K., Application of static liquid-phase microextraction to the analysis of organochlorine pesticides in water. *J. Chromatogr. A* 2001, *919* (2), 381–388.

390. Liu, Y.; Zhao, E.; Zhou, Z., Single-drop microextraction and gas chromatographic determination of fungicide in water and wine samples. *Anal. Lett.* 2006, *39* (11), 2333–2344.

391. Pan, H. J.; Ho, W. H., Determination of fungicides in water using liquid phase microextraction and gas chromatography with electron capture detection. *Anal. Chim. Acta* 2004, *527* (1), 61–67.

392. Sanagi, M. M.; See, H. H.; Ibrahim, W. A.; Naim, A. A., Determination of pesticides in water by cone-shaped membrane protected liquid phase microextraction prior to micro-liquid chromatography. *J. Chromatogr. A* 2007, *1152* (1–2), 215–219.

393. Sandahl, M.; Mathiasson, L.; Jönsson, J. Å., On-line automated sample preparation for liquid chromatography using parallel supported liquid membrane extraction and microporous membrane liquid–liquid extraction. *J. Chromatogr. A* 2002, *975* (1), 211–217.

394. Lüthje, K.; Hyötyläinen, T.; Rautiainen-Rämä, M.; Riekkola, M. L., Pressurised hot water extraction–microporous membrane liquid–liquid extraction coupled on-line with gas chromatography–mass spectrometry in the analysis of pesticides in grapes. *Analyst* 2005, *130* (1), 52–58.

395. Chia, K. J.; Huang, S. D., Analysis of organochlorine pesticides in wine by solvent bar microextraction coupled with gas chromatography with tandem mass spectrometry detection. *Rapid Commun. Mass Spectrom.* 2006, *20* (2), 118–124.

396. Schellin, M.; Hauser, B.; Popp, P., Determination of organophosphorus pesticides using membrane-assisted solvent extraction combined with large volume injection-gas chromatography–mass spectrometric detection. *J. Chromatogr. A* 2004, *1040* (2), 251–258.

397. Zhao, E.; Han, L.; Jiang, S.; Wang, Q.; Zhou, Z., Application of a single-drop microextraction for the analysis of organophosphorus pesticides in juice. *J. Chromatogr. A* 2006, *1114* (2), 269–273.

398. Zhu, L.; Ee, K. H.; Zhao, L.; Lee, H. K., Analysis of phenoxy herbicides in bovine milk by means of liquid–liquid–liquid microextraction with a hollow-fiber membrane. *J. Chromatogr. A* 2002, *963* (1–2), 335–343.

399. Zuin, V. G.; Schellin, M.; Montero, L.; Yariwake, J. H.; Augusto, F.; Popp, P., Comparison of stir bar sorptive extraction and membrane-assisted solvent extraction as enrichment techniques for the determination of pesticide and benzo[*a*]pyrene residues in Brazilian sugarcane juice. *J. Chromatogr. A* 2006, *1114* (2), 180–187.

400. Ulrich, S., Solid-phase microextraction in biomedical analysis. *J. Chromatogr. A* 2000, *902* (1), 167–194.

401. Deng, C.; Li, N.; Wang, X.; Zhang, X.; Zeng, J., Rapid determination of acetone in human blood by derivatization with pentafluorobenzyl hydroxylamine followed by headspace liquid-phase microextraction and gas chromatography/mass spectrometry. *Rapid Commun. Mass Spectrom.* 2005, *19* (5), 647–653.

402. Dong, L.; Shen, X.; Deng, C., Development of gas chromatography–mass spectrometry following headspace single-drop microextraction and simultaneous derivatization for fast determination of the diabetes biomarker, acetone in human blood samples. *Anal. Chim. Acta* 2006, *569* (1–2), 91–96.

403. Li, N.; Deng, C.; Yao, N.; Shen, X.; Zhang, X., Determination of acetone, hexanal and heptanal in blood samples by derivatization with pentafluorobenzyl hydroxylamine followed by headspace single-drop microextraction and gas chromatography-mass spectrometry. *Anal. Chim. Acta* 2005, *540* (2), 317–323.

404. Li, N.; Deng, C.; Yin, X.; Yao, N.; Shen, X.; Zhang, X., Gas chromatography–mass spectrometric analysis of hexanal and heptanal in human blood by headspace single-drop microextraction with droplet derivatization. *Anal. Biochem.* 2005, *342* (2), 318–326.

405. Pranaitytė, B.; Jermak, S.; Naujalis, E.; Padarauskas, A., Capillary electrophoretic determination of ammonia using headspace single-drop microextraction. *Microchem. J.* 2007, *86* (1), 48–52.

406. Jonsson, O. B.; Nordlöf, U.; Nilsson, U. L., The XT-tube extractor: a hollow fiber–based supported liquid membrane extractor for bioanalytical sample preparation. *Anal. Chem.* 2003, *75* (14), 3506–3511.

407. Nozal, L.; Arce, L.; Simonet, B. M.; Ríos, Á.; Valcárcel, M., In-line liquid-phase microextraction for selective enrichment and direct electrophoretic analysis of acidic drugs. *Electrophoresis* 2007, *28* (18), 3284–3289.

408. Shah, F. U.; Barri, T.; Jönsson, J. Å.; Skog, K., Determination of heterocyclic aromatic amines in human urine by using hollow-fibre supported liquid membrane extraction and liquid chromatography–ultraviolet detection system. *J. Chromatogr. B* 2008, *870* (2), 203–208.

409. Fang, H.; Zeng, Z.; Liu, L.; Pang, D., On-line back-extraction field-amplified sample injection method for directly analyzing cocaine and thebaine in the extractants by solvent microextraction. *Anal. Chem.* 2006, *78* (4), 1257–1263.

410. Fiamegos, Y. C.; Nanos, C. G.; Stalikas, C. D., Ultrasonic-assisted derivatization reaction of amino acids prior to their determination in urine by using single-drop microextraction in conjunction with gas chromatography. *J. Chromatogr. B* 2004, *813* (1–2), 89–94.

411. Ma, M.; Kang, S.; Zhao, Q.; Chen, B.; Yao, S., Liquid-phase microextraction combined with high-performance liquid chromatography for the determination of local anaesthetics in human urine. *J. Pharm. Biomed. Anal.* 2006, *40* (1), 128–135.

412. Shrivas, K.; Wu, H. F., Single drop microextraction as a concentrating probe for rapid screening of low molecular weight drugs from human urine in atmospheric-pressure matrix-assisted laser desorption/ionization mass spectrometry. *Rapid Commun. Mass Spectrom.* 2007, *21* (18), 3103–3108.

413. Shrivas, K.; Wu, H. F., Modified silver nanoparticle as a hydrophobic affinity probe for analysis of peptides and proteins in biological samples by using liquid–liquid microextraction coupled to AP-MALDI-ion trap and MALDI-TOF mass spectrometry. *Anal. Chem.* 2008, *80* (7), 2583–2589.

414. Sobhi, H. R.; Yamini, Y.; Esrafili, A.; Abadi, R. H. H. B., Suitable conditions for liquid-phase microextraction using solidification of a floating drop for extraction of fat-soluble vitamins established using an orthogonal array experimental design. *J. Chromatogr. A* 2008, *1196–1197*, 28–32.

415. Sudhir, P.-R.; Wu, H.-F.; Zhou, Z.-C., Identification of peptides using gold nanoparticle-assisted single-drop microextraction coupled with AP-MALDI mass spectrometry. *Anal. Chem.* 2005, *77* (22), 7380–7385.

416. de Jager, L.; Andrews, A. R. J., Development of a screening method for cocaine and cocaine metabolites in saliva using hollow fiber membrane solvent microextraction. *Anal. Chim. Acta* 2002, *458* (2), 311–320.

417. Yang, X. L.; Luo, M. B.; Ding, J. H., Rapid determination of nicotine in saliva by liquid phase microextraction–high performance liquid chromatography. *Chin. J. Anal. Chem.* 2007, *35* (2), 171–175.

418. Bjørhovde, A.; Halvorsen, T. G.; Rasmussen, K. E.; Pedersen-Bjergaard, S., Liquid-phase microextraction of drugs from human breast milk. *Anal. Chim. Acta* 2003, *491* (2), 155–161.

419. Gjelstad, A.; Andersen, T. M.; Rasmussen, K. E.; Pedersen-Bjergaard, S., Microextraction across supported liquid membranes forced by pH gradients and electrical fields. *J. Chromatogr. A* 2007, *1157* (1–2), 38–45.

420. Gjelstad, A.; Rasmussen, K. E.; Pedersen-Bjergaard, S., Simulation of flux during electro-membrane extraction based on the Nernst–Planck equation. *J. Chromatogr. A* 2007, *1174* (1–2), 104–111.

421. Gjelstad, A.; Rasmussen, K. E.; Pedersen-Bjergaard, S., Electrokinetic migration across artificial liquid membranes: tuning the membrane chemistry to different types of drug substances. *J. Chromatogr. A* 2006, *1124* (1–2), 29–34.

422. Balchen, M.; Gjelstad, A.; Rasmussen, K. E.; Pedersen-Bjergaard, S., Electrokinetic migration of acidic drugs across a supported liquid membrane. *J. Chromatogr. A* 2007, *1152* (1–2), 220–225.

423. Gupta, P. K.; Manral, L.; Ganesan, K.; Dubey, D. K., Use of single-drop microextraction for determination of fentanyl in water samples. *Anal. Bioanal. Chem.* 2007, *388* (3), 579–583.

424. Sarafraz-Yazdi, A.; Mofazzeli, F.; Es'haghi, Z., Directly suspended droplet three liquid phase microextraction of diclofenac prior to LC. *Chromatographia* 2008, *67* (1), 49–53.

425. Raich-Montiu, J.; Krogh, K. A.; Granados, M.; Jönsson, J. Å.; Halling-Sørensen, B., Determination of ivermectin and transformation products in environmental waters using hollow fibre-supported liquid membrane extraction and liquid chromatography–mass spectrometry/mass spectrometry. *J. Chromatogr. A* 2008, *1187* (1–2), 275–280.

426. Ma, M.; Cantwell, F. F., Solvent microextraction with simultaneous back-extraction for sample cleanup and preconcentration: preconcentration into a single microdrop. *Anal. Chem.* 1999, *71* (2), 388–393.

427. Wang, G.-Q.; Zhang, R.-J.; Sun, Y.-A.; Xie, K.; Ma, C.-Y., Characterization and semiquantitative analysis of volatile compounds in orange juice by use of headspace-solvent microextraction and gas chromatography. *Chromatographia* 2007, *65* (5–6), 363–366.

428. Xiao, Q.; Yu, C.; Xing, J.; Hu, B., Comparison of headspace and direct single-drop microextraction and headspace solid-phase microextraction for the measurement of volatile sulfur compounds in beer and beverage by gas chromatography with flame photometric detection. *J. Chromatogr. A* 2006, *1125* (1), 133–137.

429. Tankeviciute, A.; Kazlauskas, R.; Vickackaite, V., Headspace extraction of alcohols into a single drop. *Analyst* 2001, *126* (10), 1674–1677.

430. Tankeviciute, A.; Rozkov, A.; Kazlauskas, R.; Vickackaite, V., Extraction of alcohols into a xylene drop. *Chem. Anal. (Warsaw)* 2002, *47* (6), 935–944.

431. Tankeviciute, A.; Vickackaite, V.; Kazlauskas, R., Gas chromatographic determination of alcohols using microextraction with xylene drop. *Chemija (Vilnius)* 2002, *13* (4), 227–231.

432. Shrivas, K.; Wu, H. F., Rapid determination of caffeine in one drop of beverages and foods using drop-to-drop solvent microextraction with gas chromatography/mass spectrometry. *J. Chromatogr. A* 2007, *1170* (1–2), 9–14.

433. Yu, C.; Liu, Q.; Lan, L.; Hu, B., Comparison of dual solvent–stir bars microextraction and U-shaped hollow fiber-liquid phase microextraction for the analysis of Sudan dyes in food samples by high-performance liquid chromatography–ultraviolet/mass spectrometry. *J. Chromatogr. A* 2008, *1188* (2), 124–131.

434. Hyötyläinen, T.; Tuutijärvi, T.; Kuosmanen, K.; Riekkola, M. L., Determination of pesticide residues in red wines with microporous membrane liquid–liquid extraction and gas chromatography. *Anal. Bioanal. Chem.* 2002, *372* (5), 732–736.

435. González-Peñas, E.; Leache, C.; Viscarret, M.; Pérez de Obanos, A.; Araguás, C.; López de Cerain, A., Determination of ochratoxin A in wine using liquid-phase microextraction combined with liquid chromatography with fluorescence detection. *J. Chromatogr. A* 2004, *1025* (2), 163–168.

436. Valcárcel, M.; Cárdenas, S.; Simonet, B. M.; Carrillo-Carrión, C., Principles of qualitative analysis in the chromatographic context. *J. Chromatogr. A* 2007, *1158* (1–2), 234–240.

437. Batlle, R.; Nerin, C., Application of single-drop microextraction to the determination of dialkyl phthalate esters in food simulants. *J. Chromatogr. A* 2004, *1045* (1–2), 29–35.

438. Wang, X.; Jiang, T.; Yuan, J.; Cheng, C.; Liu, J.; Shi, J.; Zhao, R., Determination of volatile residual solvents in pharmaceutical products by headspace liquid-phase microextraction gas chromatography–flame ionization detector. *Anal. Bioanal. Chem.* 2006, *385* (6), 1082–1086.

439. Wood, D. C.; Miller, J. M.; Christ, I.; Majors, R. E., Headspace liquid microextraction. *LC•GC Eur.* 2004, *17* (11), 573–579.

440. Hylton, K.; Mitra, S., Automated, on-line membrane extraction. *J. Chromatogr. A* 2007, *1152* (1–2), 199–214.

441. Barri, T.; Bergström, S.; Norberg, J.; Jönsson, J. Å., Miniaturized and automated sample pretreatment for determination of PCBs in environmental aqueous samples using an on-line microporous membrane liquid–liquid extraction–gas chromatography system. *Anal. Chem.* 2004, *76* (7), 1928–1934.

442. Melwanki, M. B.; Huang, S. D., Extraction of hydroxyaromatic compounds in river water by liquid–liquid–liquid microextraction with automated movement of the acceptor and the donor phase. *J. Sep. Sci.* 2006, *29* (13), 2078–2084.

443. Trtic-Petrovic, T.; Jönsson, J. Å., Determination of drug–protein binding using supported liquid membrane extraction under equilibrium conditions. *J. Chromatogr. B* 2005, *814* (2), 375–384.

444. Jeannot, M. A.; Cantwell, F. F., Solvent microextraction as a speciation tool: determination of free progesterone in a protein solution. *Anal. Chem.* 1997, *69* (15), 2935–2940.

445. Jeannot, M. A.; Cantwell, F. F., Mass transfer characteristics of solvent extraction into a single drop at the tip of a syringe needle. *Anal. Chem.* 1997, *69* (2), 235–239.

446. Qin, X. Y.; Meng, J.; Li, X. Y.; Zhou, J.; Sun, X. L.; Wen, A. D., Determination of venlafaxine in human plasma by high-performance liquid chromatography using cloud-point extraction and spectrofluorimetric detection. *J. Chromatogr. B* 2008, *872* (1–2), 38–42.

447. Guo, Y. G.; Zhang, J.; Liu, D. N.; Fu, H. F., Determination of *n*-octanol–water partition coefficients by hollow-fiber membrane solvent microextraction coupled with HPLC. *Anal. Bioanal. Chem.* 2006, *386* (7–8), 2193–2198.

448. Liu, J. F.; Hu, X. L.; Peng, J. F.; Jönsson, J. Å.; Mayer, P.; Jiang, G. B., Equilibrium sampling of freely dissolved alkylphenols into a thin film of 1-octanol supported on a hollow fiber membrane. *Anal. Chem.* 2006, *78* (24), 8526–8534.

449. Sangster, J., *Octanol–Water Partition Coefficients: Fundamentals and Physical Chemistry*. Chichester, UK, Wiley, 1997.

450. Liu, J. F.; Jönsson, J. Å.; Mayer, P., Equilibrium sampling through membranes of freely dissolved chlorophenols in water samples with hollow fiber supported liquid membrane. *Anal. Chem.* 2005, *77* (15), 4800–4809.

CHAPTER 7

SME EXPERIMENTS

7.1 INTRODUCTION

The experiments in this chapter were chosen to be representative of the types of experiments likely to be attempted by a new user of the SME technique. They have either been adapted from published methods and modified for this book or developed in our laboratories. The single-drop (SDME and HS-SDME) experiments can be conducted with equal success using either a manual or a computerized autosampler. Typical useful autosampler parameters are included for each experiment where appropriate. The experiments that follow include direct-immersion extractions from water using single-drop microextraction (SDME), direct-immersion extraction from soil dispersed in water using hollow fiber–protected two-phase microextraction (HF(2)ME), and extraction from water using dispersive liquid–liquid microextraction (DLLME). Also included are headspace extractions from water using single-drop microextraction (HS-SDME) and HF(2)ME. The experiments are rounded off by a comparison between static solvent extractions and dynamic solvent extractions. A complete list of experiments is presented in Table 7.1.

Solvent Microextraction: Theory and Practice, By John M. Kokosa, Andrzej Przyjazny, and Michael A. Jeannot
Copyright © 2009 John Wiley & Sons, Inc.

TABLE 7.1   SME Experiments

Experiment	Section
Determination of gasoline diluents in motor oil by HS-SDME	7.3
Determination of BTEX in water by HS-SDME	7.4
Analysis of halogenated disinfection by-products by SDME and HS-SDME	7.5
Analysis of volatile organic compounds by SDME and HS-SDME	7.6
Analysis of residual solvents in drug products by HS-SDME	7.7
Analysis of arson accelerant by HS-SDME	7.8
Analysis of PAHs by SDME and HS-SDME	7.9
Determination of acetone in aqueous solutions by derivatization HS-SDME	7.10
Determination of pesticides in soil by HF(2)ME	7.11
Determination of PAHs and HOCs by DLLME	7.12
Dynamic headspace and direct immersion extractions (DY-SME)	7.13

SME methods that are most readily adaptable to the use of an autosampler include direct-immersion SDME and headspace extraction HS-SDME. Direct-immersion SDME involves positioning the syringe needle tip in the sample solution (usually, water) and exposing a 1- to 3-μL drop of extracting solvent to the sample for a period of 1 to 30 min. HS-SDME involves positioning the needle tip and drop in the headspace over a sample in a vial. A number 2 point 10-μL syringe (such as a Hamilton curved bevel tip) has been found to have the greatest retention of the solvent drop. Salt can be added, and the sample can be stirred or heated to increase extraction efficiency. Stirring is done most conveniently in a computer-controlled heating-stirring module, but can be performed using a standard stirrer or stir-hot plate. When used with an autosampler, the stirrer can be defined for the computer software as a tray. A third SME method, hollow fiber–protected two-phase micro-extraction [(HF(2)ME), often referred to as liquid-phase microextraction (LPME)] has also been adapted successfully for use with an autosampler. However, most extractions using this mode are conducted with manual extraction, as will be done here.

Direct-immersion (DI) extractions are usually slow compared to headspace (HS) extractions, in part due to the slower diffusion rates for analytes in water than in air. If an autosampler system is available, extraction times can be decreased success-fully for very dilute solutions using dynamic SDME, in which the drop is moved in and out of the syringe needle continuously during the extraction. This process continually exposes a new surface of the drop to the water, decreasing diffusion times from the water to the drop. An analogous technique, where the extracting solvent is moved in and out of the syringe barrel along with the water sample is even more effective. However, this in-syringe dynamic method should be avoided for SDME unless very clean water samples are used, to avoid contaminating the chromatographic system.

One must be careful to account for a decrease in drop size during the extraction due to evaporation or solubility in the sample, as well as small losses of the drop due to creep of the drop up the outside surface of the needle. Typically, even for a high-boiling solvent, it is advisable to use a drop size 0.1 to 0.5 µL larger than the final pull-up/injection volume. Larger drop/injection ratios are needed for volatile solvents. For instance, when using toluene as the extracting solvent, if a drop size of 3 µL were used, calculations indicate that about 0.65 µL might be lost due to solubility in water (even when using headspace), and about 0.1 µL might be lost due to solvent creep. Therefore, a maximum injection volume of 2 µL would be prudent, to avoid injecting air or water, which might contain salt or other matrix components. Extraction solvent loss can be minimized by addition of NaCl (100 to 300 mg/mL water) or Na_2SO_4 (100 to 300 mg/mL water) to an aqueous sample.

These drop effects become more crucial when using elevated extraction temperatures or long extraction times, even for high-boiling extraction solvents such as 1-octanol or tetradecane, unless a needle cooling device is used for the extraction. It is therefore recommended that extraction temperatures be limited to 30 to 50°C and extraction times limited to 3 to 10 min for headspace and 5 to 15 min. for direct extractions when first developing a method. It is also recommended that, if possible, a nearly saturated salt solution (300 mg NaCl/mL water or 300 mg Na_2SO_4/mL water) be used to increase extraction efficiency from water solutions and to decrease extraction solvent depletion.

Chapter 5 describes a more detailed methodology for choosing an appropriate extraction mode, solvent, and extraction conditions. You may also find the material in Chapter 6 useful if you are attempting to locate a reference procedure for extraction of a particular analyte.

The following applications have been chosen to represent SME alternatives to several traditional sample preparation methods: (1) direct analysis, (2) liquid–liquid extraction, (3) purge-and-trap extraction, and (4) static headspace extraction. Many of these methods have been challenged in recent years by new techniques such as SPME and SPE. SME is not expected to supplant these techniques, but will be an additional powerful weapon in the arsenal available to analysts.

7.2 RECOMMENDED EXPERIMENTAL CONDITIONS

Table 7.2 lists the ranges of important experimental parameters which are recommended starting points for an experiment. Refer to Chapters 2, 4, 5, and 6 for recommended solvents and extraction modes for specific analytes.

Whenever possible, use headspace SME for sample preparation. This technique works well for most nonpolar, low- to medium-molecular-weight compounds. The extraction solvents of choice for low- to medium-boiling-range analytes are alkanes with boiling points 50 to 100°C higher than the highest-boiling analyte of interest. These include dodecane, tridecane, tetradecane, and pentadecane. Hexadecane may be used with caution, since it melts at 18°C. 1-Octanol is also a useful solvent, especially for polar analytes such as aldehydes, ketones, phenols, and alcohols. The GC inlet must be used in split mode (10:1 to 50:1) to yield sharp, resolvable peaks with these solvents. Lower-boiling solvents such as toluene, the xylenes, or hexane

TABLE 7.2   **Experimental Parameters Affecting Extraction Efficiency**

Parameter	Ranges Recommended
Sample amount	1–5 mL
Headspace volume	0.5–1 mL
Extraction solvent volume SDME	1–2 μL
Extraction solvent volume HF(2)ME	3–10 μL
Extraction solvent volume HS-SDME	1–2 μL
Stirring-rate DI modes	600 rpm
Stirring-rate HS modes	600–1600 rpm
Salt concentration	30%
Incubation and extraction time (SDME)	1–5 and 5–20 min
Incubation and extraction time (HF(2)ME)	0 and 15–60 min
Incubation and extraction time (HS-SDME)	1–5 and 5–10 min
Temperature (DI modes)	20–40°C
Temperature (HS modes)	20–60°C
pH of sample and acceptor (acidic analytes)	3 and 10
pH of sample and acceptor (basic analytes)	10 and 3

may be used to extract high-boiling analytes such as the PAHs. These solvents can be injected with either split or splitless modes.

Analytes at concentrations higher than 50 μg/L in water may be extracted efficiently without the addition of NaCl or Na_2SO_4. Salt addition at near saturation is recommended, however, for analytes at lower concentrations and to minimize extraction solvent loss to sample water.

Sample extraction temperatures for volatile analytes are best between 20 and 35°C. Polar and higher-boiling analytes may require higher extraction temperatures, but is suggested that the extraction temperature be limited to a maximum of 50 to 60°C.

Efficient agitation or stirring of water samples is necessary to minimize extraction time and increase extraction efficiency. A stir rate of 600 to 1200 rpm is suggested. Drop loss is possible for the SDME methods if too high a stirring rate is used, however, which normally limits stirring rates to a maximum of 600 rpm. Use of an orbital agitator requires the method to turn off the mixer during extraction. If a stirrer or orbital mixer is not available, vortex the sample for at least 1 min or until all salt is dissolved. This will shorten the time necessary for equilibration with the headspace for HS modes. Next, allow the sample to sit to equilibrate for an hour before extraction.

The maximum practical SDME extraction solvent drop volume is 3 μL. It is recommended that the drop volume be 0.1 to 0.5 μL larger than the injection volume, however. This will compensate for loss of the solvent through evaporation, dissolution, and creep up the outside of the needle.

Direct-extraction SDME is used when HS-SDME would give poor results, such as with high-boiling or polar analytes when extracting from water. Any water-insoluble solvent can be used, but the solvent used most commonly is 1-octanol.

To avoid the injection of water sample, especially if salting out has been used or the drop partially dissolves in the water, use a drop larger than the injection volume. For instance, if a 3-μL drop is used for the extraction, an injection volume of 2 to 2.5 μL should be considered.

To monitor the integrity of the microdrop and to account for small variations in the size of the drop withdrawn into the syringe after the extraction, an internal standard in the extracting solvent is needed. This can either be an added chemical, similar in boiling point to the extracted analytes, or an impurity originally present in the extracting solvent. Added standards have included decafluorobiphenyl (for GC-MS or GC-ECD analysis) or naphthalene or biphenyl (for GC-FID work). Examples of impurities that can be used as standards include tridecane in 1-octanol and decane in tetradecane. The absolute amount of standard does not need to be known; it simply has to be the same for each analysis.

When extracting analytes present at very low concentrations, SME may require purification of even the highest-purity commercial solvents. Solvents absorb chemicals from the atmosphere and even when freshly opened may contain minor amounts of homologs and oxidation products which may co-elute with analytes of interest. For example, hexadecane contains very small amounts of decane and dodecane as well as aldehydes. These can be removed by single or double vacuum distillation. The impurities may, however, be useful, for they can be used as in-drop internal standards to monitor the stability of the drop and minor variations in drop size from run to run.

The hollow fiber extraction modes [HF(2)ME and HF(3)ME] require individual preparation of each fiber. The fibers should be rinsed with acetone and air dried prior to use to remove manufacturing impurities. One end of the fiber may be sealed to encapsulate the extracting solvent. This is often done by crimping and heat sealing with a device such as a soldering iron, or just wrapping with a wire. However, some procedures leave both ends open. In one variation, the tube is bent into a U-shape and the ends are not immersed in the sample. One end can also be left open, with the water-immiscible solvent exposed to the sample. It is also possible to fill the tube and then seal both ends. The sealed tube is then stirred with the sample for extraction, removed, and punctured with a syringe needle to remove the extraction solvent. Normally, the fiber is immersed in the extraction solvent to saturate the polymer pores, wiped of excess solvent if necessary, and then a syringe used to fill the interior with solvent. In a few cases the presaturation step is omitted. The prepared fiber is attached to the sample vial through a hole in the septum cap, attached directly to the tip of the syringe needle, or attached to a micropipette tip which is, in turn, inserted through the vial cap septum.

Because the extraction solvent is protected by the hollow fiber membrane, the solvent cannot be lost, as can happen with an SDME extraction. As a result, much higher stirring rates are generally used, ranging from 1000 to 1600 rpm. However, solvent loss by evaporation and solubility can also occur. Thus, similar rules apply for choosing a solvent using a hollow fiber method as when using SDME.

Finally, the hollow fiber mode may also use a three-phase extraction. The pores of the fiber are saturated with a water-immiscible solvent while the lumen can be filled with an aqueous solution. This mode is generally used to extract acidic or basic analytes. To extract an acidic analyte, the sample solution is made acidic, to minimize ionization and allow the analyte to pass through the nonpolar fiber and organic solvent within its pores. The lumen of the fiber is filled with a basic solution, which ionizes the acidic analyte and accumulates it in the acceptor solution. Basic analytes are extracted by reversing the pH gradients.

Since the analytes must pass through the polymer and the organic solvent in the pores to enter the bulk of the extracting solvent, HF extractions are usually much slower than single-drop extractions. However, HF modes have the potential for exhaustive extractions, especially HF(3)ME extractions.

Hints specific to various extractions are noted in the experimental procedures recommended below.

7.3 DETERMINATION OF GASOLINE DILUENTS IN MOTOR OIL BY HS-SDME

ASTM method D 3525-93 for determining the concentration of gasoline present in motor oil involves injecting a 50% hexadecane solution of the oil directly onto a packed GC column, using tetradecane as an internal standard.[1] Although this works well, the method also involves injecting all the components of the oil, including viscous polymers, antioxidants, water, dirt, carbon, and other materials that would degrade the column. The method would not be appropriate for a capillary column system. The alternative would be to use static headspace extraction. Static headspace is simple and straightforward, but low-boiling components dominate in the headspace, leading to a chromatogram pattern (Figure 7.1) very different from that

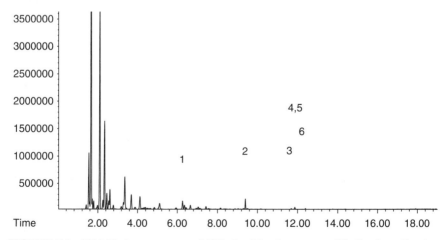

FIGURE 7.1. GC-MS chromatogram of 100 μL of headspace over 50 μL of gasoline in a 2-mL crimp-top vial. Temperature: 50°C. Inlet split: 25:1. Peak identification: (1) benzene, (2) toluene, (3) ethylbenzene, (4) *p*-xylene, (5) *o*-xylene, (6) *m*-xylene.

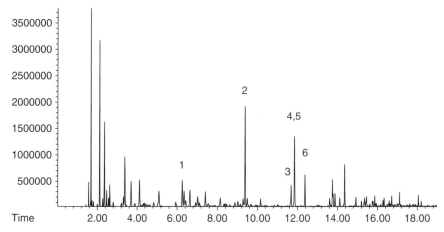

FIGURE 7.2. GC-MS chromatogram of 1 µL of 2.2% gasoline solution in hexadecane. Inlet split: 25:1. Peak identities same as in Figure 7.1.

of a GC of 1 µL of gasoline (Figure 7.2) dissolved in hexadecane.[2] HS-SDME of gasoline in motor oil using 1 µL of tetradecane or hexadecane as the extracting solvent, on the other hand, yields a chromatogram (Figure 7.3) that is nearly identical with that of gasoline. A calibration curve is prepared by the extraction of gasoline solutions in motor oil samples with either hexadecane or tetradecane. Figure 7.4 is an example of the HS-SDME extraction with hexadecane of 0.5 mL of used automotive motor oil contaminated with heat-degraded gasoline, followed by GC/FID analysis.

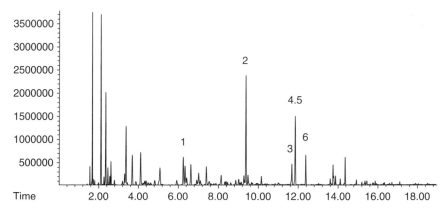

FIGURE 7.3. HS-SDME GC-MS, 1.8% gasoline in engine oil standard. Sample volume: 0.5 mL in a 2-mL vial; extracting solvent: hexadecane (1 µL); extraction time: 3 min; extraction temperature 50°C; inlet split: 25:1. Peak identities same as in Figure 7.1.

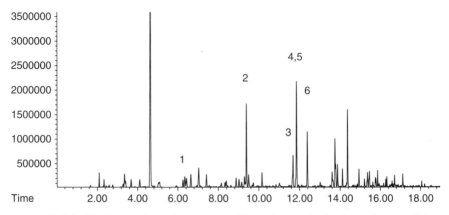

FIGURE 7.4. HS-SDME GC-MS, gasoline diluent in used oil. Extraction conditions same as in Figure 7.3 and peak identities same as in Figure 7.1. Surrogate standard: ethyl acetate.

7.3.1   Experimental

A four-point calibration curve is prepared by extraction of samples of gasoline in motor oil (0.5 to 4%). Used oil samples (0.5 mL, accurately weighed) are then extracted. Optimal extraction conditions are 0.5-mL samples of oil in a crimped 2-mL vial, heated for 30 minutes at 50°C, and extracted with 1.0 µL of hexadecane for 3 minutes. The extract is injected into an Agilent 6890 GC equipped with a HP-5, 30 m length × 0.25 mm diameter × 1 µm film at a 2-mL/min flow rate, with a 25:1 split. The injector is set to 250°C, the FID detector to 300°C, and the oven is programmed from 35 to 135°C at a rate of 10°C/min, followed by a second ramp at 25°C/min to 260°C.

7.3.2   Results and Discussion

The amounts of gasoline in motor oil samples using this technique agree well with results obtained using the ASTM method. Minor differences (well within ASTM requirements) would be minimized if weathered gasoline rather than new gasoline were used to prepare the standards. The technique also enables the quantification of BTEX components (benzene, toluene, ethylbenzene, and *o*-, *m*-, and *p*-xylene) in an analogous fashion. Use of an inlet split is necessary to ensure sharp peaks and to avoid overloading the column and detector. Samples were found to be too viscous to stir effectively, so they should be equilibrated at 50°C for 30 minutes. Higher temperatures are undesirable, due to possible losses through the septum during sampling, to overheating, and to loss of the drop. A 3-minute extraction time was found to be optimal, since extracted amounts approach equilibrium at that point. The literature procedure used ethyl acetate as a surrogate to monitor the extraction efficiency, but an internal standard added to the extracting solvent is more important to monitor the drop and ensure that it is not lost during the extraction, as well as to correct for minor drop-size variances. Since tetradecane is an impurity present in commercial 99% hexadecane, this provides a good internal standard in the drop. The hexadecane cannot, of course, be used to monitor the drop since the peak would saturate the detector.

7.3.3 Additional Experimental Recommendations

Instead of hexadecane, pentadecane or tetradecane may be used as the extraction solvent. Hexadecane impurity in the tetradecane may be used as the internal standard. Hexadecane melts at 18°C and may have to be heated in a cold laboratory. Tetradecane melts at 5.5°C and pentadecane at 8 to 10°C and are less of a problem. All high-boiling solvents, however, tend to coat or passivate the pneumatic system of the GC. This does not result in carryover of sample between runs but does necessitate injection of one or two blanks of solvent before making a sample run, to ensure that the GC is passivated. It also then requires at least one or two blank runs following a series of samples with no solvent after using these solvents to bleed off any solvent remaining in the GC. In addition, it is imperative to use a split flow that will remove solvent from the pneumatic lines and valves and to monitor the charcoal split vent trap to make sure that it does not plug. Lower-boiling solvents result in similar problems but do not affect the pneumatic electronics by plugging the filter, even if it has been saturated.

The autosampler settings should include a slow needle pull-up speed (0.5 μL/s), and a delay of 1 s following pull-up due to the viscosity of the extracting solvent, a drop size of 1.2 μL, and an injection drop size of 1 μL. Wash 1 and wash 2 should contain the extracting solvent. Since the concentrations of gasoline are in the percent range, there is no need to distill the solvent if is at least 99% pure. However, it may have to be vacuum distilled if decane or other impurities interfere with the chromatogram.

7.4 DETERMINATION OF BTEX IN WATER BY HS-SDME

BTEX components (benzene, toluene, ethylbenzene, and o-, m-, p-xylenes) are major components of gasoline and are frequently monitored in water to detect leakage of underground gasoline storage facilities. Present popular extraction and concentration methods widely in use include purge-and-trap and SPME. In a classic set of experiments, the interrelationships of drop size, sample size, temperature, extraction time, agitation, and analyte octanol–water partition coefficient (log K_{ow}) for the extraction of BTEX components in water by HS-SME was studied and compared to theoretical results.[3] Ionic effect (salt concentration) was not studied.

7.4.1 Experimental

Calibration standards are prepared at 870 ppm in methanol. Stock solutions are stored at 4°C and freshly prepared every 2 weeks. The stock solution is spiked into water daily at levels ranging from 8.7 to 870 ppb. Samples of 1.5 mL in 2-mL or 20 mL in 40-mL crimped vials containing Teflon-coated stir bars are extracted with hexadecane for various times, stirring rates, temperatures, and drop sizes ranging from 1 to 3 μL. The extract is injected into an Agilent 6890 GC equipped with a HP-5, 30 m length × 0.25 mm diameter × 1 μm film at a 2-mL/minute flow rate, with a 10:1 split. The injector is set to 250°C, the FID detector to 300°C, and the oven temperature is programmed from 35 to 260°C as follows: 35°C for 5 min, then raised to 135°C at 10°C/min, and to 260°C at 25°C/min, then held at 260°C for 5 min.

The autosampler settings should include a slow needle pull-up speed (0.5 μL/s) and a delay of 1 s following pull-up due to the viscosity of the solvent, a drop size of 1.2 μL, and an injection drop size of 1 μL. Wash 1 and wash 2 should contain the extraction solvent.

7.4.2 Results and Discussion

The results agree closely with theory. Increasing the drop size is directly proportional to the amount of material extracted (up to exhaustive extraction). However, the maximum viable drop size is 3 μL, since larger drops are not stable and fall off the needle. A drop size of 1 μL was found to be adequate in this study. Temperature was found to affect the water/headspace and headspace/drop equilibria. An increase in temperature was found to decrease the extraction efficiency into the drop for the components studied, and a final temperature of 23°C (room temperature) was found to give a maximum extraction.

The amount of sample extracted is not directly proportional to sample size. For analytes with smaller log K_{ow} values, a greater percentage of the analyte is extracted with a 1.5-mL sample size than for a 20-mL sample size. Analytes with large log K_{ow} values, however, can be extracted in larger amounts when using larger sample volumes than when using small sample volumes.

In most cases, extraction efficiency increases to a maximum with extraction time, and then levels off or even decreases with additional time. In this study, a 6-minute extraction time was found to be adequate.

Finally, extraction efficiency is directly proportional to stirring rate at a given extraction time. Higher stirring rates decrease the equilibrium time for the transfer of analyte to the headspace. Thus, the stirring rate should be maximized, as practical, short of causing dislodgement of the drop from the needle.

All of these results are predicted from the theory governing SME, as presented in Chapter 3.

7.4.3 Additional Experimental Recommendations

Lower detection limits can be obtained by the addition of 300 mg NaCl/mL of sample. Dynamic extraction, discussed later, may also be useful. Larger volumes of water samples will not yield significantly greater amounts of extracted material. See Chapter 4 for example calculations.

7.5 ANALYSIS OF HALOGENATED DISINFECTION BY-PRODUCTS BY SDME AND HS-SDME

SME has been used successfully to extract the trihalomethane chlorination disinfection by-products from drinking water using both HS-SDME and SDME. The procedures yielded better results than the liquid–liquid EPA extraction procedure (method 551.1).[4]

7.5.1 Experimental: HS-SDME

In one study, 25-mL water samples are treated with 7.5 g of NaCl (300 mg NaCl/mL water) and extracted with 1 μL of 1-octanol in 40-mL vials (15 mL headspace) for 10 min at 20°C while stirring at 800 rpm.[5] The extract is then analyzed by GC/ECD. A 25:1 inlet split is utilized. Limits of detection reported range from 0.15 to 0.4 μg/L, with RSDs of 8 to 11%.

In a second study, 1-mL samples are treated with 300 mg of Na_2SO_4 and extracted with 1 μL of either hexadecane or tetradecane in 2-mL vials (0.7 mL headspace) for 3 min at 23°C, either with no stirring during extraction or while stirring at 750 rpm.[6] The extract is analyzed by GC/ECD with a 25:1 inlet split. An internal standard (1,4-bromofluorobenzene or decafluorobiphenyl) is added to the extracting solvent to monitor the stability of the drop and to account for minor variations in the volume of the drop injected, run to run. These experiments can be conducted using both manual and automated HS-SDME, which yields nearly identical limits of detection ranging from 0.05 to 0.10 μg/L and RSDs ranging from 8 to 9% for manual extraction to 7 to 12% for automated extraction. Figure 7.5 is a

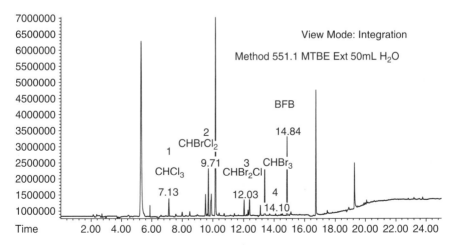

FIGURE 7.5 GC-ECD chromatogram of tap water extracted using EPA method 551.1; extracting solvent: MTBE (3 mL); sample volume: 50 mL; injection volume: 1 μL (splitless). Peak identification: (1) chloroform, (2) bromodichloromethane, (3) dibromochloromethane, (4) bromoform, (BFB) bromofluorobenzene (I.S.).

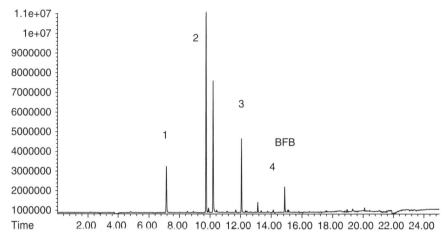

FIGURE 7.6 GC-ECD chromatogram of tap water extracted using HS-SDME; extracting solvent: hexadecane (1 µL); sample volume: 1 mL; extraction time: 3 min; extraction temperature: 25°C; inlet split: 25:1; salt concentration: 300 mg Na_2SO_4/mL. Peak identities same as in Figure 7.5.

chromatogram of a city tap water sample extracted using method 551.1 [a 3-mL methyl *t*-butyl ether (MTBE) extract of 50 mL of water], and Figure 7.6 is a chromatogram of 1 mL of the same tap water extracted by 1 µL of hexadccanc in 3 min using an autosampler. Note the large number of impurities in the MTBE extract.

7.5.2 Experimental: SDME

The two SDME procedures published used hexane as the extraction solvent. The earlier study did not use added salt or stirring and did not report the water sample volume.[7] Samples are extracted with 2 µL of hexane, at the tip of a syringe immersed in the water, for 15 min at 25°C. Samples are analyzed by GC/ECD with a splitless inlet. Limits of detection were not reported.

In the second study, 5-mL water samples in 10-mL vials (5 mL headspace) are treated with 900 mg of NaCl (180 mg/mL) and extracted with 2 µL of hexane at 25°C for 5 min while stirring at 600 rpm.[8] One microliter of the extract is drawn up into the syringe and analyzed by GC/ECD with a splitless inlet. Limits of detection range from 0.23 to 0.45 µg/L. RSD values were below 6%.

7.5.3 Results and Discussion

These results indicate a number of proper and improper experimental conditions. First, EPA method 551.1 suggests that water samples be salted out with sodium sulfate rather than NaCl, to avoid potential halide exchange. The salting-out characteristics of 350 mg/mL NaCl and 300 mg/mL Na_2SO_4 are very similar. The main difficulty in working with Na_2SO_4 is that a nearly saturated solution, on standing, tends to crystallize after 24 hours or so. Many investigators use low salt concentrations (10%), not realizing that the salting-out effect is exponentially dependent on salt concentrations.

Many investigators also use large volumes of sample and often large headspace volumes. Calculations clearly show that this is not necessary. A 1-mL sample volume may be more than adequate, as seen in the results above, and large headspace volumes are deleterious. Extraction times should also be kept as short as possible. SME is not usually an exhaustive extraction technique: rather, is an equilibrium or intermediate extraction technique. Once equilibrium is approached (3 to 10 min for HS-SDME and 5 to 15 min for SDME), additional extraction time is counterproductive. It should be noted, as expected from diffusion theory, that it does take longer for equilibrium to be established for SDME than HS-SDME and that using large volumes of sample in any case increases equilibration time significantly, without substantially increasing the amount of analyte extracted. The exception would be for analytes with large octanol–water partition coefficients such as PAHs.

Finally, the two SDME experiments were conducted using a low-boiling solvent (hexane, b.p. = 69°C) and splitless injection, which can give larger-area chromatographic peaks than split injections. However, low-boiling solvents tend to evaporate and dissolve significantly during extraction. The second SDME experiment deals with this issue by extracting with 2 μL of hexane and then drawing only 1 μL of the solvent back into the syringe after the extraction. It is strongly recommended that all SME experiments use an extraction drop slightly larger than the sample to be injected, to account for solvent loss due to evaporation, dissolving, or wicking up the outside of the needle. This is mandatory for SDME, to avoid the possibility of drawing up water into the syringe and ultimately contaminating the chromatographic column. In the second SDME experiment, one additional technique was used to ensure that the hexane solvent would not interfere with the analyte peaks. Use of a DB-624 chromatographic column allowed hexane to elute before the trihalomethane peaks.

7.5.4 Additional Experimental Recommendations

HS-SDME using dodecane or tetradecane as the extracting solvent is recommended. The solvents may need to be vacuum distilled if oxidized impurities in the solvents interfere with the analyte peaks. It is recommended to use a 1-mL sample to which 250 to 300 mg of Na_2SO_4 has been added. Samples should be stirred continuously and extracted at a maximum temperature of 35°C for 5 minutes. An extracting volume of 1.2 to 2.2 μL, with an injected volume of 1 to 2 μL, should be sufficient. Actual injection volumes and split ratios will depend on the GC column requirements.

7.6 ANALYSIS OF VOLATILE ORGANIC COMPOUNDS BY SDME AND HS-SDME

Traditional analysis of a wide variety of chemicals present in drinking water, wastewater, and soils has used macro liquid–liquid extraction and purge-and-trap (PT), and more recently, solid-phase extraction (SPE) and solid-phase microextraction (SPME). SME is, in most cases, capable of replicating these extraction methods, often with equivalent or better limits of detection and reproducibility, and certainly in simplicity.

7.6.1 Experimental: HS-SDME

There have been numerous papers dealing with extraction of volatile chemicals from soil and water using SME. In three similar papers, the authors addressed extraction of chlorinated hydrocarbons from water using HS-SDME.[9–11] The analytes extracted ranged from dichloromethane to tetrachloroethene. Sample sizes ranged from 2 to 40 mL, with headspace volumes ranging from 5 to 25 mL. Extracting solvents included 1-octanol, tridecane, and benzyl alcohol. Optimal extraction temperatures ranged from 20 to 25°C, extraction times from 10 to 11 min, and stirring rates from 400 to 1000 rpm. In one experiment, the ionic strength effect was studied and, as expected, extraction efficiency increased exponentially with increasing salt concentrations.

In another study, encompassing halogenated and nonhalogenated analytes from the EPA 524 list, HS-SDME, and SDME were compared.[12] Analytes ranged from chloroform and ethylbenzene to *o*-dichlorobenzene to naphthalene. A mass selective detector was used, in the full scan mode, as well as an ECD. Samples were extracted with either 1.3 µL of tetradecane or hexadecane containing decafluorobiphenyl (1.0 µL withdrawn into the syringe and injected). Extraction temperatures ranged from 23 to 50°C, with a final compromise temperature of 30°C. Salt concentrations varied from 0 to 350 mg/mL, and extraction efficiencies were found to increase with increasing ionic strength, with 350 mg/mL (concentration of saturated solution) chosen for the final experiments. Stirring rates were varied from 0 to 1500 rpm, higher stirring rates yielding better extractions, other factors being constant. The GC-MSD was also interfaced with an autosampler and single magnet mixer (SMM), which was limited to 750 rpm by the SMM firmware available at the time. The extraction solvents were injected with inlet split ratios varying from 10:1 to 25:1. A 30 m x 0.25 mm x 1.0 µm HP-5 column was used. The column was ramped from 35 to 260°C at 10°C/min and held for 5 min.

7.6.2 Results and Discussion

Limits of detection ranged from 0.002 to 0.4 µg/L for an ECD detector to 2 to 9 µg/L when an FID detector was used. Linear ranges varied with the analytes but were between 0.01 and 250 µg/L with an ECD detector and 10 to 80 µg/L with an FID detector. Limits of detection were estimated to be below 0.5 µg/L, with the MSD in full scan mode, and 0.05 µg/L with an MSD in single-ion monitoring mode. RSDs ranged from 4 to 7%.

A few examples will illustrate these findings.[12] Figure 7.7 shows the results of headspace SDME extraction of 1 mL of sample containing 0.400 µg/L of each analyte with 350 mg of salt added at 30°C and 750 rpm for 10 min. Figure 7.8

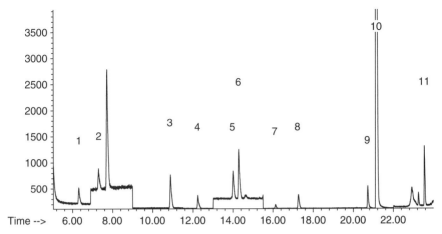

FIGURE 7.7 GC-MS SIM chromatogram of EPA method 524.2 volatiles (400 ng/L of each analyte) extracted using headspace SDME from 1 mL of water containing 350 mg/mL NaCl. Sample vial: 2 mL with a micro Teflon-coated stir bar; stirring rate: 600 rpm; extraction time: 10 min; extraction temperature: 30°C; extraction solvent: tetradecane (1.3 μL); volume injected: 1 μL; inlet split: 10:1. Peak identification: (1) chloroform, (2) benzene, (3) toluene, (4) tetrachloroethylene, (5) chlorobenzene, (6) ethylbenzene, (7) bromoform, (8) bromobenzene, (9) o-dichlorobenzene, (10) decafluorobiphenyl (I.S.), (11) naphthalene.

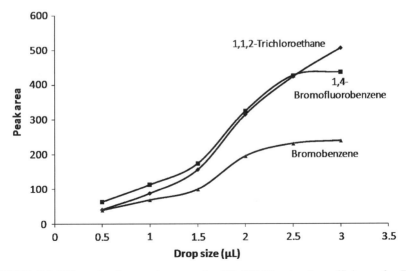

FIGURE 7.8 Effect of solvent volume on the HS-SDME extraction efficiency for EPA method 524.2 volatiles determined by GC-ECD.

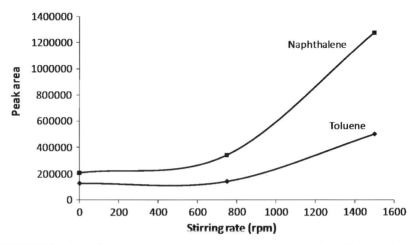

FIGURE 7.9 Effect of stirring rate (rpm) on the HS-SDME extraction efficiency for EPA method 524.2 volatiles determined by GC-MS SIM.

shows the extraction efficiency dependence on drop size, Figure 7.9 the dependence on stirring rate, Figure 7.10 the dependence on salt concentration, Figure 7.11 the dependence on extraction time, and Figure 7.12 the dependence on extraction temperature.

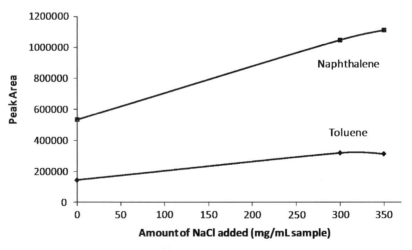

FIGURE 7.10 Effect of ionic strength on the HS-SDME extraction efficiency for EPA method 524.2 volatiles determined by GC-MS SIM.

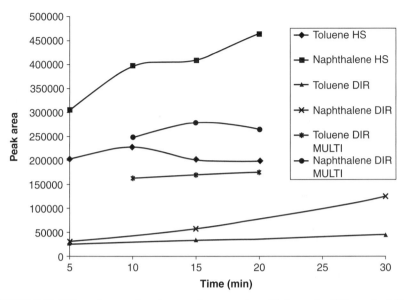

FIGURE 7.11 Effect of extraction time on the extraction efficiency for EPA method 524.2 volatiles for various extraction modes: direct, headspace, and direct dynamic in-needle (multiple). GC-MS SIM analysis.

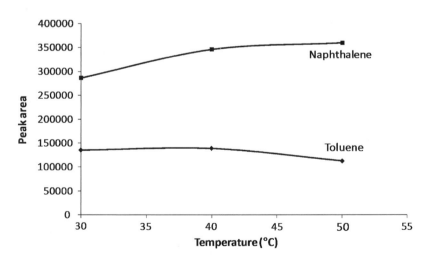

FIGURE 7.12 Effect of extraction temperature on the HS-SDME extraction efficiency for EPA method 524.2 volatiles determined by GC-MS SIM.

7.6.3 Experimental: SDME

One-microliter samples of EPA method 524 analytes containing 350 mg/mL salt were similarly extracted using SDME with 1.3 μL of tetradecane under conditions identical to the HS-SDME extractions, but yielded much poorer results.[12] Refer to Chapter 4 for the appropriate figures. Figure 4.4 illustrates the results for a 4-μg/L sample using HS-SDME and Figure 4.3 for a 4-μg/L sample using SDME. Extraction using SDME requires significantly longer extraction times and may require the use of dynamic extraction, which is discussed later.

7.6.4 Additional Experimental Recommendations

An alkane such as dodecane or tetradecane is recommended for extraction of most volatile organics normally extracted by PT. More polar analytes may be better extracted with 1-octanol. A 1-mL sample size should be sufficient in most cases, and enough salt should be added to give a nearly saturated solution (300 mg/mL). The saturated salt solutions (350 mg/mL) used above may result in small but varying amounts of undissolved salt remaining in the sample, depending on the extraction temperature. If halide exchange is a concern, Na_2SO_4 should be used instead (250 mg/mL). Extraction temperatures will be a compromise between optimal temperatures for low-boiling components and higher-boiling components. Temperatures in the range 20 to 35°C are appropriate. Extraction times ranging from 5 to 15 minutes will be adequate. Stirring should be at the highest rate possible short of affecting the drop. An internal standard such as decafluorobiphenyl must be added to the extracting solvent or a naturally occurring impurity in the solvent used to account for drop variations. The drop size should range from 1.2 to 3 μL, with 1 to 2.5 μL injected, and a split ratio of 10:1 to 50:1 used, depending on the chromatography system. For samples with very low analyte concentration, extracting solvents should be singly or doubly vacuum distilled, to remove interfering impurities.

SDME extractions are not usually necessary for most volatiles, except for the more polar analytes, for which 1-octanol may be a good choice as the extracting solvent. Longer extraction times are necessary for SDME compared to HS-SDME extractions, and dynamic extraction may be useful in decreasing the extraction times.

7.7 ANALYSIS OF RESIDUAL SOLVENTS IN DRUG PRODUCTS BY HS-SDME

At the present time, residual solvents in pharmaceuticals are analyzed using the USP 467 method, in which the product is dissolved in water, DMSO, or DMF and extracted using static headspace analysis.[13] The method essentially uses a single-point calibration and requires either a pass (lower concentrations than allowed) or fail (higher concentrations than allowed). Several very polar solvents cannot be analyzed by this method, due to their water solubilities and high boiling points. Allowed concentrations vary widely, from 5000 ppm to less than 1 ppm.

7.7.1 Experimental: Manual HS-SDME

Two studies of the use of HS-SME are available. In the first study, a drug is spiked with four solvents, and then a water solution is prepared.[14] Five-milliliter samples (a 15-mL vial) including tetrahydrofuran, methanol, dichloromethane, and ethanol, are extracted manually with 3 μL of 1-octanol for 20 min at 600 rpm. An extraction temperature was not indicated. Salt is added at 100 mg/mL concentration. The sample is analyzed by GC/FID.

7.7.2 Results and Conclusions

The limits of detection found range from 0.2 mg/L (for a 0.5-g drug sample) for tetrahydrofuran to 60 mg/L for methanol. These values are well within the required concentration levels used for method USP 467. The RSDs range from 7 to 12%, and the linear dynamic range varies from 1 to 100 mg/L for tetrahydrofuran to 10 to 1000 mg/L for methanol.

7.7.3 Experimental: Automated HS-SME

In a second approach, extraction is performed using an autosampler and a single magnet mixer (SMM), and either an MSD (scan mode) or an FID detector.[15] One-milliliter samples containing 300 mg of NaCl are prepared (in a 2-mL vial) at the USP 467 test concentrations for 12 analytes, including methanol, acetonitrile, methylene chloride, chloroform, cyclohexane, dioxane, trichloroethylene, pyridine, toluene, chlorobenzene, and *m*-xylene. Samples are prepared by dissolving analytes in DMSO and then dilution with water according to the USP method. The samples are extracted at 30 to 50°C for 5 to 7 min at 750 rpm with 1.3 to 3 μL of 1-octanol. Final conditions chosen include an extraction temperature of 30°C, an extraction time of 7 min, and 1.3 μL of 1-octanol (1 μL injected, 25:1 split ratio, 30 m × 0.25 mm × 1 μm film, HP-5 column). Samples are mixed by either stirring at 750 rpm with a micro stir bar or by vortexing prior to loading the samples in the tray.

Calibration curves with standards prepared at one-fifth to four times the USP test concentrations are used to determine the LOD and linearity for the method.

7.7.4 Results and Conclusions

Figures 7.13 and 7.14 show the chromatograms obtained for samples prepared at the USP method concentrations. Dioxane is below acceptable detection limits.

The effect of extraction temperature is shown clearly in Figure 7.15 for three analytes: methylene chloride, trichloroethylene, and pyridine. Increasing temperature decreases extraction efficiency for volatile, nonpolar analytes, but increases extraction efficiency dramatically for polar, less volatile analytes. Thus, a compromise extraction temperature is necessary. Figure 7.16 shows the effect of extraction time on extraction efficiency. It can be seen that at a particular extraction temperature, each analyte has an optimal extraction time. This is due to the fact that

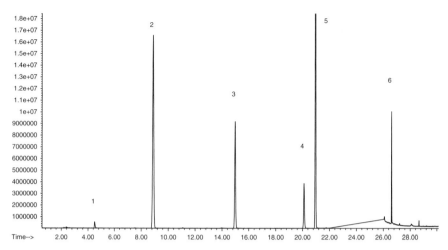

FIGURE 7.13 GC-MS SIM chromatogram of HS-SDME extract of USP 467 residual solvents in pharmaceuticals; high concentration of nonpolar analytes; extracting solvent: 1-octanol (1.3 μL); extraction time: 7 min; extraction temperature: 30°C; injection volume: 1 μL; inlet split: 20:1. Peak identification: (1) methylene chloride, (2) cyclohexane, (3) toluene, (4) chlorobenzene, (5) *m*-xylene, (6) tridecane solvent impurity.

analytes are in equilibrium between the water solution, the headspace, and the extracting solvent. As the temperature of the drop increases, analytes may have a smaller drop/headspace partition coefficient, and this temperature rise is time dependent.

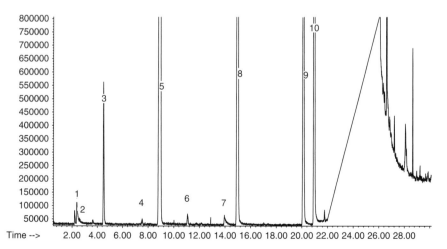

FIGURE 7.14 GC-MS chromatogram of HS-SDME extract of USP 467 residual solvents in pharmaceuticals; low concentration and polar analytes; conditions as in Figure 7.13. Peak identification: (1) methanol, (2) acetonitrile, (3) methylene chloride, (4) chloroform, (5) benzene, (6) trichloroethylene, (7) pyridine, (8) toluene, (9) chlorobenzene, (10) *m*-xylene.

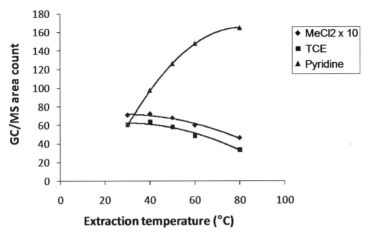

FIGURE 7.15 Effect of extraction temperature on the HS-SDME extraction efficiency for USP 467 residual solvents in pharmaceuticals. Stirred sample solution with salt added. Other conditions as in Figure 7.13.

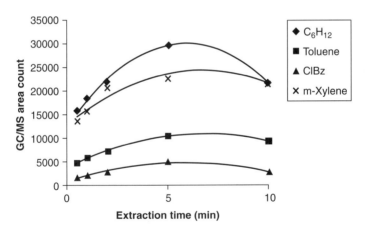

FIGURE 7.16 Effect of extraction time on the HS-SDME extraction efficiency for USP 467 residual solvents in pharmaceuticals. Stirred sample solution with no salt added. Other conditions same as in Figure 7.13.

Figure 7.17 illustrates the effect of using sample stirring or vortexing prior to sampling, as well as the dependence of the ionic strength on the extraction efficiencies of individual analytes. As can be seen, vortexing can be a very efficient method for bringing the sample to equilibrium, although not as efficient as continuous stirring. Addition of salt is most effective for relatively water-soluble compounds such as pyridine and polarizable compounds such as the halides.

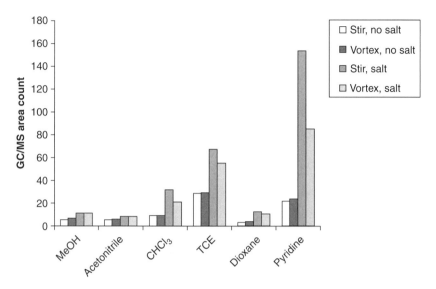

FIGURE 7.17 Effect of agitation method and ionic strength on the HS-SDME extraction efficiency for USP 467 residual solvents in pharmaceuticals. Extraction conditions as in Figure 7.13.

Finally, Figures 7.18 and 7.19 illustrate that some compounds, such as hydrocarbons, not only extract efficiently but may actually reach a saturation point in the microdrop, leading to a nonlinear calibration curve at higher concentrations. This can be overcome in some cases by increasing the drop size, but may be handled

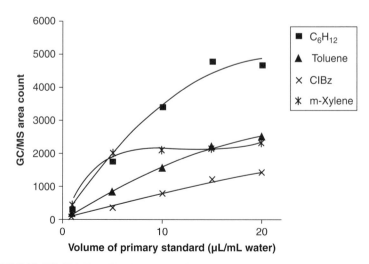

FIGURE 7.18 HS-SDME calibration curves for USP 467 analytes at method required concentrations. Conditions as in Figure 7.13.

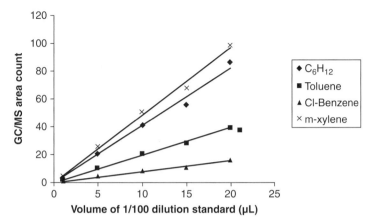

FIGURE 7.19 HS-SDME calibration curves for USP 467 analytes diluted to 1:100 method required concentrations. Conditions as in Figure 7.13.

more effectively by diluting the solution to bring the concentration of the analytes into a linear range.

7.7.5   Additional Experimental Recommendations

1-Octanol is a good choice as an extraction solvent unless it is known that the residual solvents are halogenated compounds or hydrocarbons. Dodecane or tetradecane may then be a preferable solvent if an ECD detector is used. A larger microdrop yields larger amounts of extracted analytes, but the ultimate drop size will depend on the chromatographic system and column used as well. For the solutions to be within the linear calibration range of the detector if USP standard solutions are used, it will be necessary to dilute the analytes present at high concentrations.

7.8   ARSON ACCELERANT ANALYSES BY HS-SDME

Residual arson accelerants, including gasoline, kerosene, and isopropyl alcohol, are commonly analyzed using static headspace analysis, PT, or SPME. HS-SDME is a useful alternative to these techniques. The following work is performed using a 2-mL vial, a computer-interfaced autosampler, and a single magnet mixer (SMM).[16]

7.8.1   Experimental

Samples are prepared by diluting hydrocarbon accelerants in methanol and then adding this solution (containing 0.005 μL of the accelerant) to an empty 2-mL vial, a vial containing a 1-cm piece of a matchstick or a vial with the stick and 200 μL of water or water plus 200 μL of 35% salt solution, and the vial is then crimped. Alcohol analytes are dissolved in pentane and added to the vial. Gasolines, charcoal lighter fluid, paint thinner, and Coleman camp stove fluid (white gas) samples are then extracted with 1.3 μL of tetradecane (1 μL drawn into the syringe and injected) after heating for 7 min at 50°C. A 20:1 inlet split is used.

Chromatographic peaks of the higher-boiling accelerants overlap with tetradecane, so a lower-boiling extraction solvent (heptane) is used. A total of 0.05 μL of kerosene or No.2 fuel oil (diesel fuel) is added to the vial. Since heptane evaporates rapidly at 50°C, a drop size of 2.5 μL is used, with only 1 μL remaining at the end of the extraction for analysis. Injection is performed in the splitless mode (0.5 min).

Methanol, ethanol, and 2-propanol (0.05 μL each) are extracted using 1.3 μL of 1-octanol and injected with a split mode (20:1 split).

7.8.2 Results and Discussion

Three different extraction solvents are required to accommodate the wide scope of boiling ranges and polarities of the accelerants. Lower-boiling hydrocarbons, up through gasoline and charcoal lighter fluid, are extracted well with tetradecane. Higher-boiling hydrocarbons (kerosene and No.2 fuel oil) require a lower-boiling solvent for extraction. The very polar alcohols require 1-octanol for efficient extraction. The alcohols and high-boiling accelerants are not extracted as efficiently by HS-SDME as are the lower-boiling hydrocarbons, as might be expected.

Matrix effects are very important factors in the extractions. Addition of the accelerant to a vial containing wood results in a significant decrease in the extraction efficiency. Figure 7.20 is a chromatogram for an extracted sample of 87-octane gasoline prepared by dissolving 5 μL in 1 mL of pentane and then injecting 1 μL of the solution into an empty 2-mL vial. Extracts of the same amount of gasoline

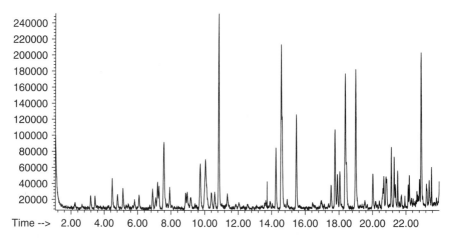

FIGURE 7.20 HS-SDME extraction of 0.005 μL of unleaded gasoline (5 μL of unleaded gasoline in 1 mL of methanol; 1 μL of the solution added to a 2-mL empty vial. Extraction temperature: 50°C; solvent: tetradecane (1.3 μL); extraction time: 7 min; volume injected: 1 μL; inlet split: 10:1.

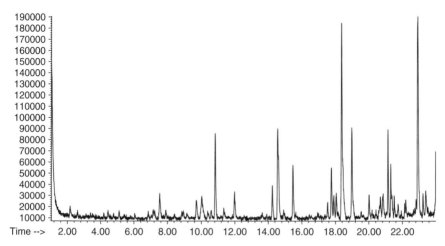

FIGURE 7.21 HS-SDME of 0.005 μL of gasoline deposited on wood. Vial volume: 2 mL; other conditions as in Figure 7.20.

deposited on a 1-cm wood match section are shown in Figure 7.21. Addition of water, a common condition for accelerant samples collected in the field, has an even more dramatic effect, as shown in Figure 7.22, which is at least partially reversed by the addition of saturated salt solution, as shown in Figure 7.23. The peaks at 18.2 and 22.6 min are decane and dodecane, present in the tetradecane as impurities.

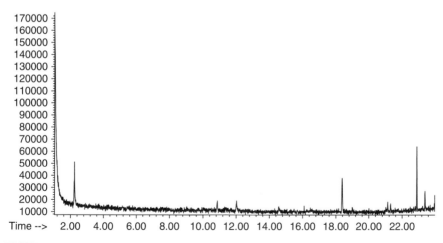

FIGURE 7.22 HS-SDME of 0.005 μL of gasoline deposited on water-soaked wood. Vial volume: 2 mL; other conditions as in Figure 7.20.

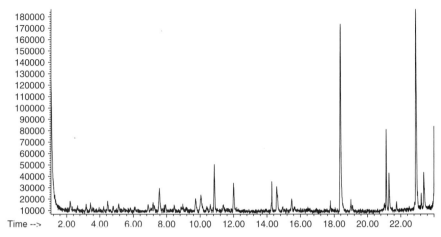

FIGURE 7.23 HS-SDME of 0.005 μL of gasoline deposited on water-soaked wood to which 100 μL of saturated NaCl solution was added. Other conditions as in Figure 7.20.

7.8.3 Additional Experimental Recommendations

Extraction temperatures higher than 50 to 60°C, often used for static headspace extractions, can be counterproductive for HS-SDME, as are long extraction times. Saturated salt solution must be added whenever water is present in the sample. It is possible to extract a sample sequentially, using tetradecane, then heptane, then 1-octanol, since the extractions are not exhaustive. A lower-boiling alcohol insoluble in water, such as hexanol, might also be useful for testing for both the high-boiling accelerants and alcohols during the same extraction.

7.9 ANALYSIS OF PAHs BY SDME

PAHs up to a molecular weight of 202 (pyrene) have been extracted from water efficiently using both SDME and HS-SDME. The resulting extracts have then been analyzed using GC or the extracting solvent evaporated, analytes redissolved in the mobile phase, and then analyzed by HPLC. Higher-molecular-weight PAHs (such as benzo[a]pyrene) are extracted with greater difficulty, especially at low concentrations, and require additional extraction time and higher sample extraction temperatures.

7.9.1 Experimental: SDME Extractions of PAHs from Aqueous Samples

In one approach, PAHs present in soil samples (1 g) are triturated with acetone (7 mL) in a 22-mL vial to release the PAHs, water added (14 mL) to decrease their solubility, and then extracted with 2 μL of octane for 11 min.[17] Soil samples are spiked with a standard mixture of 17 PAHs. The unique added device that makes extraction from the soil work is a filter placed in the vial to separate the soil particles from the microdrop. The filter is prepared from two pieces of window screening with sandwiched glass fiber in between. The screen is positioned 8 mm from the top of the vial and the syringe tip positioned just above it during the extraction. The extracted sample is then analyzed by GC with a pulsed-discharge helium ionization detector.

In a second approach, 3-mL water solutions of six PAHs, ranging from fluoranthene to benzo[*a*]pyrene, are extracted with 3 μL of toluene for 20 min at 40°C while stirring at 700 rpm using SDME. Following extraction, the solvent is evaporated and changed to acetonitrile, to allow the sample to be analyzed by HPLC.

7.9.2 Results and Conclusions

The particulate matter in the soil slurry will contaminate and ultimately dislodge an unprotected solvent drop. The specially designed device used for the soil sample above works well, but there are two simpler solutions to the problem. First, the sample can be extracted using HS-SDME. This avoids all direct contact with the soil particles and, as will be see in Section 7.9.3, allows analysis of PAHs with molecular weights up to that of pyrene. A second solution would be to protect the extraction solvent by encasing it in a microporous polymer, the HF(2)ME extraction mode. See section 7.11 for an example of extraction of analytes from soil using this mode. The limits of detection for the soil extracts in the procedure above range from 0.1 to 0.4 mg/kg soil (0.1 to 0.4 μg/sample), and RSDs range from 1 to 35%. The large RSDs are probably due to irreproducible drop sizes and not allowing extraction to proceed until it approaches equilibrium.

The calibration curves are linear over a range of 2 to 100 μg/L for fluoranthene and 8 to 100 μg/L for benzo[*a*]pyrene, with correlation coefficients between 0.9878 and 0.9921. The RSD values range from 5 to 9% and limits of detection from 1 to 3.5 μg/L.

In both cases, there is a dramatic decrease in the amount of high-molecular-weight PAHs (such as benzo[*a*]pyrene) extracted compared to lower-molecular-weight compounds (such as fluoranthene). These results are confirmed by findings in our laboratories. To maximize the extractions of high-molecular-weight PAHs, it is necessary to use elevated extraction temperatures (40 to 60°C) and longer extraction times (10 to 20 min) than those used for smaller molecules. This is possibly due to the lower diffusion coefficients in water for larger molecules. Figure 7.24 is an example of extraction of 1 mL of an aqueous solution of 16 PAHs in a 2-mL crimp cap vial with 2 μL of *o*-xylene at 60°C for 15 min. After extraction, 1.5 μL of the extraction solvent is withdrawn into the syringe and analyzed by GC/MS, using a splitless injection.

7.9.3 Experimental: HS-SDME Extractions of PAHs from Aqueous Solutions

In one procedure, 6 mL of water solutions of PAHs, ranging from naphthalene to pyrene, are extracted in a 10-mL vial at 40°C for 12 min while stirring at 400 rpm

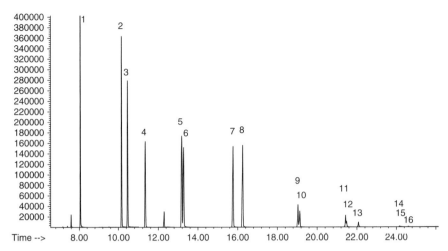

FIGURE 7.24 GC-MS SIM chromatogram of SDME extract of 1 mL of water containing 20 μg/L of 16 PAHs. Extraction solvent: *o*-xylene (2 μL); extraction time: 15 min; extraction temperature: 60°C; volume injected: 1.5 μL (splitless). Peak identification: (1) naphthalene, (2) acenaphthylene, (3) acenaphthene, (4) fluorene, (5) phenanthrene, (6) anthracene, (7) fluoranthene, (8) pyrene, (9) benzo[*a*]anthracene, (10) chrysene, (11) benzo[*b*]fluoranthene, (12) benzo[*k*]fluoranthene, (13) benzo[*a*]pyrene, (14) indino[1,2,3-*cd*]pyrene, (15) dibenzo [*ah*]anthracene, (16) benzo[*ghi*]perylene.

using HS-SDME.[18] Sodium chloride (0.135 g/mL) is used to increase extraction efficiency. A 3-μL drop of 1-butanol is used to extract the PAHs, and an in-house cooling jacket is used to cool the needle to −6°C. The extracts are then analyzed by GC-FID.

In a second procedure, 10-mL aqueous samples containing five PAHs, ranging from naphthalene to pyrene, are extracted with a 10-μL drop of saturated aqueous β-cyclodextrin solution for 10 min at 40°C.[19] A 10-μL drop is obtained by using a micropipette tip, inserted through the vial cap septum to hold the extraction solvent, which is in turn inserted through the pipette tip using an HPLC syringe. After extraction, the β-cyclodextrin is withdrawn into the syringe, the syringe removed from the pipette tip, and the extract analyzed by HPLC.

7.9.4 Results and Conclusions

The limits of detection for the HS-SDME extraction using 1-butanol range from 4 to 41 μg/L, and the RSD values range from 6 to 20% for spiked tap water. The limits of detection using β-cyclodextrin range from 1.5 to 28 μg/L. The RSD values are 6 to 7%, and the linear dynamic range extends from 0.01 to 50 μg/L.

HS-SDME results for solutions of 16 PAHs in our laboratories were close to the foregoing results. Maximum extraction efficiencies for 1-mL samples (0.2 to 40 μg/L) in water are obtained by extraction with *o*-xylene (2 μL) for 15 min at 80°C. PAHs with molecular weights higher than pyrene (MW = 202) are poorly extracted. An example of HS-SDME extraction is shown in Figure 7.25.

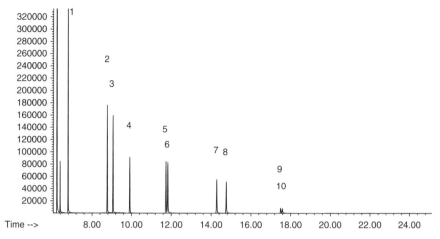

FIGURE 7.25 GC-MS SIM chromatogram of HS-SDME extract of 1 mL of water containing 8 μg/L of 16 PAHs. Extraction solvent: *o*-xylene (2 μL); extraction time: 5 min; extraction temperature: 80°C; volume injected 1 μL. Peak identities same as in Figure 7.24.

7.9.5 Additional Experimental Recommendations

Octane, decane, toluene, or xylene would be extraction solvents of choice, due to their limited solubilities in water compared to 1-butanol. In these experiments, samples are injected with a split inlet. Since the extraction solvents are lower boiling than the analytes, a standard splitless injection is possible, with resulting enhanced sensitivity. Unlike lower-boiling analytes, high-boiling analytes may have to be extracted at temperatures up to 80°C. As a result of these higher extraction temperatures, a drop size 1 μL larger than the amount of extraction solvent must be used for the analysis. Extracts can be analyzed directly by GC or could also be injected into a vial, concentrated, and acetonitrile or methanol added to enable HPLC analysis. Using β-cyclodextrin aqueous solution as the extraction solvent has the added advantage that the solution is directly compatible with reversed-phase HPLC solvents.

7.10 DETERMINATION OF ACETONE IN AQUEOUS SOLUTIONS BY DERIVATIZATION HS-SDME

Carbonyl compounds are often determined by derivatizing with a reagent such as 2,4-dinitrophenylhydrazine (DNPH) and then analyzed by HPLC with UV detection. Acetone is of particular interest in biological media (blood and urine), since it is a metabolic product indicative of diabetes. A fast, inexpensive method for small sample volumes that contain large percentages of salts and protein is required. To this end, a method has been developed to derivatize acetone in 1 mL of biological samples by adding pentafluorobenzyl hydroxylamine to the aqueous sample.[20] The resulting oxime is volatile and can be extracted by HS-SDME. This analysis can also be applied to other carbonyl compounds present in aqueous solutions. Alternatively, if the carbonyl compound is volatile enough to partition into the headspace, the derivatizing agent can also be incorporated into the extraction solvent. The following procedure demonstrates the headspace extraction of derivatized acetone from water.

7.10.1 Experimental

To a 2-mL crimp vial containing a PTFE-coated micro stir bar are added 1 mL of aqueous sample and 100 μL of a 50 mM solution (12.5 mg/mL of *O*-(2,3,4,5,6-pentafluorobenzyl)hydroxylamine hydrochloride (PFBHA) in water. The cap is crimped and the vial stirred at a constant rate of 600 to 1000 rpm at 40°C for 10 min. A 10-μL No. 2 point syringe is filled with 2 μL of *o*-xylene and the syringe inserted through the septum 1 cm into the vial. The *o*-xylene is slowly expelled to the tip of the needle and the headspace extracted for 5 min. The *o*-xylene extract (1.5 μL) is withdrawn and analyzed by GC-MS or GC-ECD using splitless injection.

7.10.2 Results and Conclusions

The extraction was developed originally to analyze acetone in whole blood using 2 μL of toluene as the extracting solvent. *o*-Xylene is suggested as a replacement for toluene since it is less soluble in water and has a higher boiling point. Even so, it is recommended that only 1.5 μL of the solvent be withdrawn into the syringe for injection. This takes into account loss of the solvent by evaporation and solubility in water. The original procedure also used in-syringe dynamic sampling to increase extraction efficiency. At the levels of acetone typically found in the blood of diabetics (approximately 10 to 20 μg/mL), a static extraction is sufficiently sensitive. The published procedure resulted in a linear range of 60 ng to 300 μg/mL (0.001 to 5.0 mM) acetone ($r^2 > 0.97$) using selective ion monitoring (*m/z* 181). Electron capture detection should give similar results.

7.10.3 Additional Experimental Recommendations

Whole blood may have to be diluted with distilled water if it is too viscous to stir effectively. The original publication's authors did not stir the sample, which avoids contamination of the stir bar. High concentrations of acetone in water solutions may have to be diluted to give a linear calibration curve. Addition of salt to clean water samples may be beneficial to promote partitioning of the derivative into the headspace for very dilute samples. If an internal standard is desired for the sample, 10 μL of a 5-mg/mL (50 mM) solution of benzaldehyde in methanol was found to be effective.

7.11 DETERMINATION OF PESTICIDES IN SOIL BY HF(2)ME

Hollow fiber–protected microextraction has been used most frequently and successfully in a three-phase mode [HF(3)ME] for extraction of acidic or basic analytes from aqueous sample, especially for extraction from biological media, such as blood, serum, and urine. The two-phase mode [HF(2)ME] has been used for the extraction of analytes such as PAHs and pesticides from aqueous samples. However, this technique has several disadvantages compared to SDME, including the inability to automate the extraction easily and the need to prepare each extraction fiber manually. HF(2)ME has one major advantage over SDME, however, in that the fiber protects the solvent from particles or dissolved macromolecules present in the sample, which would contaminate or cause dislodgement of a solvent drop suspended from a syringe needle. Thus, extraction from a soil sample slurry can be accomplished with HF(2)ME.[21]

7.11.1 Experimental

Soil sample standards are prepared by air drying, pulverizing, and sifting soil samples. Methanol solutions (1.00 mg/mL) of pesticides are added to 30 g of soil to obtain concentrations ranging from 0.05 to 5.0 μg pesticide/g of soil. The samples are mixed, air dried, and stored refrigerated. Soil (150 mg) is added to a 4-mL vial containing a stir bar, followed by 1 mL of acetone and 2 mL of water. A cap containing a 2-mm hole is attached and the sample stirred at 1000 rpm and 30°C for 10 min.

The fiber (Accurel Q 3/2 polypropylene, 600 μm i.d., 200 μm wall thickness) is cut to a length of 1.5 cm and washed with acetone to remove polymer processing impurities. Sonication is preferred, but extracting twice with 5 mL of acetone in a small beaker works well. The fiber is air dried and 1 mm of one end is tightly crimped with a small pair of needlenose pliers. The crimped flap is folded over onto the fiber and the fold is recrimped. The resulting length of uncrimped fiber is 1.3 cm. A No. 2 point 10-μL syringe is filled with 5 μL of *o*-xylene (remove all air bubbles) and the needle is pierced through the crimped fold of the fiber. This process ensures a tight fit of the syringe needle on the fiber and prevents the crimp fold from opening during extraction.

The fiber, attached to the syringe, is dipped into 100 μL of *o*-xylene for 2 s to saturate the polymer pores. The stirrer is turned off, the fiber then inserted through the vial cap hole into the sample, and the syringe clamped on a ring stand, aligning it so that the top of the fiber is immersed 0.5 cm in the water. The *o*-xylene is slowly injected into the fiber, the stirrer turned on, and extraction begun. Following extraction (10 min), 2 μL of the *o*-xylene is withdrawn into the syringe, the syringe withdrawn from the vial, the fiber removed and discarded; the syringe plunger pulled back an additional 0.5 μL, the needle wiped with a tissue moistened with *o*-xylene, and the extract injected for analysis by GC-MS or GC-ECD, depending on the analytes present.

7.11.2 Results and Discussion

Published HF(2)ME extractions from soil have obtained recoveries ranging from 90 to 100%, with linear ranges of 0.1 to 2 μg pesticide/g soil, good linearity ($r^2 = 0.98$ to 0.99), and RSDs, ranging from 6 to 9%. Acetone must be added to the soil and water for effective extraction. Up to 1 g of sample can be extracted, with water–acetone volumes ranging from 4 to 20 mL. As with SDME, high-boiling analytes should be extracted with low-boiling solvents, if possible, and GC analysis performed using splitless injection. Low-boiling analytes should be extracted with higher-boiling solvents such as 1-octanol or tetradecane. In these cases, soil samples cannot be dried, but must be analyzed wet, with added acetone and water.

7.11.3 Additional Experimental Recommendations

Many of the published procedures using HF(2)ME ignore or omit necessary experimental details required for a successful analysis. First, it should be remembered that the polymer is about 70% porous, which means that the lumen of a 1.3-cm fiber has a volume of 3.6 μL and the fiber wall holds an additional 4.6 μL of the solvent. Thus, if 2 μL of the solvent is analyzed, only 25% of the extracted analytes is analyzed, presuming that none of the solvent evaporates or dissolves in the water. If a solvent as volatile as toluene, o-xylene, or hexane is used, it will very quickly evaporate from the fiber after dipping it into the solvent to saturate the pores, so the transfer into the sample must be accomplished within 10 s. Adding an additional 5 μL from the syringe will nearly fill the fiber. Experimentation will be necessary to determine how much of a particular solvent must be added with the syringe to fill the fiber without overfilling it. If a low-boiling solvent is used, the fiber must be immersed in the sample completely to prevent evaporation. The procedure above uses o-xylene rather than the more commonly used toluene, since it is higher boiling and less soluble in water. If a higher-boiling solvent such as 1-octanol or tetradecane is used, the solvent present in the pores and interior of the fiber after dipping in the solvent will not evaporate to any great extent, and less must be added with the syringe.

A second crucial experimental point is that a standard 26-gauge syringe needle is 0.43 mm in diameter, but the fiber has a 0.6-mm internal diameter, so the fiber will not stay on the syringe needle during extraction unless the crimping method discussed above is used. As an alternative, a 22 (0.72 mm)-gauge needle can be used, but this may cause GC inlet septum degradation. If a much longer fiber is used (as is often the case in HF(3)ME), an autosampler double-gauge needle (22 to 26 gauge) can be used. Another alternative is to place the fiber into a 1-mm hole in the vial cap. The syringe needle then fits loosely into the fiber. There is less evaporation of volatile solvents with this technique, due to the tight fit of the fiber into the septum hole. A third technique is to attach the fiber to a micropipette tip (0.1 to 10 μL), which has been, in turn, inserted through the vial septum. The top of the pipette tip must be cut off so that the needle can be inserted into the fiber. This technique has been adapted to allow the use of an autosampler for the HF(2)ME mode.

The end of the fiber immersed in the sample has been left open in this example. In some cases [HF(3)ME] the end must be sealed. This can be accomplished by crimping and then heat sealing with a soldering iron, tying with thin wire, or plugging with an appropriate-diameter wire or polymer fiber.

Finally, addition of NaCl or Na_2SO_4, as with SDME, may be helpful to decrease the solubility of extracting solvent and analyte(s) in the aqueous solution. This is expecially useful when employing solvents such as toluene and 1-octanol and lower-molecular-weight nonpolar analytes. As with SDME, very water-soluble solvents such as chloroform cannot be used by themselves for extraction.

7.12 DETERMINATION OF PAHs AND HOCs BY DLLME

Polycyclic aromatic hydrocarbons (PAHs) of low to medium molecular weight and halogenated organic compounds (HOCs) are efficiently extracted from water by SDME or HS-SDME. High-molecular-weight compounds are less efficiently extracted within a reasonable experimental period (15 min or less), however. This is due to a combination of factors, including slower diffusion rates through the sample and into the solvent for large molecules. One simple way to overcome this difficulty is with dispersive liquid–liquid microextraction. Two procedures are used for DLLME. Both methods involve dissolving the extracting solvent in a small amount of solvent miscible with water, such as acetone, and injecting the solution into the aqueous sample in a centrifuge tube or conical-bottomed vial. The extracting solvent is dispersed throughout the sample as microdroplets, with a resulting very large surface area, which effectively extracts high-boiling nonpolar analytes. The sample is centrifuged to separate the extracting solvent from the sample and analyzed.

Two techniques have been used to separate the extract from the sample. The first involves using a solvent with a density greater than that of water (CS_2, CCl_4, C_2Cl_4, chlorobenzene), which is removed from the bottom of the centrifuged sample with a syringe for injection and analysis. The second technique uses a near room temperature-melting extraction liquid (1-undecanol, 1-dodecanol, tetradecane, or hexadecane), which is solidified by cooling in an ice bath, removed with tweezers, added to a vial, allowed to melt, and analyzed.

7.12.1 Experimental: Extraction of PAHs from Water

A solution of 8 μL of extracting solvent (tetrachloroethene, C_2Cl_4) in 0.5 mL of acetone is added rapidly with a syringe to 5 mL of aqueous sample in a centrifuge tube. The dispersion is immediately centrifuged (6000 rpm) for 5 min. A 10-μL syringe is used to sample 2 μL of the extract from the bottom of the tube and analyzed by GC-FID or GC-MS using splitless injection. Figure 6.12 shows the chromatogram for the extraction of 16 PAHs from water using DLLME.[22] Comparing this chromatogram with Figure 7.24, it can be seen that the higher molecular weight PAHs are more efficiently extracted by DLLME than by SDME or HS-SDME.

7.12.2 Experimental: Extraction of HOCs from Water

A solution of 10 μL of tetradecane in 0.5 μL of acetone is added rapidly with a syringe to 5 mL of aqueous sample in a centrifuge tube. The dispersion is centrifuged immediately for 5 min (6000 rpm) and the tube immersed in an ice–salt bath to solidify the tetradecane. A pair of tweezers is chilled in an ice-water bath, wiped with a tissue, and the floating solidified tetradecane quickly removed and added to a sample vial. The solvent is allowed to melt and 1–3 μL analyzed by

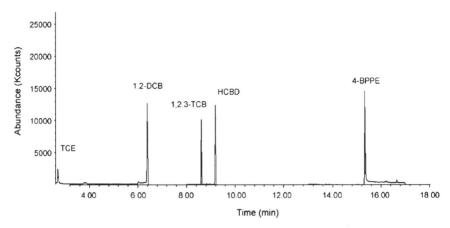

FIGURE 7.26 GC-MS chromatogram of 5 mL of lake water spiked with 0.2 to 1 μg/L of halogenated organic compounds after DLLME extraction with 10 μL of 2-dodecanol; volume injected: 3 μL. Peak identification: TCE, trichloroethylene; 1,2-DCB, 1,2-dichlorobenzene; 1,2,3-TCB, 1,2,3-trichlorobenzene; HCBD, hexachlorobutadiene; 4-BPPE, 4-bromodiphenyl ether. (Reprinted with permission from Ref. 23; copyright © 2008 Elsevier.)

GC-MS or GC-ECD using a split injection. Figure 7.26 shows a chromatogram for the DLLME extraction of halogenated organic compounds.[23]

7.12.3 Results and Conclusions

Either technique can be used to extract the two types of analytes. This technique is most useful for extraction of high-molecular-weight nonpolar chemicals. The major disadvantage of this extraction mode is that it cannot easily be fully automated, although an autosampler could be used to inject the separated solvent if it is transferred to a reduced-volume vial insert. Unlike SDME extraction, all PAHs are effectively extracted with this mode, including PAHs larger than pyrene.

7.12.4 Additional Experiment Recommendations

Care must be taken not to remove any water with the extract, especially if salt is added to aid in solvent separation and analyte extraction. Small amounts of water will tend to cling to the walls of the sample vial, however, and not pose a problem. This method cannot be used for samples that contain significant amounts of solids, such as soils. Little work has been done using this mode of extraction for low-boiling analytes. This may be a fruitful area of research.

7.13 DYNAMIC HEADSPACE AND DIRECT IMMERSION EXTRACTIONS (DY-SME)

Dynamic SME was developed by Lee and co-workers, who performed the process manually. Additional work was done using in-house constructed autosamplers and commercial syringe pumps and, more recently, using fully computer-controlled autosamplers. Dynamic SME can be performed using either headspace or direct extractions, but is most useful for SDME, due to the prolonged times needed for analyte diffusion through water to the drop, compared to diffusion through the headspace to the drop. Because headspace extraction occurs faster than direct-immersion extraction, dynamic extraction for HS-SDME is usually unnecessary.

7.13.1   Experimental

There are essentially two modes for DY-SME: in-needle (exposed) and in-syringe (unexposed). The in-needle technique is identical with normal HS-SDME or SDME except that the drop is moved continually into and out of the needle tip during the extraction. The in-syringe technique pulls water or air (5 to 7 μL), along with 1 to 3 μL of extraction solvent, into and out of the syringe barrel during the extraction. The in-syringe technique can be more effective in extracting analytes, but there is a serious disadvantage when used for direct-immersion SDME: Water that contains any suspended or soluble macrosolids or salts may contaminate the chromatographic system if it enters the GC inlet. Therefore, this technique should not be used for SDME unless using clean, filtered water samples with no added salt. The method may be effective in some cases for HS-SDME.

When the in-needle mode is used, the drop size should be 0.3 to 1.0 μL larger than the volume pulled back into the syringe so that contaminants in the water (dissolved and undissolved solids, salt) are prevented from entering the syringe needle and contaminating the chromatograph.

In the original studies, the in-needle dynamic SDME and HS-SDME experiments were conducted by manipulating the syringe plunger manually. The extraction was performed by taking up 1 μL of isooctane (with 1,4-dibromobenzene as an internal standard) into a 10-μL syringe, placing the needle into a water solution containing 10 chlorobenzenes, pulling up 3 μL of water quickly, holding for a dwell time of 0.5 to 5 s, and repeating the process 15 times.[24] The water was expelled and the extract analyzed by GC-MS. Limits of detection varied from 0.02 to 0.05 μg/L, percent recoveries from 86 to 105%, and RSDs from 2 to 7%.

HS-SDME was manually performed in a similar fashion, using a 2-μL volume of toluene and a 5-μL volume of sample headspace pulled up into and out of the syringe up to 25 times to extract chlorobenzenes from soil (1 g of soil in 1.5 mL of water).[25] Detection limits ranged from 6 to 14 ng/g soil (4 to 9 μg/L water) and RSDs from 6 to 18%.

In later studies involving extraction of alcohols with 1-octanol and BTEX in water samples with dodecane, syringe plunger manipulation was controlled by an in-house constructed device capable of controlling the plunger movement rate and dwell time (the time for the plunger to remain at the maximum withdrawal).[26,27] Samples were analyzed by GC-MS (alcohols) and GC-FID (BTEX). Detection limits ranged from 1 μg/L (2-pentanol) to 97 μg/L (methanol) and RSDs from 6 to 16% for the alcohols and 0.2 to 0.4 μg/L and RSDs from 7 to 9% for the BTEX.

In an in-syringe dynamic SDME study extraction of PAHs (ranging from pyrene to benzo[b]fluoranthene) with toluene, the syringe plunger was manipulated with a syringe pump.[28] The technique was compared with static direct-immersion SDME. In these experiments, the toluene extract was evaporated to dryness, dissolved in acetonitrile, and the solution analyzed by HPLC-UV. The detection limits for the static method ranged from 1 to 4 µg/L, and the dynamic method detection limits ranged from 0.4 to 0.6 µg/L.

In a second study involving in-syringe dynamic SDME, phthalates were extracted from water using 2 µL of hexane and a 8-µL syringe headspace for 30 extraction repetitions.[29] Detection limit values ranged from 0.4 to 4 µg/L, and relative standard deviations were in the 5 to 6% range.

In a comprehensive study, a computer-controlled autosampler was used for a series of experiments involving most common SME modes.[30] A comparison was made between in-needle dynamic HS-SDME, in-syringe dynamic HS-SDME, in-syringe dynamic SDME, static HS-SDME, and SDME. In-needle dynamic SDME was not studied. This study also examined the potential of hollow fiber–protected two-phase microextraction [HF(2)ME].

The headspace experiments were conducted with BTEX in water and the direct-immersion SDME and HF(2)ME experiments with PAHs (acenaphthene, fluorene, fluoranthene, and pyrene), and the samples were analyzed with a Saturn GC-MS.

7.13.2 Results and Discussion

Beyond the standard variables: stirring rate, extraction time, extraction solvent, drop size, and ionic strength, three additional variables are important for dynamic extractions: number of cycles, dwell time (the time the syringe plunger is held at the maximum volume), and plunger speed. For in-syringe dynamic SDME, a plunger speed of 1 to 1.5 µL/s is optimal, but for in-syringe dynamic HS-SDME, a plunger speed of 10 µL/s was found to be optimal. A dwell time of 5 s is optimal for both techniques. The optimal number of cycles for in-syringe dynamic HS-SDME and dynamic SDME is between 30 and 60 times, depending on the solvent.

In-needle dynamic HS-SDME requires only 10 cycles for optimal extraction and is twice as effective as static exposed-drop HS-SDME. Dynamic HS-SDME and static HS-SDME are both more effective than in-syringe dynamic HS-SDME.

The last study also compared hollow-fiber SME [HF(2)ME] to SDME. HF(2)ME (often referred to in the literature as LPME) involves placing 10 to 20 µL of solvent in an end-capped porous polyethylene tube 1 to 1.5 cm long and extracting analytes from water through the polymer into the solvent. Due to the larger volume of the extracting solvent, this technique is often an exhaustive extraction technique. However, only 1 to 2 µL of the extract is analyzed by GC, although the solvent can be removed, concentrated, and nonvolatile extracts analyzed by HPLC. The HF(2)ME method in the study was automated. Although this technique has great utility for extraction of analytes from aqueous samples with a complex matrix such as blood, serum, and soil samples, since these contaminants cannot pass through the polymer barrier, the study showed it to be less effective than SDME. A major disadvantage of HF(2)ME at the present time is that the special fiber construction needed to automate the method must be handmade, leading to considerable manual effort. This may be the best alternative, however, for extraction of biological samples.

Although none of these studies examined in-needle dynamic SDME, this technique has been shown to be an effective alternative to in-syringe dynamic SDME and HF(2)ME.[12] As an example, Figure 7.7 shows a 1-mL sample containing 400 pg/mL of volatile analytes extracted by static HS-SDME for 10 min at 30°C with 350 mg/mL NaCl added. When the experiment was repeated using direct-immersion SDME, very little was extracted, even after 30 min. However, when in-needle dynamic SDME was used with 30 cycles (during a 10-min period), an extraction efficiency similar to that obtained by static HS-SDME (Figure 4.5) was obtained, as shown in Figure 4.6.

7.13.3 Additional Experimental Recommendations

The studies described above indicate that the first choice for extraction of volatile chemicals is static HS-SDME. Water samples should be saturated with salt and temperatures kept close to 30°C for lower-boiling analytes and as high as 50 to 60°C for high-boiling analytes. Nonpolar analytes are best extracted with alkane or aromatic solvents such as toluene, xylene, dodecane, or tetradecane. Polar analytes are best extracted with a solvent such as 1-octanol. Extraction times should range from 5 to 10 min. In-needle dynamic HS-SDME may be necessary for lower concentrations of analytes if an autosampler is available.

Although in-syringe dynamic SDME is most effective for direct extractions, contamination of the syringe and analytical equipment may be a problem unless extractions are from clean filtered, water samples. If this method is used, a dwell time of 5 s and a plunger motion rate of 1 to 1.5 μL/s are optimal. It may be possible to use a volatile solvent such as chloroform with this technique if extraction temperatures are kept low.

In-needle dynamic SDME is faster and more effective than static SDME and should be tried if an autosampler is available. A drop size 0.3 to 0.5 μL larger than the injection volume is needed to prevent water uptake into the syringe.

REFERENCES

1. *Standard Test Method for Gasoline Diluent in Used Gasoline Engine Oils by Gas Chromatography*, ASTM Test Method D 3525–93, American Society for Testing and Materials, West Conshohocken, PA.
2. Kokosa, J. M.; Przyjazny, A., Headspace microdrop analysis: an alternative test method for gasoline diluent and benzene, toluene, ethylbenzene and xylenes in used engine oils. *J. Chromatogr. A* 2003, *983* (1–2), 205–214.
3. Przyjazny, A.; Kokosa, J. M., Analytical characteristics of the determination of benzene, toluene, ethylbenzene and xylenes in water by headspace solvent microextraction. *J. Chromatogr. A* 2002, *977* (2), 143–153.
4. Munch, D. J.; Hautman, D. P., CFR Method 551.1, Methods for the Determination of Organic Compounds in Drinking Water, Supplement 3, 1995.
5. Zhao, R. S.; Lao, W. J.; Xu, X. B., Headspace liquid-phase microextraction of trihalomethanes in drinking water and their gas chromatographic determination. *Talanta* 2004, *62* (4), 751–756.

6. Kokosa, J. M.; Przyjazny, A.; Christ, I., Automated headspace analysis of drinking water disinfection by-products. Paper 1970–1, presented at PittCon 2003, March 13, 2003, Orlando, FL.

7. Buszewski, B.; Ligor, T., Single-drop extraction versus solid-phase microextraction for the analysis of VOCs in water. *LC·GC Eur.* 2002, *15* (2), 92–97.

8. Tor, A.; Aydin, M. E., Application of liquid-phase microextraction to the analysis of trihalomethanes in water. *Anal. Chim. Acta* 2006, *575* (1), 138–143.

9. Yamini, Y.; Hosseini, M. H.; Hojaty, M.; Arab, J., Headspace solvent microextraction of trihalomethane compounds into a single drop. *J. Chromatogr. Sci.* 2004, *42* (1), 32–36.

10. Li, X.; Xu, X.; Wang, X.; Ma, L., Headspace single-drop microextraction with gas chromatography for determination of volatile halocarbons in water samples. *Int. J. Environ. Anal. Chem.* 2004, *84* (9), 633–645.

11. Zhang, T.; Chen, X.; Li, Y.; Liang, P., Application of headspace liquid-phase microextraction to the analysis of volatile halocarbons in water. *Chromatographia* 2006, *63* (11–12), 633–637.

12. Kokosa, J. M.; Przyjazny, A., A comparison of direct and headspace solvent microextraction for the determination of selected volatile organic compounds. Paper 1680–4, presented at PittCon 2007, February 28, 2007, Chicago.

13. Chapter 467, Residual solvents, *Pharmacop. Forum*, 2007, *33* (3), 1.

14. Wang, X.; Jiang, T.; Yuan, J.; Cheng, C.; Liu, J.; Shi, J.; Zhao, R., Determination of volatile residual solvents in pharmaceutical products by headspace liquid-phase microextraction gas chromatography–flame ionization detector. *Anal. Bioanal. Chem.* 2006, *385* (6), 1082–1086.

15. Kokosa, J. M.; Christ, I., Residual solvents in drug products using liquid phase microextraction/gas chromatography. Paper 2900–8, presented at PittCon 2008, March 6, 2008, New Orleans, LA.

16. Kokosa, J. M; Lemke, F.; Christ, I.; Watts, V.; Sindy, E., Arson accelerant analysis using liquid phase microextraction gas/chromatography. Paper 1170–4, presented at PittCon 2008, March 4, 2008, New Orleans, LA.

17. Zhang, H.; Andrews, A. R. J., Preliminary studies of a fast screening method for polycyclic aromatic hydrocarbons in soil by using solvent microextraction–gas chromatography *J. Environ. Monit.* 2000, *2* (6), 656–661.

18. Shariati-Feizabadi, S.; Yamini, Y.; Bahramifar, N., Headspace solvent microextraction and gas chromatographic determination of some polycyclic aromatic hydrocarbons in water samples. *Anal. Chim. Acta* 2003, *489* (1), 21–31.

19. Wu, Y,; Xia, L.; Chen, R.; Hu, B., Headspace single drop microextraction combined with HPLC for the determination of trace polycyclic aromatic hydrocarbons in environmental samples. *Talanta* 2008, *74* (4), 470–477.

20. Deng, C.; Li, N.; Wang, X; Zhang, X.; Zang, J., Rapid determination of acetone in human blood by derivatization with pentafluorobenzyl hydroxylamine followed by headspace liquid-phase microextraction and gas chromatography/mass spectroscopy. *Rapid Commun. Mass Spectrom.* 2005, *19* (5), 647–653.

21. Hou, L; Lee, H. K., Determination of pesticides in soil by liquid-phase microextraction and gas chromatography–mass spectrometry. *J. Chromatogr. A* 2004, *1038* (1–2), 37–42.

22. Rezaee, M.; Assadi, Y.; Hosseini, M. M.; Aghaee, E.; Ahmadi, F.; Berijani, S., Determination of organic compounds in water using dispersive liquid–liquid microextraction. *J. Chromatogr. A* 2006, *1116* (1–2), 1–9.

23. Leong, M.; Huang, S, Dispersive liquid–liquid microextraction method based on solidification of floating organic drop combined with gas chromatography with electron-capture or mass spectrometry detection. *J. Chromatogr.* A 2008, *1211* (1–2), 8–12.

24. Wang, Y.; Kwok, Y. C.; He, Y.; Lee, H. K., Application of dynamic liquid-phase microextraction to the analysis of chlorobenzenes in water by using a conventional microsyringe. *Anal. Chem.* 1998, *70* (21), 4610–4614.

25. Shen, G.; Lee, H. K., Headspace liquid-phase microextraction of chlorobenzenes in soil with gas chromatography–electron capture detection. *Anal. Chem.* 2003, *75* (1), 98–103.

26. Saraji, M., Dynamic headspace liquid-phase microextraction of alcohols. *J. Chromatogr.* A 2005, *1062* (1), 15–21.

27. Mohammadi, A.; Alizadeh, N., Automated dynamic headspace organic solvent film microextraction for benzene, toluene, ethylbenzene and xylene: renewable liquid film as a sampler by a programmable motor. *J. Chromatogr.* A 2006, *1107* (1–2), 19–28.

28. Hou, L.; Lee, H. K., Application of static and dynamic liquid-phase microextraction in the determination of polycyclic aromatic hydrocarbons. *J. Chromatogr.* A 2002, *976* (1–2), 377–385.

29. Xu, J.; Liang, P.; Zhang, T., Dynamic liquid-phase microextraction of three phthalate esters from water samples and determination by gas chromatography. *Anal. Chim. Acta* 2007, *597* (1), 1–5.

30. Ouyang, G.; Zhao, W.; Pawliszyn, J., Automation and optimization of liquid-phase microextraction by gas chromatography. *J. Chromatogr.* A 2007, *1138* (1–2), 47–54.

ACRONYMS AND ABBREVIATIONS

AAS	atomic absorption spectrometry
APDC	ammonium pyrrolidine dithiocarbamate
BBP	butylbenzyl phthalate
BFB	p-bromofluorobenzene
b.p.	boiling point
BSA	N,O-bis(trimethylsilyl)acetamide; bovine serum albumin
BSTFA	bis(trimethylsilyl)trifluoroacetamide
BTEX	benzene, toluene, ethylbenzene, xylenes
BZA	benzoylacetone
CCD	central composite design
CE	capillary electrophoresis
CF	continuous flow
CFME	continuous-flow solvent microextraction
CPE	cloud-point extraction
CRM	certified reference material
CWA	chemical warfare agent
DAD	diode array detector
DBP	di-n-butyl phthalate
DD	drop-to-drop
DDTC	diethyldithiocarbamate
DEHP	bis-2-ethylhexyl phthalate
DFBP	decafluorobiphenyl
DHE	dihexyl ether
DI	direct immersion

Solvent Microextraction: Theory and Practice, By John M. Kokosa, Andrzej Przyjazny, and Michael A. Jeannot
Copyright © 2009 John Wiley & Sons, Inc.

DLLME	dispersive liquid–liquid microextraction
DMF	dimethylformamide
DMSO	dimethyl sulfoxide
DNPH	2,4-dinitrophenylhydrazine
DSD	directly suspended droplet
DY-SME	dynamic solvent microextraction
ECD	electron capture detector
8-HQ	8-hydroxyquinoline
EME	electromembrane extraction
EPA	Environmental Protection Agency
ET	electrothermal (or flameless)
ETAAS	electrothermal atomic absorption spectrometry
FIA	flow injection analysis
FID	flame ionization detector
FLD	fluorescence detector; (fluorimetric detector)
FPD	flame photometric detector
GC	gas chromatography
HD	hydrodistillation
HF(2)	hollow fiber–protected two-phase
HF(3)	hollow fiber–protected three-phase
HFME	hollow fiber–protected microextraction
HLLE	homogeneous liquid–liquid extraction
HOC	halogenated organic compound
HPLC	high-performance liquid chromatography
HS	headspace
HS-SDME	headspace single-drop microextraction
Htfa	1,1,1-trifluoroacetylacetone
ICH	International Conference on Harmonization
ICP	inductively coupled plasma
i.d.	inside diameter
IDL	instrument detection limit
IR	infrared
IS	internal standard
LC	liquid chromatography
LDR	linear dynamic range
LGLME	liquid–gas–liquid microextraction
LLE	liquid–liquid extraction
LLLME	liquid–liquid–liquid microextraction
LOD	limit of detection
LOQ	limit of quantitation
LPME	liquid-phase microextraction
LVI	large-volume injection
MAE	microwave-assisted extraction
MALDI-MS	matrix-assisted laser desorption ionization mass spectrometry
MASE	membrane-assisted solvent extraction

MCL	maximum contamination level
MD	microwave distillation
MDL	method detection limit
MEKC	micellar electrokinetic chromatography
MEPS	microextraction in packed syringe
MESI	membrane extraction with a sorbent interface
MMLLE	microporous membrane liquid–liquid extraction
m.p.	melting point
MS	mass spectrometry
MSD	mass selective detector
MTBE	methyl *tert*-butyl ether
MTBSTFA	*N*-methyl-*N*-(*tert*-butyldimethylsilyl)trifluoroacetamide
MW	molecular weight
NPD	nitrogen–phosphorus detector; (thermionic detector)
NSAIDs	nonsteroidal anti-inflammatory drugs
OCPs	organochlorine pesticides
OPPs	organophosphorus pesticides
OVAT	one variable at a time
PAHs	polycyclic aromatic hydrocarbons
PAN	1-(2-pyridylazo)2-naphthol
PBBs	polybrominated biphenyls
PBDEs	polybrominated diphenyl ethers
PCBs	polychlorinated biphenyls
PCP	personal care product
PDHID	pulsed-discharge helium ionization detector
PEEK	poly(ether ether ketone)
PET	poly(ethylene terephthalate)
PFBAY	pentafluorobenzaldehyde
PFBHA	*O*-2,3,4,5,6-(pentafluorobenzyl)hydroxylamine hydrochloride
PHWE	pressurized hot-water extraction
PLE	pressurized liquid extraction
PP	polypropylene
ppb	parts per billion
ppm	parts per million
ppt	parts per trillion
PT	purge-and-trap
PTC	phase-transfer catalysis
PTFE	polytetrafluoroethylene
rpm	revolutions per minute
RSD	relative standard deviation
RSM	response surface methodology
RTIL	room-temperature ionic liquid
SBME	solvent bar microextraction
SD	single drop
SDME	single-drop microextraction

SE-FIA	solvent extraction–flow injection analysis
SFE	supercritical fluid extraction
SIM	single-ion monitoring; (selected ion monitoring)
SLM	supported liquid membrane
SME	solvent microextraction
SMM	single magnet mixer
SPDE	solid-phase dynamic extraction
SPE	solid-phase extraction
SPME	solid-phase microextraction
SRM	standard reference material
TBA$^+$	tetrabutylammonium cation
TBAB	tetrabutylammonium bromide
TBT	tributyltin
TCMs	traditional Chinese medicines
TCPH	2,4,6-trichlorophenylhydrazine
THM	trihalomethane
TOPO	trioctylphosphine oxide
TPT	triphenyltin
TSCC4P	tetraspirocyclohexylcalix[4]pyrrole
UPLC	ultraperformance liquid chromatography
USAEME	ultrasound-assisted emulsification–microextraction
USE	ultrasound-assisted extraction
USP	*United States Pharmacopeia*
UV	ultraviolet
VIS	visible region or visible spectrophotometry
VOCs	volatile organic compounds

SME MODES: CLASSIFICATION AND GLOSSARY

In this appendix we present a classification of the many different modes of solvent microextraction that have been developed and described in the analytical literature. At present, there are 19 different SME modes, and they were introduced in this book in Chapters 2, 5, and 6. To bring some order into confusion, we try to classify all these SME techniques based on the number of phases that exist during the extraction process (Figure A.1). It should be kept in mind, however, that following extraction there have to be at least two phases present, sample and extract, to be able to separate the enriched analytes from the donor phase.

In one-quasi-one-phase SME, during extraction the sample and the solvent either form a homogeneous solution—one-phase SME (homogeneous, temperature-controlled, and cloud-point SME)—or a very fine suspension (emulsion)—quasi-one-phase SME (dispersive and ultrasound-assisted SME) (Figure A.2). Following extraction, the sample and the extract are separated by centrifugation, change in temperature, ionic strength, or pH.

There are two major modes of two-phase SME: direct immersion and membrane assisted (Figure A.3). In the former mode, an aqueous sample is in direct contact with an extracting solvent, which is insoluble or very slightly soluble in water (has low polarity or is nonpolar). In the latter case, an aqueous sample is separated from an extracting organic solvent by a polymeric membrane, which is porous in all membrane-assisted two-phase SME modes. Following extraction, the entire extract

Solvent Microextraction: Theory and Practice, By John M. Kokosa, Andrzej Przyjazny, and Michael A. Jeannot
Copyright © 2009 John Wiley & Sons, Inc.

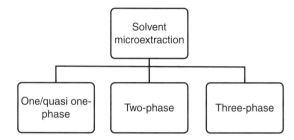

FIGURE A.1. Classification of SME modes based on the number of phases during the extraction process.

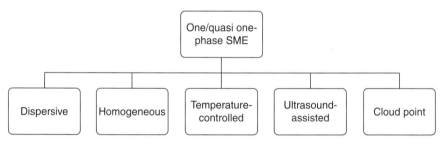

FIGURE A-2. Modes of one-quasi one-phase-solvent microextraction.

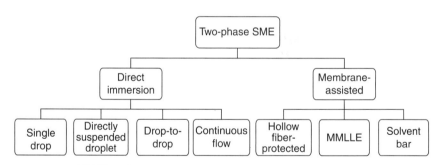

FIGURE A-3. Modes of two-phase solvent microextraction.

or an aliquot of it is introduced into an analytical instrument used for final determination.

In three-phase solvent microextraction, the donor solution (sample) is separated from the acceptor solution (extractant) by a barrier (third phase). The barrier can be a gas (headspace), a liquid, or a nonporous polymeric membrane (MASE) (Figure A.4). The gas can be present as a headspace over the sample solution (headspace, hollow fiber–protected headspace microextraction), or it can be trapped in the pores of a microporous polymeric membrane (liquid–gas–liquid

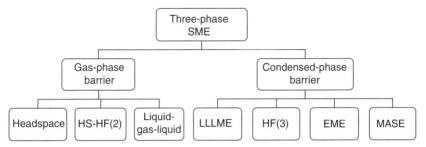

FIGURE A-4. Modes of three-phase solvent microextraction.

microextraction). The barrier liquid can float on the surface of a sample solution, or it can be trapped in the pores of a microporous polymeric membrane. The SME mode involving the organic solvent floating on the surface of an aqueous sample solution is called liquid–liquid–liquid microextraction. When the barrier organic liquid is trapped in the pores of a microporous polymeric membrane, separating two aqueous solutions: sample (donor) and acceptor, the technique is often called supported liquid membrane. In four of the three-phase SME modes (hollow fiber–protected, liquid–liquid–liquid, liquid–gas–liquid, and electromembrane extraction), the extracting solvent is an aqueous solution. In the other three modes (headspace, hollow fiber–protected headspace, and MASE), the extractant is frequently an organic solvent, although pure water and aqueous solutions have also been used in HS-SDME.

A.1 GLOSSARY OF ONE-QUASI-ONE-PHASE SME MODES

Cloud-Point Extraction (CPE) is a sample preparation technique based on the fact that when heated above a certain temperature (cloud point), aqueous solutions of several nonionic surfactants separate into two phases: a surfactant-rich phase (a small volume), and a supernatant aqueous phase (bulk amount) that has a concentration of surfactant close to the critical micellar concentration. The hydrophobic analytes from the sample are concentrated in the surfactant-rich phase. Compared to the initial sample volume, the surfactant-rich solution volume is very small; thus, a high enrichment factor can be obtained. The phase separation is accomplished by centrifugation. The surfactant-rich phase is then cleaned up and/or diluted, followed by final determination of the analytes.[1]

Dispersive Liquid–Liquid Microextraction (DLLME) is a solvent microextraction technique in which an extracting organic solvent, insoluble or very slightly soluble in water (nonpolar or with a very low polarity) and having a density higher than that of water, is first dissolved in a small volume of dispersing solvent, which is miscible with water (methanol, acetone, or acetonitrile), and the solution is next added to the donor solution (sample). The extracting solvent forms a fine suspension (emulsion) with the sample, having a very large interfacial area (see Figure 2.2). This results in rapid extraction of analytes. Following extraction, the suspension is separated by centrifugation, and the organic extract layer is analyzed.

Homogeneous Liquid–Liquid extraction (HLLE) is a one-phase SME mode in which the analyte originally present in a true homogeneous solution is extracted into a very small volume of sedimented phase formed from the solution by the phase separation phenomenon. The ternary component solvent system and the perfluorinated surfactant system are the two usual modes of homogeneous liquid–liquid extraction. In HLLE, the initial state is a homogeneous solution, so that the surface area of the interface between the aqueous phase and the water-miscible organic solvent phase is infinitely large. The difference between DLLME and HLLE is that in the former technique the extracting solvent is dispersed in the form of fine droplets throughout the sample and can be separated from the aqueous phase solely by centrifugation, whereas in the latter technique a true homogeneous solution is formed, which requires the addition of a reagent to increase the ionic strength or adjust the pH of solution prior to centrifugation to separate the sample from the acceptor.

Temperature-Controlled Ionic Liquid-Dispersive Liquid-Phase Micro-extraction In this a solvent microextraction technique, a small volume of ionic liquid (extracting solvent) is first completely dissolved in an aqueous sample by raising the temperature, followed by separation of the two phases (sample and extract) through lowering the temperature and centrifugation.

Ultrasound-Assisted Emulsification–Microextraction (USAEME) is a quasi-one-phase SME mode in which an emulsion of organic solvent insoluble or very slightly soluble in water and having a density higher than water is formed by the addition of a small volume of the solvent to an aqueous sample, followed by sonication. The extract is then separated from the sample by centrifugation. In contrast to DLLME, this technique avoids the use of a disperser solvent and is thus more environmentally friendly.

A.2 GLOSSARY OF TWO-PHASE SME MODES

A.2.1 Direct-Immersion Modes

Continuous-Flow Microextraction (CFME) is a dynamic SME technique in which a drop of organic solvent immiscible with water is held at the outlet tip of poly (ether ether ketone) (PEEK) tubing that is immersed in a continuously flowing sample solution in the extraction chamber.[2] The presence of both diffusion and convection ensures rapid mass transfer of analytes from the sample into the extractant. A microsyringe can also be used to hold a microdrop of the extracting solvent (see Figure 6.13).[3,4] In a modification of CFME called cycle flow microextraction, the sample is recycled by putting the waste outlet of tubing through which the sample flows into the sample reservoir.[5, 6]

Directly Suspended Droplet Microextraction (DSDME) is a two-phase SME mode in which a free droplet of organic solvent immiscible with water and having a density lower than that if water is delivered to the surface of an aqueous sample, which is then agitated by stirring. Following extraction, the extract can be separated from the sample in one of two ways: either an aliquot of the extract is collected by a microsyringe and analyzed,[7, 8] or the system is placed in an ice bath, the organic extract freezes, forming a pellet, which is collected with tweezers, transferred to a different vial, and melted. The latter method calls for a solvent with a melting point close to ambient temperature (e.g., 1-undecanol, m.p. 19?C).[9–14] In a variation of this technique called *directly suspended droplet liquid–liquid–liquid microextraction*, a freely suspended droplet of an aqueous solvent is delivered to the top-center position of a layer of immiscible organic solvent floating on top of an aqueous sample while being stirred by a stirring bar placed on the bottom of the sample vial.[15]

Drop-to-Drop Solvent Microextraction (DDME) is a direct–immersion two-phase SME mode in which both the sample solution and the drop of organic solvent suspended at the tip of a microsyringe needle both have very small volumes, on the order of microliters (see Figure 2.1). As a result, the need for agitation is eliminated and the mass transfer rate of analytes into the extractant is high. The technique is particularly useful when the sample volume available is limited (blood, plasma, or serum).

Single-Drop Microextraction (SDME) is the most common form of two-phase SME, in which a drop of organic solvent immiscible with water, typically 1 to 4 μL in volume and hanging from the tip of the needle of a microsyringe, is immersed into an aqueous sample solution (usually 2 to 40 mL). The solution is stirred to improve mass transfer of the analytes into the drop. The technique is best suited for analytes of low volatility (semivolatiles) and polarity, and relatively clean sample matrices such as tap water or groundwater. Its limitations include partial dissolution of the drop, the danger of drop dislodgment, and a relatively long time of extraction compared to dispersive liquid–liquid microextraction. SDME can be fully automated.

A.2.2 Membrane-assisted modes

Hollow Fiber–Protected Two–Phase SME [HF(2)ME] is a two-phase SME mode in which the extracting organic solvent fills the pores and the lumen of a microporous polymeric fiber, typically made of polypropylene. The fiber is immersed in an aqueous sample solution. The fiber can be U-shaped with two guiding needles taken from a conventional medical syringe supporting the two ends of a hollow fiber to facilitate filling and emptying the lumen.[16] This variant is sometimes called *liquid-phase microextraction*. One end of the fiber can also be fixed on a GC microsyringe needle, and this design is called *hollow fiber–protected liquid–phase microextraction* (see Figure 2.3).[17] Hollow fiber–protected two-phase SME is used most often to extract analytes from complex, dirty matrices such as physiological fluids, because in addition to analyte enrichment it provides substantial sample cleanup. It also permits more vigorous sample agitation than does single-drop SME.

Microporous Membrane Liquid–Liquid Microextraction   (MMLLE) is an SME mode in which a microporous membrane, typically made of flat sheet polypropylene, separates a flowing aqueous sample solution from an extracting organic solvent, which usually remains stagnant but can also flow. Sometimes the term *MMLLE* has also been used for hollow fiber–protected two-phase SME, although it is commonly reserved for systems with flowing donor solutions. MMLLE has been used both in an *off-line* and an *online* configuration with gas chromatography.

Solvent Bar Microextraction   (SBME) is a modification of hollow fiber–protected two-phase solvent microextraction in which the organic solvent is confined in a short piece of hollow fiber that is sealed at both ends, called a solvent bar. The solvent bar is then placed in the aqueous solution for extraction. The solution is stirred and the solvent bar tumbles in it freely (see Figure 2.5).[18] This results in faster mass transfer and higher enrichment factors then those for syringe-based hollow fiber–protected two-phase SME.[16]

A.3   GLOSSARY OF THREE-PHASE SME MODES

A.3.1   Three-Phase SME Modes with a Gas-Phase Intermediate (Barrier)

Headspace Solvent Microextraction   (HSM) is also called *headspace single-drop microextraction* (HS-SDME), a three-phase technique in which the droplet of an organic solvent is hanging from the tip of the needle of a GC microsyringe in the headspace above a condensed (liquid or solid) sample in a closed container (see Figure 6.4). The solvent is usually a high-boiling organic compound, although in some cases pure water or aqueous solutions have also been used as the extractants. The technique is suitable for preconcentration of volatile and semivolatile analytes. It can be used for very complex liquid sample matrices (wastewater, physiological fluids) as well as solid samples. HS-SME can be fully automated.

Hollow Fiber–Protected Headspace Solvent Microextraction   [HS-HF(2) ME] is a three-phase SME mode in which a length of microporous polypropylene hollow fiber, impregnated with an organic solvent, is fitted onto the needle of a microsyringe filled with a volume of the same organic solvent. The solvent is extruded from the syringe barrel into the lumen of the fiber. The fiber with the solvent (still attached to the microsyringe) is then exposed to the headspace above a liquid or solid sample or to a gaseous sample. The use of the fiber increases the interfacial area, resulting in a faster transport of the analytes into the solvent. It also prevents the solvent from being dislodged from the tip of the syringe. At the end of the extraction period, the solvent with the extracted analytes is withdrawn from the fiber lumen into the microsyringe and analyzed.

Liquid–Gas–Liquid Microextraction   (LGLME) is a three-phase SME mode in which the hydrophobic membrane as well as the air inside the pores of the membrane forms the barrier between aqueous sample solution and aqueous extraction (acceptor) solution. The analytes pass through the barrier by gas diffusion.[19] The technique completely eliminates the need for organic solvent and provides a substantial sample cleanup in addition to analyte enrichment.

A.3.2   Three-Phase SME Modes with a Condensed-Phase Intermediate (Barrier)

Liquid–Liquid–Liquid Microextraction   (LLLME) is a three-phase SME mode in which a microdrop of an aqueous solution (acceptor solution) hanging from the tip of the needle of a GC microsyringe is immersed in a layer of immiscible organic solvent floating on top of an aqueous sample (donor solution) (see Figure 6.14). A small volumetric flask or Teflon ring placed in a headspace vial can be used for LLLME, which makes the experimental setup very simple. The organic solvent used in LLLME must meet two requirements: It has to be immiscible with water, and its density must be lower than that of water.

Hollow Fiber–Protected Three–Phase Solvent Microextraction   [HF(3)ME] is an SME mode in which a microporous polypropylene hollow fiber with an organic solvent immiscible with water immobilized in the pores of the fiber is used to extract analytes from an aqueous sample (donor solution) into another aqueous solution that fills the lumen of the fiber (acceptor solution) (see Figure 2.6). In some papers this technique is also called liquid–liquid–liquid microextraction, but in this book the term LLLME is reserved for the SME mode with a layer of organic floating on top of an aqueous sample (donor solution). HF(3)ME is used for the enrichment of analytes having acid or base properties. By an appropriate choice of pH of donor and acceptor solutions the analytes can be extracted from the donor solution into the organic solvent and then into the acceptor solution. A modification of HF(3)ME called *carrier-mediated HF(3)ME* can be used to extract highly polar hydrophilic analytes that would otherwise not be extracted by the organic solvent. In this approach, a relatively hydrophobic ion pairing reagent (called a carrier) with acceptable water solubility forms ion-pairs with the analytes. The ion-pair complexes are then extracted into the organic phase, followed by back-extraction into an aqueous acceptor phase (see Figure 3.7).

Electromembrane Extraction   (EME) is a modification of hollow fiber–protected three-phase SME in which the driving force for extraction is an electrical potential difference applied across the supported liquid membrane.[20–27] The equipment for EME is the same as that for HF(3)ME, except for the addition of two electrodes and a dc power supply (see Figure 3.8). The analytes in the EME system are ionized in both the sample and acceptor solutions, and this promotes electro-kinetic migration across the organic liquid membrane. For basic analytes, a positive electrode is placed in the sample and the negative electrode is located in the acceptor solution; the potential is reversed for acidic compounds. The use of an electrical potential difference as the driving force shortens the extraction time to typically 5 min per extraction while retaining large enrichment factors and high extraction efficiency.

Membrane-Assisted Solvent Extraction   (MASE) is a three-phase SME technique in which an organic solvent fills a nonporous polypropylene membrane bag, which is placed in an aqueous sample solution. The solution is agitated (see Figure 6.8). Since the membrane is nonporous, it acts as the real barrier. To speed up the mass transfer kinetics, the membrane thickness is reduced (typically to 30 µm) and the extraction temperature is increased. The technique can be fully automated.

REFERENCES

1. Han, F.; Yin, R.; Shi, X.; Jia, Q.; Liu, H.; Yao, H.; Xu, L.; Li, S., Cloud point extraction–HPLC method for determination and pharmacokinetic study of flurbiprofen in rat plasma after oral and transdermal administration. *J. Chromatogr. B* 2008, *868* (1–2), 64–69.
2. Liu, W.; Lee, H. K., Continuous-flow microextraction exceeding 1000-fold concentration of dilute analytes. *Anal. Chem.* 2000, *72* (18), 4462–4467.
3. Liu, X. J.; Chen, X. W.; Yang, S.; Wang, X. D., Continuous-flow microextration coupled with HPLC for the determination of 4-chloroaniline in *Chlamydomonas reinhardtii*. *Bull. Environ. Contam. Toxicol.* 2007, *78* (5), 368–372.
4. Liu, Y.; Hashi, Y.; Lin, J.-M., Continuous-flow microextraction and gas chromatographic–mass spectrometric determination of polycyclic aromatic hydrocarbon compounds in water. *Anal. Chim. Acta* 2007, *585* (2), 294–299.
5. Xia, L.; Hu, B.; Jiang, Z.; Wu, Y.; Chen, R., 8-Hydroxyquinoline–chloroform single drop microextraction and electrothermal vaporization ICP-MS for the fractionation of aluminium in natural waters and drinks. *J. Anal. At. Spectrom.* 2005, *20* (5), 441–446.
6. Xia, L.; Li, X.; Wu, Y.; Hu, B.; Chen, R., Ionic liquids based single drop microextraction combined with electrothermal vaporization inductively coupled plasma mass spectrometry for determination of Co, Hg and Pb in biological and environmental samples. *Spectrochim. Acta B* 2008, *63* (11), 1290–1296.
7. Sarafraz-Yazdi, A.; Raouf-Yazdinejad, S.; Es'haghi, Z., Directly suspended droplet microextraction and analysis of amitriptyline and nortriptyline by GC. *Chromatographia* 2007, *66* (7–8), 613–617.
8. Yangcheng, L.; Quan, L.; Guangsheng, L.; Youyuan, D., Directly suspended droplet microextraction. *Anal. Chim. Acta* 2006, *566* (2), 259–264.
9. Dadfarnia, S.; Salmanzadeh, A. M.; Shabani, A. M. H., A novel separation/preconcentration system based on solidification of floating organic drop microextraction

for determination of lead by graphite furnace atomic absorption spectrometry. *Anal. Chim. Acta* 2008, *623* (2), 163–167.

10. Farahani, H.; Ganjali, M. R.; Dinarvand, R.; Norouzi, P., Screening method for phthalate esters in water using liquid-phase microextraction based on the solidification of a floating organic microdrop combined with gas chromatography–mass spectrometry. *Talanta* 2008, *76* (4), 718–723.

11. Farahani, H.; Yamini, Y.; Shariati, S.; Khalili-Zanjani, M. R.; Mansour-Baghahi, S., Development of liquid phase microextraction method based on solidification of floated organic drop for extraction and preconcentration of organochlorine pesticides in water samples. *Anal. Chim. Acta* 2008, *626* (2), 166–173.

12. Khalili-Zanjani, M. R.; Yamini, Y.; Yazdanfar, N.; Shariati, S., Extraction and determination of organophosphorus pesticides in water samples by a new liquid phase microextraction–gas chromatography–flame photometric detection. *Anal. Chim. Acta* 2008, *606* (2), 202–208.

13. Sobhi, H. R.; Yamini, Y.; Esrafili, A.; Abadi, R. H. H. B., Suitable conditions for liquid-phase microextraction using solidification of a floating drop for extraction of fat-soluble vitamins established using an orthogonal array experimental design. *J. Chromatogr. A* 2008, *1196–1197*, 28–32.

14. Khalili Zanjani, M. R.; Yamini, Y.; Shariati, S.; Jönsson, J. Å., A new liquid-phase microextraction method based on solidification of floating organic drop. *Anal. Chim. Acta* 2007, *585* (2), 286–293.

15. Sarafraz-Yazdi, A.; Mofazzeli, F.; Es'haghi, Z., Directly suspended droplet three liquid phase microextraction of diclofenac prior to LC. *Chromatographia* 2008, *67* (1), 49–53.

16. Barri, T.; Jönsson, J. Å., Advances and developments in membrane extraction for gas chromatography: techniques and applications. *J. Chromatogr. A* 2008, *1186* (1–2), 16–38.

17. Shen, G.; Lee, H. K., Hollow fiber–protected liquid-phase microextraction of triazine herbicides. *Anal. Chem.* 2002, *74* (3), 648–654.

18. Jiang, X.; Lee, H. K., Solvent bar microextraction. *Anal. Chem.* 2004, *76* (18), 5591–5596.

19. Farajzadeh, M. A.; Vardast, M. R.; Jönsson, J. Å., Liquid–gas–liquid microextraction as a simple technique for the extraction of 2,4-di- *tert*-butyl phenol from aqueous samples. *Chromatographia* 2007, *66* (5–6), 415–419.

20. Balchen, M.; Reubsaet, L.; Pedersen-Bjergaard, S., Electromembrane extraction of peptides. *J. Chromatogr. A* 2008, *1194* (2), 143–149.

21. Basheer, C.; Tan, S. H.; Lee, H. K., Extraction of lead ions by electromembrane isolation. *J. Chromatogr. A* 2008, *1213* (1), 14–18.

22. Gjelstad, A.; Andersen, T. M.; Rasmussen, K. E.; Pedersen-Bjergaard, S., Microextraction across supported liquid membranes forced by pH gradients and electrical fields. *J. Chromatogr. A* 2007, *1157* (1–2), 38–45.

23. Gjelstad, A.; Rasmussen, K. E.; Pedersen-Bjergaard, S., Simulation of flux during electro-membrane extraction based on the Nernst–Planck equation. *J. Chromatogr. A* 2007, *1174* (1–2), 104–111.

24. Kjelsen, I. J. O.; Gjelstad, A.; Rasmussen, K. E.; Pedersen-Bjergaard, S., Low-voltage electromembrane extraction of basic drugs from biological samples. *J. Chromatogr. A* 2008, *1180* (1–2), 1–9.

25. Pedersen-Bjergaard, S.; Rasmussen, K. E., Electrokinetic migration across artificial liquid membranes: new concept for rapid sample preparation of biological fluids. *J. Chromatogr. A* 2006, *1109* (2), 183–190.

26. Pedersen-Bjergaard, S.; Rasmussen, K. E., Extraction across supported liquid membranes by use of electrical fields, *Anal. Bioanal. Chem.* 2007, *388* (3), 521–523.

27. Xu, L.; Hauser, P. C.; Lee, H. K., Electro membrane isolation of nerve agent degradation products across a supported liquid membrane followed by capillary electrophoresis with contactless conductivity detection. *J. Chromatogr. A* 2008, *1214* (1–2), 17–22.

INDEX

Abbreviations: t = table; f = figure

313